SUSTAINABLE FOOD SECURITY
IN WEST AFRICA

SUSTAINABLE FOOD SECURITY IN WEST AFRICA

Edited by

W. K. ASENSO-OKYERE

University of Ghana

GEORGE BENNEH

University of Ghana

and

WOUTER TIMS

Center for World Food Studies,
Free University of Amsterdam

KLUWER ACADEMIC PUBLISHERS

BOSTON / DORDRECHT / LONDON

A C.I.P. Catalogue record for this book is available from the Library of Congress

ISBN 0-7923-9919-6

Published by Kluwer Academic Publishers,
P.O. Box 17, 3300 AA Dordrecht, The Netherlands.

Sold and distributed in the U.S.A. and Canada
by Kluwer Academic Publishers,
101 Philip Drive, Norwell, MA 02061, U.S.A.

In all other countries, sold and distributed
by Kluwer Academic Publishers Group,
P.O. Box 322, 3300 AH Dordrecht, The Netherlands.

Printed on acid-free paper

This book was typeset by Ms. Hellen P. Agbesi,
ISSER, University of Ghana, Legon.

Printed in the Netherlands

CONTENTS

PREFACE

The initiative under which this book was prepared is the Reseau SADAOC (Securité Alimentaire Durable Africain Occidental Centrale — Sustainable Food Security in Central West Africa) which is a network of researchers and policy makers in Central West Africa (Burkina Faso, Cote d'Ivoire, Ghana, Mali and Togo) and the Netherlands which has been seeking to achieve food security in West Africa through research and policy dialogue since 1992.

The SADAOC Programme arose out of the frustrations of donors as to how to tackle the issue of food security in West Africa and the dissatisfaction of policy makers with research methods and the channelling of research results for effective policy making to enhance food security in the sub-region. The main goal of the SADAOC programme was therefore to develop a new and structural approach to help to improve food security in West Africa by strengthening local research capacity and stimulating high quality research that supports policy design. The initiative is unique in that it defines a joint responsibility of researchers and policy makers for research findings, the application and implementation of the research recommendations as well as the incorporation of the feed back in the on-going research.

The SADAOC programme was motivated from the recognition that complementarities exist between the coastal and sahelian countries in West Africa and that food security in the sub-region can greatly improve if these complementarities were exploited through trade, harmonisation of sub-regional policies and general economic cooperation. Although it was not possible to undertake both research and policy dialogue activities in all the SADAOC countries of West Africa due to resource limitations, it is anticipated that the development of methodologies for research and some of the research results obtained and presented in this book would be applicable to problems and issues in most of the countries of West Africa.

While the book does not cover all the possible areas that can be discussed under sustainable food security, its composition of the natural components of food security in terms of availability (production and trade), accessibility (incomes and poverty status) and utilisation (health and nutrition) can be appealing to all those interested in improving food security in the developing world. It is hoped that any gaps in coverage will be covered by future work under the SADAOC programme and other initiatives in the sub-region.

Many thanks go to the Netherlands government for providing the funds for the SADAOC programme which made it possible to publish this book and to the many individuals both in the Netherlands and West Africa whose hard work and perseverance contributed to the realisation of most of the objectives of Reseau SADAOC.

W. Kwadwo Asenso-Okyere, University of Ghana
George Benneh, University of Ghana
Wouter Tims, Free University of Amsterdam

Section 1
INTRODUCTION

1

CHAPTER 1

THE STATUS OF FOOD SECURITY IN WEST AFRICA

W.K. Asenso-Okyere, G. Benneh and *W. Tims*

The Dimensions of Food Security

Food security is an important component of human welfare; as an indicator of development it tells us a good deal about changes taking place in human life. It can be applied to analyse a wide range of issues pertaining to the food and nutritional situation at the worldwide, national, household and individual levels. Food security is defined as the ability of countries, regions or households to meet their required levels of food consumption at all times (Siamwalla and Valdes, 1981). Even though the wider ranges of food security are important, its ultimate focus is on food security at the household and individual levels, where food security is defined as a situation in which a household has both physical (i.e. supply) and economic (i.e. effective demand) access to adequate food for all its members and it is not at undue risk of losing such access (Shama, 1992).

Food insecurity may be transitory or chronic in nature. Transitory food insecurity is temporary or transient, arising from temporary shortfalls in food supply relative to requirements, or because of a temporary loss of adequate effective purchasing power for food. Seasonal and annual fluctuations in food production, natural catastrophes, the temporary loss of employment and similar factors often lead to transitory food insecurity. Chronic food insecurity is, on the other hand, persistent and almost intractable. According to recent FAO estimates (FAO, 1995), the total number of people around the world suffering from chronic undernutrition — or chronic food insecurity — was between 800 and 900 million people in the last 20 years, but apparently declining slowly. Sub-saharan Africa was however the most notable exception from this worldwide trend, showing an increase over that same period from about 100 million to about 200 million. It demonstrates the difficult problems of Africa in past years and the appropriateness of renewed interest in food security research and other programmes.

One may conveniently divide the factors which influence or determine food security into two broad groups: those on the supply-side and those on the demand-side. Whereas the supply side factors are concerned with food availability, the demand-side factors determine the degree of access to available food. Supply of food is affected by the natural resource endowments of the community, available technology and its dissemination (for food production, storage and preservation), prices, market opportunities and ability to augment own production with external supplies when the need arises.

A large part of West Africa is in the sahelian ecological zone where agro-climatic conditions are harsh and the capacity for food production is low. Large areas of land cannot support food crop production in some countries while in some others, rainfall is so low and erratic that crop yields are not only low but also highly unstable (Olayemi, 1996). On the demand side, food security is affected by household incomes and economic assets (including stocks of animals), prices, demographic factors such as number, gender and age composition of households, and socio-cultural factors like health and sanitation status, educational level, cultural norms, and food consumption habits.

Considering that on the average agriculture accounts for up to 80 percent of the labour force of the active population, 36 percent of GDP,[1] and 60 percent of export revenues, the need to put due emphasis on the sector when setting national development priorities is necessary. Overall economic growth and improvement in the standard of living of the people can hardly be achieved without it (Dapaah, 1996). Policies taking account of both technical and social research are increasingly needed to achieve a long term economic growth and an end to food insecurity. Still, many governments in the West African sub-region do not give the sector the attention it needs in terms of policy commitment and budgetary outlays. Available information indicates that West African countries on average allocate no more than 10 percent of their national public expenditure to agriculture annually (Dapaah, 1996).

Despite low public investments in agriculture, technological research efforts have been successful for a number of crops in the sub-region. According to the World Bank (1995), such research has made it possible to raise yields of traditional or local varieties significantly (Table 1.1).

TABLE 1.1 Yield Increases for Main Crops in
Sub-Saharan Africa (Mt/ha)

Crop	Traditional Yield	Improved Yield
Palm Oil	1.0	2.0-3.0
Coconut	1.5	2.7
Cocoa	0.4	2.0-3.0
Coffee (robusta)	0.25	2.0-3.0
Rubber	0.6-0.8	2.0-3.5
Maize	1.0	3.0-4.0
Rice	0.9	3.0
Cassava	7.0	25.0

Source: World Bank, 1995

Unfortunately, the full benefits of the research results were not achieved because the research agenda in West Africa was concentrated mainly on biological aspects and neglected the equally important policy, storage, agro-processing and marketing problems.

Post-harvest crop care is important: it lowers marketing margins and ensures remunerative prices to producers. However, the evidence indicates that food marketing is inefficient in many West African countries. A detailed marketing margin analysis done in Ghana found that about 70 percent of the average consumer's expenditure on food is represented by marketing costs (Ministry of Agriculture, 1988). Studies carried out in Burkina Faso, Mali and Senegal similarly estimated that transport and other transfer costs ranged from the equivalent of 50 to 75 percent of final cost of the goods at terminal markets (Badiane and Delgado, 1995).

The capacity of many West African countries to import adequate amounts of food to meet increasing deficits in domestic food production has been increasingly constrained by stagnating exports, worsening balance of payment problems and increasing debt burdens. With the supplies of food aid decreasing around the world, reliance on food aid to supplement domestic supplies in West Africa is becoming an increasingly risky policy.

---

[1] This disparity between the population share compared to the share in GDP indicates that the rural (agricultural) population lags behind in terms of income compared to their compatriots dwelling in cities. This does not necessarily imply lesser productivity, but also policy-induced low prices for their products in past years.

On the demand side, access to food and therefore to purchasing power is closely linked to the issue of poverty. Food security cannot be improved without reducing the incidence of poverty which unfortunately is increasing at alarming levels in West Africa. At the national level, only two countries (Cote d'Ivoire and Senegal) have an average per capita GNP which is above the African average (of $615); the West African average is much lower, at $365 per capita and only about 60 percent of the African average (World Bank, 1995). Over the period 1980-93, West African countries recorded an average per capita decrease of their GNP by 0.85 percent.

Using a poverty line of one U.S. dollar per day in terms of international purchasing power, the World Bank (1996) estimates the incidence of poverty in selected West African countries in 1990 to be 15.4 percent in Cote d'Ivoire, 20.4 percent in Ghana, and 34.1 percent in Nigeria. The average for six countries (the three above, including Guinea-Bissau, Mauritania and Senegal) is 32.3 percent which exceeds the corresponding averages of 2.5 percent in the Middle East/North Africa sub-region and 14.7 percent in East Asia. Still, this average is lower than the 52.9 percent estimated for all of Sub-saharan Africa (Olayemi, 1996).

Poverty tends to be more pervasive in the rural areas of most West African countries, although urban poverty is also becoming a major problem, with increasing rural-urban migration but without a concomitant expansion in urban employment or social amenities. Part of poverty in some regions is cyclical or seasonal, while in other regions with similar characteristics it is not. It would be useful to investigate under what circumstances such seasonality is exhibited, and how measures to alleviate seasonal poverty may differ from those to alleviate chronic poverty.

The high population growth rate coupled with a high rate of urbanisation poses a challenge to the efforts to reduce poverty in the sub-region. The West Africa Perspective Study (WALTPS) has estimated that the population of West Africa is likely to reach 430 million by 2020 and the proportion of town dwellers to increase from 40 percent in 1990 to 60 percent by 2020. This phenomenal increase in the urban population will bring with it new challenges for food security.

Adequacy of food consumption by all people is the ultimate goal of food security policies, and this adequacy must be reflected in the adequacy of food nutrient intake. Olayemi (1996) argues that West Africa's average calorie intake of 2243 kilo calories per capita per day on average falls short of the minimum requirement recommended by the Food and Agriculture Organisation. Only in six West African countries that the average calorie intake appears to be higher than the African average of 2289 kilo calories; only Cote d'Ivoire seems to meet the international standard of calorie intake.

Regional Food Security

Due to differences in vegetation and climate, the food production mix of the northern parts of West Africa is different from that of the southern parts of the sub-region. Whereas the northern parts produce cereals, legumes and livestock, the southern parts are suited for roots and tubers, oil-bearing tree crops and fish. Thus food security can be enhanced in West Africa if the complementarities that exist between the coastal and the sahelian countries can be exploited. For instance, livestock produced in the sahelian parts of the sub-region can be exchanged for roots and tubers and fish can be obtained from the coastal parts through trade.

For example, during some parts of the year onions produced in Ghana are sold in Burkina Faso and at other times of the year onions produced in Burkina Faso are sold in Ghana. These flows are made attractive because the onset of the seasons is slightly different in the two countries.

There is a lot of duplication in national agricultural research systems in West Africa which can be eliminated if the countries cooperate more intensively. With agro-climatic conditions almost the same in Ghana and Cote d'Ivoire, substantial economies can be realised if the two countries cooperate in the area of agricultural research. Similarly, the Savanna Agricultural Research Institute in Ghana can collaborate with Burkina Faso, whereas Burkina Faso and Mali can cooperate with each other. There is already some research cooperation under the auspices of the CILSS in the area of food security and natural resources among the sahelian countries and it is hoped that the SADAOC initiative will stimulate this cooperation by bringing in the coastal countries.

The existence of a large monetary union covering the francophone countries (except Guinea Conakry) and the presence of other monetary systems pose some constraints but also opportunities to trade among West African countries. It has been easier to conduct trade under the one monetary system of the CFA franc. The exchange rate policies under the CFA franc zone do have an impact on trade relationships in the sub-region; so are the policies under the other monetary systems. For instance, after the devaluation of the CFA franc in January 1993, livestock and meat imports from Burkina Faso, Mali and Niger into Ghana became cheaper than those from Europe and Australia and therefore generated a lot of trade in livestock from the sahelian to the non-CFA franc zone coastal countries.

The collapse of the West Africa Clearing House which was based in Freetown, Sierra Leone, has been a substantial loss for economic cooperation among West African countries. The Clearing House served as the avenue along which trade reconciliations took place, using the West Africa Unit of Currency which had firmly established conversion rates with the currencies in the sub-region.

Another issue that has hampered trade among the countries of West Africa is the continued dependence on colonial ties which tends to focus on trade between former colonies and the colonial power, to the detriment of regional trade. As a result, most West African countries trade mostly with European countries, whereas official intra-West Africa trade is only marginal. At the same time, however, due to similarities in ethnic groups in border areas, and therefore similarities in consumption patterns, there exists a substantial amount of informal trade at the borders that does not enter into the national trade statistics of the countries concerned. For example, the Ewe ethnic group stretches from South-East Ghana through Togo to Benin. It is therefore not surprising that cassava dough moves from Ghana to Togo and to Benin, to be used to prepare gari which is a staple food for the Ewes.

The cassava dough is head loaded across the border and no recording of the transaction takes place, which implies that cassava trade statistics for the countries involved will definitely understate these flows.

To facilitate trade and other forms of cooperation, the ECOWAS Treaty makes provision for free movement of people and goods. It is unfortunate that this provision has not been implemented within the spirit of the Treaty. Although a West African citizen does not need a visa to visit another West African country, the existence of a large number of road checkpoints and the extortions of money that occur at these checkpoints discourage people from enjoying the free movement which is provided under the Treaty. During a recent trip from Abidjan to Accra by road, as many as 27 checkpoints could be counted, which not only tended to lengthen travel time unnecessarily but also added considerably to costs.

One of the threats to food security in West Africa is conflict and war. The wars in Liberia and Sierra Leone compromise the contribution these two countries could have made to the food security of their own and of the sub-region as a whole. Food production has declined drastically and in some cases it has not been possible to get food aid to the people because of armed bandits who often ambush personnel of aid agencies. Where there is no outright war, conflicts have made it impossible for farmers to achieve anything like their full production potentials, as also occurred during the ethnic conflict in the northern part of Ghana in 1994. Political differences and suspicions of interference in other's internal affairs have led to the closure of the borders between Ghana and Togo a few times in the past. Any time the borders are closed, the free movement of people and goods which otherwise occurs every day is halted and the informal border trade between the two countries is severely reduced. People use unauthorised routes, at the risk of getting their goods confiscated when they are apprehended by security agents.

National Food Security

Many countries aspire to achieve food security which is not synonymous with food self-sufficiency. The latter connotes the internal capacity of a country or household to supply all the food it requires, whereas food security does not require self-sufficiency in food supply. Food security can be achieved by importing food, as long as a country has the capacity to import additional food from abroad without involving undue supply risks. Country governments may try to minimise these risks by promoting domestic food production and thus reducing dependence on external supplies. Risks may include inadequate foreign exchange earnings, possibilities of untimely deliveries and interference of foreign suppliers for political and other reasons. When the import programme cannot be implemented reliably, the ensuing domestic turmoil can be politically disastrous. This is the reason why many governments have attempted to achieve a high level of food self-sufficiency, sometimes at almost any cost. In fact, these costs have in a number of cases become so high that the budget problems became uncontrollable and forced governments to restructure their policies. The maintenance costs of strategic levels of public food stocks as part of those policies have also come in for review in many countries.

Notwithstanding these policies aimed at self-sufficiency, the growth of food imports into West Africa was quite high between 1986 and 1992, reflecting a poor domestic production record. Olayemi (1996) reports that food imports increased in all countries over the period, except for Benin, Niger and Sierra Leone. Virtually all countries, except Benin and Nigeria, also received substantial amounts of food aid during the period. Overall food imports into the sub-region increased at an annual rate of 9.7 percent per annum over the 1986-92 period, compared to 3.3 percent for Africa.

Food aid can have a dampening effect on prices and therefore reduce the incentive for domestic production. When food aid is distributed in kind and particularly when given out free of charge, it can have a deleterious effect on food markets. The food-for-work programme, designed to promote labour intensive infrastructural development projects as a food security enhancing strategy for the poorest people, is one of the food aid programmes that turned out to be detrimental to agricultural development in countries like Ghana. Invariably, the recipients sold some of the food received and diverted the money to purchase non-food items. A direct income supplement would have been more useful.

Instead of food aid, developed countries might focus their aid programmes on ways to expand domestic food production in developing countries, both by increasing the agricultural area and by increasing yields, as well as more efficient post-harvest management of storage and processing. Assistance to the strengthening of research institutes and to the expansion of their capacity for research may have a higher pay-off in the long term than food aid. Food aid may then be used to meet the transient shortfalls in food supplies that may be due to unforeseen circumstances.

National food security policies in West Africa have been complicated by rapid population growth. In most of the countries of the sub-region, the growth in food production has lagged behind the rate of growth of their populations. At more than 3 percent population growth, the rate of growth of food production has to be higher; otherwise deficits in food supplies are bound to occur and will need to be met from imports. Although the sub-region as a whole increased its imports of food over the period 1986-92, the capacity to sustain increasing food imports is limited for many of the countries. Export earnings hardly increased over the period and most countries of West Africa incurred structural deficits on current account of their balance of payments ranging from 3 to 18 percent of their GNP, with a weighted average of around 6-7 percent.

Many of the West African countries have only paid lip service to family planning. Ghana had a population policy in 1969 but it has not been able to effectively implement this programme which could have significantly reduced its population growth rate of some 3.2 percent per annum. Thus, contraceptive use in Ghana is still very low at about 13 percent among the sexually active population; still lower achievement is registered in most of the other West African countries.

Pragmatic macroeconomic and sectoral policies are essential as conditions for a strong and vibrant agricultural sector, as there needs to be a proper enabling environment for the sector to function efficiently. Macroeconomic policies have to be guided by an understanding of motivations which operate at the micro level, but these have been largely missing until recently. Most current analysis (through models or otherwise) has been either one or the other separately, with the link between the two often ignored. For instance, Orbeta (1994) found that the protein and calorie intake of low income households is better served by an across-the-board tariff reduction than by an equivalent percentage of currency devaluation. Clearly, these two policies are not equivalent (cet par, tariff reduction increases imports while devaluations decrease them) but both do feature in conventional adjustment packages.

Macroeconomic policies should be conducive to a competitive agricultural sector compared to the other sectors of the economy. For example the exchange rate regime should encourage the agricultural sector to produce both for the domestic market and for exports. The exchange rate should reflect realistically the international value of the domestic currency. Before structural adjustment, policies were adopted by many West African countries to maintain over-valued currencies which tended to discourage export production. It also necessitated direct intervention by governments to husband and allocate scarce foreign exchange. This was for example the case before the CFA franc was devalued in 1993. A more realistic exchange rate should make it possible to obtain foreign exchange in a competitive market, for example to purchase needed agricultural inputs.

Interest rates should reflect the relation between the demand and supply of money, with no special favours to be given to particular sectors of the economy or to the public sector. Sound monetary policies should create the conditions where it is profitable to borrow for agricultural investments and production. To be able to keep the nominal interest rates down, steps must be taken to control inflation. This calls for effective monetary policy in the presence of a disciplined fiscal policy which relates the government's expenditure programme to its revenues. In some West African countries efforts to bring down inflation have been fruitless because governments have not exercised the necessary restraint in the fiscal programmes and have continued to allow budget deficits to get out of hand. Tax policies of governments should not discourage agricultural production.

Because of the easy ways to tax the agricultural sector, certainly when this is done through the mechanisms of price controls, some governments have overtaxed the sector to an extent which caused a negative supply response. A case in point is the high taxation of the cocoa industry by the government of Ghana, which contributed to the severe decline in cocoa production in the 1970's to early 1980's. The cocoa industry rebounded in the mid to late 1980's when incentive prices were reintroduced.

Agricultural sector policies should promote increased production. The prices must give the correct signals. This implies that markets must be developed for both inputs and outputs and that markets must be allowed to dictate prices. Only in cases of observed market failure is some intervention justified for a specific period. It has become customary under structural adjustment to remove subsidies from inputs and to liberalise the distribution of the inputs. However, these policies have been implemented without a complementary policy for output marketing development. The result is that as input prices have risen, at the same time output prices have plummeted too low to protect the profitability of adopting technologies which are essential to raise food production.

Household Food Security

The main objective of food security is for the individual to be able to obtain adequate food needed at all times and to be able to utilise the food to meet the needs of the body. Analysis of issues of food security at the individual level needs to be conducted at the household level where decisions about production and consumption converge. For about 80 percent of the active population of West Africa employed in agriculture, these decisions are taken by farm households.

Production decisions facing the household include land and other resource acquisitions, the production mix given the endowments of the household and the effective demand including own food and other needs. Meeting own food needs depended in the past on the household's ability to mobilize the labour needs associated with its agricultural activities, whereas land was usually not the constraining factor. This situation is rapidly changing. In recent times household labour is becoming increasingly short in supply due to rural-urban migration and increased schooling of the youth. During the peak labour demand periods it is possible that farm labour wages are above the urban wage rate because of its relative scarcity. Shortage of rural labour greatly hampers farm production and poses a growing constraint to meet household food needs in West Africa.

The demand for food in the urban areas not only increased rapidly, but it also caused a demand for certain new types of foods due to a shift in demand patterns. The demand for convenience foods like rice and gari and other processed food products has gone up. Rice has become a staple food for areas which hitherto consumed very little rice. As a result, rice imports have increased to supplement domestic production.

In addition to production problems the household is faced with post-harvest problems. When the weather is favourable, food is abundant at harvest time and prices fall, becoming a disincentive for production in the next season. The traditional way of storing food for subsistence is no longer adequate in a market economy where large quantities of food can be stored in order to take advantage of increased scarcities associated with time utility. New methods of processing food have to be devised to satisfy an increasingly sophisticated consumer demand and also to increase the shelf-life of food products.

Subsistence production has by and large given way to producing both for the household and for the market: profit motives feature prominently in the household's production-consumption decisions. In this kind of regime the farmer must be able to rely on efficient markets which promise the biggest returns to the resources invested and used. However, markets remain undeveloped in many parts of West Africa. It costs much to transport food from points of production to the few markets that are available. Therefore, quite often perishable food items are left to rot when at the same time these are scarce in the consuming centres. Market information is generally poor and so market integration is virtually absent, giving rise to large disparities in prices for the same commodities at different locations.

The other side of the food security problem facing the household is the consumption and utilisation of food. High fertility and a declining rate of mortality imply that the dependency ratio of households is increasing. That is, the proportion of household members who work to feed the household is decreasing. Difficulties in feeding household members, coupled with cultural factors, have been shown to cause intra-household food inequality in certain societies. Haddad and Kanbur (1991) have observed widespread intra-household food inequality in some communities in Ghana, where boys were found to be given larger portions of food than girls. Men tend to be given larger portions of meat than children who need more protein for proper physical and brain development.

Inadequate calorie intake by household members has resulted in low body mass indices for many West African adults. The children also tend to show unacceptable levels of anthropometric characteristics (low Z-scores for weight-for-height and height-for-age). In addition to low calorie intakes there are also often micro-nutrient, protein and iron deficiencies arising out of consumption of poor unbalanced diets. Some of these diets may be deficient due to poverty but at other times they are the consequence of ignorance and inadequate nutrition knowledge. The household production mix could be easily changed so that foods supplying the basic food nutrients would be obtained without additional costs. Those who purchase their food needs from the market can also obtain the basic nutrient characteristics from inexpensive sources with the requisite knowledge. Nutrition education therefore becomes important in ensuring affordable good nutrition for the household.

Due to inadequate sanitation, sub-standard health care facilities and other forms of deprivation like the absence of potable water, disease and poor health abound in many communities. Afflicted household members cannot contribute to household production and their ability to utilise food to their maximum benefit is hampered. In some cases, other household members have to take time off from their productive activities to provide care for the sick, especially when it involves children. Asenso-Okyere and Dzator (1994) have estimated that care-taking alone accounts for 64.2 percent of the cost of malaria care.

One positive contribution of migration is the relief it provides through remittances to household members left behind. In 1990, official remittances from Ghanaians living abroad amounted to more than $200 million and for some households these transfers provided for their entire livelihoods for the whole year. Seini, Nyanteng and van der Boom (1997) estimate that remittances comprise 4.9 percent of household incomes in Ghana.

Increased remittances have come at a time when the social safety net provided by the extended family system is eroding rapidly and more emphasis is shifting to the nucleus family. In the past the extended family system ensured that no family member went hungry, and health care was provided for the sick, but increased urbanisation and other factors have left these responsibilities more in the care of the nucleus family. As the social safety net provided by the extended family declines, communities and governments need to devise other ways of providing social security for the people to cater for them in times of need and destitution.

Environmental Issues in Food Security

The key to food security is sustainability. There must be assurance that production and consumption today do not jeopardise production and consumption tomorrow. With increasing population pressure, land acquisition and the maintenance of soil fertility are becoming very important issues for food security in West Africa. In many places land has become scarce and difficult to get, so that a variety of tenurial arrangements have evolved. Some of these do not ensure security of tenure and therefore discourage the kinds of (often long-term) investments needed to avoid soil degradation and to maintain its fertility. Due to increased demand for land, shifting cultivation which guaranteed adequate soil nutrients has given way to much shorter fallow periods or more permanent systems of cultivation which require more sophisticated ways of managing the fertility and physical characteristics of the soil. It has given rise to increased use of purchased inputs like fertilizers which require higher returns to make these ventures profitable. Although farming systems have evolved which tend to rejuvenate the soil with minimal use of purchased inputs (like crop rotations, mixed farming and agro-forestry), these are only sparsely applied.

In addition, current human activities are causing severe environmental degradation and these have been exacerbated by structural adjustment programmes (SAPs). The negative impact of SAPs on the environment is transmitted in two ways. First, the higher rates of interest that result from a combination of tight money and credit policies, along with measures to eliminate financial repression increase the rate of natural resource extraction through the usual time preference channel. Second, the shortage of foreign exchange due to increased demand for imports, along with incentives to utilise export opportunities more effectively means, for most developing countries, the 'mining' of their natural resources.

In West Africa, intensive and extensive cultivation, overgrazing, timber and lumbering activities and firewood exploitation are degrading the environment. The concern for wanton destruction of the environment led to the Earth Summit to come up with AGENDA 21 which is a plan of action for combating environmental problems. Some West African countries have taken the issue seriously and are taking steps to safeguard the environment in its development activities. In this regard, Ghana requires the preparation of an environmental impact assessment before an investment or development project can be approved for implementation. This may be appropriate as it is, but the problem is its enforcement at the implementation stage.

The International Economic Order and Food Security

Next to price competitiveness, the international market requires high quality standards and the ability to supply relatively large quantities in a timely manner. Some West African countries have difficulties in meeting these requirements. Due to the high cost of production and transport, together with the maintenance of overvalued currencies, it is sometimes difficult for West African countries to compete in the international agricultural market with some of its food products. The World Bank (1996) has reiterated that countries which instituted policy reforms have also gained price competitiveness with their exports of standardised products, like garments and ethnic hand-crafted merchandise compared to their Asian rivals. It was possible for Ghana to increase its exports of cassava chips only after a deliberate policy was implemented by the Ministry of Food and Agriculture to increase the adoption of improved practices, which significantly increased yields. This was important to enable Ghana lower its cost per unit so that it could compete with exports from Thailand which has been the traditional source of supply of cassava chips to Europe. Exports of cassava chips from Ghana to Europe increased from less than one metric ton in 1994 to 2001 metric tons in 1995.

Sometimes quantitative restrictions imposed by the European Union have not made it possible for West African countries to realise their full export potential for some of the few commodities like pineapples and bananas for which the sub-region seems to have a competitive edge.

The coming into force of the World Trade Organisation (WTO) in January 1995 marks a new era in international trade relationships. The WTO includes agriculture and services which hitherto did not fall under the umbrella of the parent organisation, the General Agreement on Tariffs and Trade (GATT). The Agreement on Agriculture which is designed to promote access, ensures tariff reductions and has rules to reduce market restrictions and to increase access opportunities. Specific reliefs have been provided to protect the domestic agriculture of less developed countries (LDC's) from imports. LDC's are given the opportunity to set tariff levels for their imports to replace all previous import restrictions, but are exempt from any reductions of those tariffs in coming years. Subsidies are permitted when they take the form of payments under environmental programmes, public stock-holding for food security purposes and general agricultural development services such as research, pest and disease control, extension and advisory services. Under the Lome IV Convention, some countries (including those in West Africa) enjoyed Generalised Preferences which gave them preferential access to the European Union markets. Such preferences will in future erode with the general reduction of tariffs and equal market access to all participants under the WTO. In future, all LDC's will have to compete with all others on the same score. This may create difficulties for West African countries which are less competitive than Asian and Latin American countries.

There has been a debate about developing countries standing to lose or gain from the WTO Agreement. Whereas opinions differ, it is generally assumed that the impact depends mostly on the negotiating ability of the countries involved. West African countries have to develop and make use of negotiating skills to obtain the various concessions which are possible under the WTO. They should fight for the removal of tariffs on those goods that are of export interest to them and for the elimination of tariff escalation. They may also consider to what extent they may want or need to open their own markets.

Food Security at the Crossroads

Many of the problems which beset and face food security in West Africa and the future of its peoples have been spelled out in this chapter. As partners in SADAOC we have tried to come to grips with those problems in past years, at least with some aspects and in some of the countries, trying through research to obtain better insights. The other chapters in this book may suggest to what extent we have achieved something.

The learning process is at the crossroads in several respects. Above all, there is an understanding now that it takes a comprehensive view of micro-processes before one can come up with suggestions about macroeconomic policies. There are no sweeping statements to be made about the future, or about appropriate actions in the future, without a thorough understanding about motivations of people, in all their diversity, and about their objectives in the areas of food security. Policies should have this understanding as their backbone so as to enhance the chances of their success.

In another sense, these crossroads were also reached in terms of a much stronger conviction in Africa that not all ills could be attributed to the harsh international conditions under which countries must develop, but that sound social and economic policies at the domestic level have an important role to play. The margins within which these policies can be selected are much narrower than was thought earlier. There are still heated debates about structural adjustment, its effects and where it ultimately will lead the countries engaged in it. But the turn towards a more open stance has been made, with the constant question raised whether it will also benefit those who suffer most from food insecurity.

The move on the domestic front is now being reinforced by the new principles which have been agreed worldwide to liberalise trade in the widest sense under the WTO. Its beginnings are marked by the protectionist attitudes on the part of most participants which make a true liberalisation of agricultural trade illusory, at least for the first years. But the principles stand, providing developing countries with greater access to major markets in the time to come. This may become a major underpinning of domestic efforts to further liberalise their own economies, as international benefits are in that way within reach.

New issues arise and need to be accepted as new challenges. The environmental consequences of agricultural progress cannot be ignored if food security improvements are to be lasting from one generation to the other. Also, when improvements are realised, some groups are bound to be excluded from its benefits, due to their relatively backward conditions at the outset of the process.

As a consequence, policies to meet their needs will become more specific, complex and more difficult to implement effectively. Adjustment of policies, sometimes abandoning earlier approaches on behalf of new ones will remain an essential feature of an effective stance that deals with these matters. Research which can support these moves will be needed, in the conviction that a food secure West Africa is not only a desirable but also a feasible goal.

REFERENCES

Asenso-Okyere, W.K. and Dzator, J.A. 1994. Household cost of seeking malaria care. A retrospective study of two districts in Ghana. Monograph Series No.4 Health Social Sciences Research Unit. Institute of Statistical, Social and Economic Research, University of Ghana, Legon.

Badiane, O. and Delgado, C. 1995. Behrman, J.R. 1990. The action of human resources and poverty on one another. Living Standards Measurement Study Working Paper No.74, Washington, D.C. The World Bank.

Dapaah, S.K. 1996. Research and policy interdependence for sustainable development in 21st century West Africa. A paper presented at the Regional Conference on Governance and Development: Prospectus for 21st Century West Africa, Accra, Ghana.

FAO. 1995. Food, Agriculture and Food Security: The Global Dimension, WFS 96/TECH/1 (advance unedited version), Food and Agriculture Organisation of the United Nations, Rome.

Haddad, L. and Kanbur, R. 1991. Intrahousehold Inequality and the Theory of Targeting. Policy, Research and External Affairs. Working paper. The World Bank, Washington, D.C.

Ministry of Agriculture. 1988. Marketing Margins of Selected Food Commodities.

Orbeta, A.C. Jnr. 1994. Review of Existing Household Models in the Philippines, Mimeo, Manila: PIDS.

Olayemi, J.K. 1996. Poverty, Food Security and Development in West Africa. A paper presented at the Regional Conference on Governance and Development. Prospectus for 21st Century West Africa, Accra, Ghana.

Seini, A.W., Nyanteng, V.K. and B. van den Boom. 1997. Income and Expenditure Profiles of Ghanaian Households. In: Asenso-Okyere et al. (eds.), Sustainable Food Security in West Africa, Kluwer Academic Publishers, Dordrecht: The Netherlands.

Siamwalla, A. and Valdes, A. 1984. Food Security in Developing Countries: International Issues: In: Eicher, C.K. and Staaz, J.M. (eds), Agricultural development in the third world. Baltimore: Johns Hopkins University Press.

Shama, R.P. 1992. Monitoring Access to Food and Household Security. Food Nutrition and Agriculture, 2(4).

World Bank. 1996. Nigeria: Poverty Assessment. Draft Report No. 14733-UNI.

World Bank. 1996. Africa can compete in Europe. Findings, Africa Region, No. 65.

World Bank. 1995. World Development Report, 1995.

CHAPTER 2

AGRICULTURAL PRODUCTION, FOOD SECURITY AND POVERTY IN WEST AFRICA

Matthew Okai

Introduction

The West African sub-region is divided into four broad physiographic units: the moist coastal forest belt, the transitional savanna zone, the semi-arid Sudano-sahelian, and the Sahara desert in the extreme north. Although the sub-region as a whole has a reasonably rich and potentially productive natural resource base, the development of the agriculture and rural sector has so far proved singularly difficult since it has either stagnated or is in decline. The retrogression[1] of the sector is due largely to endogenous factors, especially the outcome of the disintegration of rational resource management and conservation systems, and the consequent relative decline of the agricultural sector's contribution to national output. As agriculture is the principal link between population and the environment, and in the absence of alternative appropriate resource-conserving and/or regenerative technologies, traditional land use and conservation practices had not been able to adapt quickly enough to cope with increasing population (both human and animal) pressure on the resources. Consequently, the sub-region had not been able to sustain agricultural production and ecological stability. This chapter is in four parts. The first section reviews briefly the colonial *structural* distortions[2] of the pattern of agricultural development. Intervention that heralded the onset of ecological degradation is reviewed in the second section. The third section reviews the implication of this degradation on food security and rural poverty while the concluding section makes some recommendations.

STRUCTURAL DISTORTIONS OF AGRICULTURAL AND RURAL DEVELOPMENT

Traditional Equilibrium Production Systems

Over thousands of years the agriculturalists and nomadic pastoralists evolved *agriculture-based* and *animal-based* resource combinations and adaptations which sustained ecological stability since production remained in equilibrium with the environment. The traditional systems are also characterised by non-exploitative social production relations consisting of: (i) the right of

---

[1] The retrogression of agriculture in the context of this chapter refers to net plant nutrient depletion due to extractive and physically destructive agricultural practices incapable of sustaining ecological stability; lack of mutual beneficial interaction and inter-relationships of physico-chemical (soils, water, climate, nutrients); biological (plants, animals, pests and diseases); socio-economic (labour, capital, markets, socio-cultural (social security)); technological (tools, machines, practices); and managerial (knowledge, decision-making and experience).

[2] Structural distortion is the dichotomous pattern of development by which scientific research is concentrated on developing the export crop sub-sector rather than developing the productivity of the smallholder in general.

every individual to productive use of land (non-exclusive tenure systems) by virtue of citizenship in a given socio-cultural group; (ii) the overlapping of rights (reciprocal lien) over land and its products among individuals and groups; and (iii) rigid prohibition against individual alienations of land over which a person has specific but never absolute right. In some instances the use of competitive exclusion to manage the intensity of natural resources exploitation is practised by nomadic pastoralists.

Colonial Distortions of Production Patterns

The incorporation of the sub-region into the world economy exposed it to the hegemon forces far beyond its territorial control. With the support of the colonising power, the colonialists soon subordinated the indigenous population and their medieval empires, and successfully gained directly or indirectly control over scarce resources of land, water and labour. The material level of life was very low, due to the failure of the social organisation to motivate the people to increase production, together with an inability of technical change to significantly raise society's production frontier. In order to achieve the colonial commercial objective of promoting agricultural production for exports, emphasis was placed on the development of exportable agricultural commodities. This emphasis was reflected in the export bias in the allocation of land resources, resulting in the structural distortion of the pattern of agricultural production, and hampering a more spatially balanced development, as priorities were directed towards the trading partners.

The unequal economic relations created new integrative bonds which economically relegated the sub-region to the periphery from where agricultural raw materials were extracted to feed the growing manufacturing industry in the Americas and Europe. Such unequal relations, tied to the world economy, were soon reflected within the sub-region with resource-poor zones increasingly becoming a periphery to the better endowed and rich coastal belt. An attempt to substitute modern for indigenous technology was not successful because the colonial authorities intervened without undertaking the necessary diagnosis of complex traditional farming systems which had been the outcome of dynamic interactions and processes of adaptation to resource constraints. The super-imposition of non-integrating commercial agricultural production over non-retreating old traditional technologies merely introduced extractive and destructive agricultural practices incapable of sustaining the ecological balance. As traditional, cultural and ecological settings gradually lost their viability, the intensified utilisation of the resource base without compensatory increases in productivity, led to self-reinforcing ecological degradation.

Post-Colonial Distortions of Development

Post-colonial development efforts to accelerate the modernisation of the agricultural economy yielded the unintended result of further ecological degradation because the technologies promoted had only tangential organic relations to ecological and cultural settings. The interactive process acquired a synergistic quality where the remedy intended for environmental deterioration became part of already prevailing and hazardous mechanisms.

Reforms of Resource Tenure Systems

Post-independence development interventions introduced many innovative programmes such as reforms of the resource tenure systems, strategies for rural development covering: (i) commodity or 'export crop' projects aimed at increasing the production of export crops by smallholders; (ii) integrated rural development (IRD), comprising multi-faceted and multidimensional area-based projects which aimed to raise living standards through the introduction of diverse services; (iii) functional projects designed to remove or attenuate single critical obstacles to development; and (iv) sub-sectoral projects which were expected to stimulate the development of specific sub-sectors usually over wide geographical areas through the delivery of basic services.

The justification for reforms of resource tenure systems is premised on the argument that so long as a communally-owned resource remains abundant, the absence of the individualisation of landed property rights does not have serious inimical consequences. However, as land becomes a scarce factor of production as a result of population pressure and/or growth in demand for its products, communal ownership of land becomes unstable and produces harmful effects in the form of mismanagement and/or over-exploitation of the valuable life-support resource. The rising land scarcity value has consequently created the need to substitute private rights for communal ownership, which involves changing the pattern of ownership, usage and land control. Such changes are expected to provide sufficient security of tenure and an incentive to invest in land development and environmental protection. Security of tenure permits, it is argued, pledging land as collateral for credit needed for on-farm development. Land administration laws are progressively being reformed to take account of this element. The laws permit entrepreneurial individuals to acquire land and to have it registered in order to ensure security of tenure.

Rural Development

Area-based development projects embracing major agriculture sub-sectors were expected to provide a sustained increase in output and improvement in the quality of life of a significant proportion of the rural poor in a given area. The focus was expected to be on activities which would either raise incomes directly or at least make the potential latent resources more productive. The achievement of sustainable agriculture was expected to be within the capability of governments to design and implement policies to foster the participation of the smallholder and the poorer segments of the rural population in such a way that it would increase their productive capacity to the greatest extent possible.

Sedentarisation of the Pastoralists

Underlying the justification for the introduction of policies for the sedentarisation of the nomadic pastoralists is that both the colonial administration and the independent African governments regarded the pastoralist way of life as conflicting and incompatible with the standard of civilised behaviour, manners and values. Most African countries, therefore, enshrined in their national development plans, programmes to improve the standard of nomadic life with emphasis placed on raising their living conditions, improving the quality of their herds and integrating them into the national economy. A further impetus for the sedentarisation policy was the observation that the yield of the traditionally-managed herd reaching the commercial market remained low at between three to eight percent of potential supply (Baldus, 1978:38; Fielder, 1973:33). Furthermore, the animals being offered for sale were often predominantly either too old, lean or sick. Picardi and Seifert (1977), for instance, reported that in the Sahel during the drought year, pastoralists characteristically dumped dying animals on the market in an effort to salvage something from their starving herds. This is because pastoralists value their animals as recurrent products and sell older animals only after they have become useless for milk supply (Picardi and Seifert, 1977:300). Meanwhile, the failure of the pastoral communities to provide an adequate meat supply resulted in a growing burden of import bills of meat and meat products. After independence, there was a growing domestic demand for meat especially in the expanding urban areas of Africa. Malnutrition, especially protein malnutrition, was a source of serious concern. The increase in demand in developed countries and the emerging markets of the Middle East had the impact of altering the structural composition of the livestock population. In the past, the nomads in the Sahelian countries kept large numbers of camels. These animals are, by their nature, hardier than other livestock and can live without water for up to 28 days during the dry season and up to 3 months during the wet season.

The expanding markets for meat in the coastal countries of West Africa stimulated an expansion in cattle and small ruminants rearing to meet demand. Over the years, therefore, cattle and small ruminant populations gradually but steadily grew thereby partially replacing camels. Cattle need water and pasture and for this reason, both government-assisted and unassisted settlements grew near water points (Arecchi, 1984:2). Because of overstocking, the pastoralists near these water points started growing sorghum and millet while the children herded the family animals in the rangeland that was reduced to make place for cereal crop production. Furthermore, rangeland upon which the herds in the traditional sector were being raised was also diminishing due to a process of environmental deterioration attributed overwhelmingly to mismanagement rather than to any continuing process of climatic change (FAO, various years).

The combination of the growing imports of meat and meat products, the degradation of rangelands and the Sahel drought of 1968-73, all helped to rekindle efforts to conserve the ranges and to make them more productive to achieve the objective of self-sufficiency in basic foodstuffs. It was felt that these objectives could be met through the development of better systems of grazing, the provision of water supplies, restocking, the construction of modern abattoirs and the establishment of beef ranching and grazing reserves, all of which have elements of natural resources conservation and protection. It was the view of the authorities that the resilience of the pastoral environment to withstand drought had been diminished and with controlled range management, land and livestock population, carrying capacity would gradually improve as pasture would be able to regenerate. The governments, therefore, committed themselves to a policy of sedentarisation of the pastoral communities.

Meanwhile, earlier notions were that pastoral communities of Africa were resistant to change and were strictly bound to tradition and culture. The governments, faced with this intractable nomadic independence, introduced policies of the sedentarisation often without paying regard, or appreciation to the pastoralist mode of life. Thus, they brought about far-reaching changes in the economy, the social fabric and the political organisation of the nomadic communities. Accordingly, following the advent of colonisation, the process of integration and encapsulation of nomadic pastoralists into centralised policies and economic systems were introduced and even reinforced after independence.

The basic thrust of post-independence strategies for the development of pastoral areas, apart from technological advances (immunisation, water supply, disease control, etc.) were the encouragement or enforcement of a change in the communal tenure system to landholding, because it was argued that the non-exclusive resource tenure was inappropriate since it dissipates economic rent. It was also argued by economists that multiple owners/users of resources in non-exclusive tenure systems have a built-in incentive to compete against each other over unallocated natural benefits, resulting in the over-exploitation of the common resource. There was therefore, concern about the misuse of the earth's natural resources like pasture on which every pastoralist is entitled to graze his livestock as he desires. Since there is no control over the number of animals each individual grazes, the common pasture is ultimately destroyed. In accordance with the concept of the 'tragedy of the commons', individual self-interest dictates that an additional animal grazing on the pasture, gives the individual the total benefit from the additional livestock while all the users of the pasture share the cost of over-grazing. As in the short run each individual's benefits outweigh his costs, everyone overuses the pasture. It can therefore be concluded that common resources are over-exploited according to individual interest; thus, the resources will be over-exploited and destroyed by the same beneficiaries.

The incentive structure created by the individualisation of landed property right optimises levels of resource exploitation. Under the sedentarisation programmes, groups of individual pastoralists are organised either into cooperatives, companies, etc., principally for ranching and for the optimisation resource use while at the same time protecting and conserving the environment. Under this arrangement, non-members are excluded from competing for the use of originally non-exclusive resources and from deriving income from it. While economic rent can be maximised by excluding non-members, the cost of exclusion (civil unrest, cattle raiding, bloodshed) may outweigh the benefits, when measured in terms of the marginal cost of defending the resource which may exceed the marginal rent gains (Behnke, 1994:20).

DYNAMIC INTERLINKAGES BETWEEN THE RETROGRESSION OF AGRICULTURE AND ENVIRONMENTAL STRESS

Reforms of Resource Tenure Systems and the Marginalisation of the Rural Poor

The individualisation of landed property rights is expected to raise productivity because of the adoption of modern methods of agricultural production. Linked to rates of structurally and socially unequal pattern of development, this reform is unlikely to be an adequate vehicle for an integrated social change since productive resources could be usurped from one form of livelihood to another, often at the expense of the weak in society. In the savanna and the Sahel entrepreneurial individuals who are usually a prominent elite in society, have expropriated some tracts of land in valley bottom for the cultivation of rice. These valleys traditionally used to be reserved for dry season grazing and also for hunting. Rice cultivation precludes seasonal grazing and hunting. The pastoralists may, therefore, be forced to abandon their traditional livelihood and habitat altogether, either by seeking extra-tribal area employment or grazing, or by adopting unfamiliar agro-pastoral systems (Kohen, 1983). Also the advantages of knowledge of land laws, for instance in Senegal, favoured the urban elite where, under the law of National Domain of 1964, prospective landowners were expected to establish titles and request registration within six months from the date of passage of the law. Rural people were not aware of the law and all non-deeded lands became part of the national domain (Feder and Noronha, 1987).

Rural Development Failure

The integrated rural development approach, a form of area planning, operated within the framework of three distinct but interlinked levels of the economy. At the tertiary level, where macro-economic policies are conceived and executed, the effects of policy changes are channelled through 'institutional infrastructure' at the intermediate (meso) level where changes in economic determinants (relative prices of outputs and inputs, domestic terms of trade) occur.

18

At this level, therefore, the implementation of IRD plans are in effect multi-sectoral within the boundaries of the project area. Although some governments have established rural development institutions for coordinating sectoral plans, in practice the coordination of multi-sectoral agencies has not been possible and, therefore, adversely affected project delivery. It is at the micro-economic level that the impacts of policy changes are felt by individual enterprises and households, and thus provoke a response. At this level agricultural pricing policy plays a crucial role for farmers' response, although other delivery services are equally critical. The agricultural pricing policy, in particular, has been one of the most complex and controversial issues in development economics, especially the attempt to resolve the conflict between the interests of consumers versus producers. This problem has been aggravated by the fact that the agricultural sector, as a producer of food, has a very high public profile and thus is of the most public character in terms of policy and programme need, while at the same time it is the most private in terms of day-to-day decision-making regarding production, marketing and consumption. There is also the other problem of price discrimination against agriculture, either directly through price control, indirectly through exchange rate or terms of trade distortions. In such a situation capital moves out of agriculture, thereby retarding the necessary technological improvement which is vital for sustainable agricultural production.

The integrated rural development approach was an attempt to plan from the 'bottom up' and also an attempt to find a compromise between the development goals of the rural community and the nation. In this context, it presupposed that the government actually represents the entire nation and the dirigiste planning expresses the goals and aspirations of the rural community as a whole, just as development goals are established at the local level in a democratic process. It also presupposes that the rural community is institutionally, technically and intellectually able to plan its own future. More importantly was the non-participation of the rural community in project identification and formulation. Over 30 years of experience had indicated that the IRD is still a top-down planning process dominated by government officials. From an institutional, technical and intellectual standpoint, rural communities rely heavily on their tradition in decision-making, particularly in the form of consensus building, and reciprocal help. These traditions are tied to kinship, tribes, clans, etc. Area development projects usually demand a different sense of community, encompassing a territorial unit consisting of many ethnic groups. In this respect the primary social fabric was lost as well as the tradition of reciprocal help to be replaced by self-interest, and consensus building is replaced by regimentation.

Environmental Impacts of Sedentarising Pastoralists

In the pastoral areas the nomads are ever conscious of the fact that they are the least insulated against the ravages of a hostile environment and in the event of the decimation of their animals by drought, depredation or diseases, there is no means of survival. In response to this situation, nomadic pastoralists offset losses in a variety of ways: (i) retain as many animals as possible, (ii) raid others, (iii) arrange for spatial distribution of their animals with relations or friends thus diffusing their survival strategies through the community or (iv) keep a mixture of camels, cattle and small ruminants some of which are capable to survive under harsh conditions (Alan, 1967). They are aware of the minimum number of animals considered essential to ensure the community's survival in times of natural calamity. Dahl and Hjort, taking an ideal herd composed of 50 percent estimate that cows and heifers, reflecting an economy based on recurrent products (milk and blood), and assuming a normal off-take of eight percent, estimate that an average family of 4.9 adult equivalents requires 50 head of cattle (Dahl and Hjort, 1976).

The need to maintain recurrent products dictates the pastoralists' economic priorities;

usually other domestic requirements are sacrificed in favour of these essentials. Inadequacies of these basic necessities will force the pastoralists to sell a few cattle. In the Sahel, Picardi and Seifert (1977) reported that from 1955 to 1965, rainfall in the zone and sub-desert areas averaged significantly higher than in the preceding 55 years. The pastoralists responded by increasing the number of cattle which the improved range was able to support. When the drought of 1968-73 hit the Sahel, the animal population was not destocked fast enough to keep the herd size in equilibrium with the environment. With advances in veterinary medicine and services, after the drought the pastoralists were able to restock their herd quickly. It has, therefore, been observed that usually, following these droughts and their accompanying famines, pastoralists try to avoid the recurrence of the catastrophe by increasing the herd to the desired level for survival purposes. Picardi and Seifert concluded that as long as the pastoralist's priority concerns his own short-term survival, no combination of economic and technical programmes will succeed in preventing the destruction of the range and the associated human suffering (Picardi and Seifert, 1977:302).

It is easy to understand and appreciate the pastoralists' reluctance to sell cattle. Selling one's animals in the commercial market will be tantamount to alienating oneself from the principal means of security. Sales of animals, of course, would generate cash income which in turn would raise demand for an inflow of consumer goods, but the distributive networks for food in pastoral areas are virtually non-existent and commercial infrastructure and formal retail outlets are unavailable. Accordingly, to sell one's subsistence base is to throw a pastoralist into a state of perpetual dependency. Furthermore, with extensive mobility, it would be unreasonable to expect pastoralists to accumulate or store consumer goods which are to be carried along as they roam the rangeland.

A pastoralist, therefore, finds it imperative to keep large numbers of stock from which he obtains recurrent products and to remain within the traditional network of collective social security. Under these conditions, many scholars of nomadism have indicated that marketed off-take may be 3 to 8 percent, with local consumption, feasts and losses raising the figure to about 15 percent which is considered the ceiling for pastoral communities. The low off-take is regarded by many scholars as being an indication of a perverse reaction to market forces because in a situation where the environment is harsh, unstable and insecurity widespread, individual pastoralists face extinction if they are not part of a wider network and ingenuous strategies for survival.

Under these conditions it had been difficult to sustain equilibrium grazing. In a study of the impact of the Sahel drought, Picardi and Seifert discovered that the resilience of rangeland (the maximum sustainable yield level or balance at which it can be exploited indefinitely without reduction of productivity) was destroyed by over-grazing. The reproductive potential of rangeland vegetation was seriously impaired because the yearly cropping or grazing exceeded the annual foliage production, and the vegetation reproductive capacity diminished fast since the situation was not relieved. They further observed that the regeneration of arid land vegetation took much longer than its destruction through over-grazing; where large areas were involved, the restoration of plant cover, even on fertile soil, may take as long as a generation. The dynamics and interaction of these factors, the onset of which were self-reinforcing, was responsible for the acceleration of desertification in the Sahel, a process which, according to their simulation, began in the twenties (Picardi and Seifert, 1977:298).

Meanwhile Lo and Sene observed that the decrease in plant cover brought the onset of wind, sheet and gulley soil erosion which in turn reduced soil fertility and water retentive capacity. The added difficulty being faced in regenerating the tree stock following the sharp decline in rainfall has been the lowering of the water table as a result of the prolonged drought, and the absence of techniques for the reproduction of the tree species as well as the protection of young saplings (Lo and Sene, 1989:453). Because of all these self-reinforcing interactions, few plants are being produced in succeeding years and if herd size is not reduced, the plants will severely be overgrazed, reducing further plant cover since perennials can no longer support the large root systems necessary to reach underground water and annuals can no longer produce sufficient seeds to reproduce themselves (Picardi and Seifert, 1977:298). Meanwhile, in the Sahel, a number of aid organisations financed the sinking of permanent deep wells and allowed the further increase of stock productivity by opening up new pastures. Veterinary services and tsetse fly eradication programmes also reduced the normally high stock death rate from disease. Human diseases have also been reduced as more clean potable water from permanent deep wells were available. Thus, both human and animal population grew rapidly since the eighties because of technological intervention in the pastoral system and more favourable rainfall. Pastoralists have, therefore, been able to graze their herds since the deep wells were able to provide the needed water for their stock (Picardi and Seifert, 1977:298).

The result of the sedentarisation differed considerably from the stated intent. The sedentarisation resulted in a decline of the resource base and increased insecurity as a result of livestock concentration in marginal areas especially around water points. Following the 1966-73 drought, there was a decline in the early seventies in human and livestock populations caused by famine and the large exodus to urban areas as a result of the pastoralists fleeing the tragedy. The increase in the urban population led to indirect pressure on rangeland. The new entrants have low income, and rely on cheap energy sources, principally charcoal, which is burnt in rural areas for delivery to urban centres. Urban need for charcoal therefore, caused an irreversible demand as wood fuel became an agricultural marketable commodity. Meanwhile, in rural areas as a result of the energy shortage, the land was increasingly deprived of its fertilising nutrients which are being used for energy consumption. Cattle dung which had been used as organic manure was no longer used since it had taken the place of firewood (Lo and Sene, 1989:453). Agricultural waste like stubbles which had been used to protect the land from erosion was being used, due to scarcity of wood, for energy.

In a study of fuel wood demand and supply, Anderson indicated that available fuel wood together with the burning of animal dung and crop residues, met the entire need of rural households' energy in West Africa, and appreciably more than half of the demand of urban households. He estimated in 1988 that more than 90 percent of the population depended on fuel wood, animal dung or crop residues for heating and cooking. Although he admitted that statistics were scarce on the extent to which animal dung and crop residues were consumed, he indicated that annual fuel wood consumption was estimated to be around 0.6 cubic metres (m^3) per caput among households consuming fuelwood or charcoal in urban areas, and 0.8 or more in rural areas. The range tended to be quite broad from 0.5m to 1.0m dictated by the cooking habits and methods, thermal capacity of wood used, and the conversion efficiency of making charcoal from wood. Often the conversion losses were not fully compensated for by improvement in cooking efficiency. Broadly, in a household of six persons, the consumption rates were roughly equivalent to one ton of oil in urban areas and one and a half tons in rural areas (Anderson and Fishwick, 1984:11). Based on these estimates, Anderson and Fishwick indicate that by the early eighties, the annual rate of consumption reached the point in the Sahel where it exceeded the annual addition to supply through the growing of trees by about 30 percent. The study revealed that tree stocks were declining at an accelerating rate because consumption was increasing exponentially with population growth while the annual addition to supply through new tree growth (the mean annual increment) was declining in proportion to the volume of stocks (Anderson and Fishwick, 1984:2).

Based on these estimates, the excess consumption over the mean annual increment was expected to rise to 130 percent in the Sahelian countries by the end of the century. These estimates did not include land clearing for agricultural purposes, which in some countries contributed more to the degeneration than fuel wood consumption or to depletion in stocks arising from industrial consumption. The decline of tree stocks in the surrounding countryside is notable within a radius of 80 to 160 kilometres. Mensching reports that a particularly serious desertification problem in the Sahel has been the entirely unplanned and largely uncontrollable destruction of the natural stock of trees. As a result, there has been a vast increase in circular destruction of the savannah tree stock surrounding settlements. The circles of destruction around settlements have been widening everywhere and overlapping (Mensching, reprint:47). The consumption of fuel wood has grown directly with the growth of the rural population, and the strong urban demand is unlikely to be reversed until rising scarcity and relative prices compel people to reduce consumption, get involved in the tree planting programmes or turn to substitutes. Lo and Sene (1983) also point out that the wood crisis is being brought about by urban poverty because alternative sources of energy such as butane require a marked improvement in purchasing power in rural as well as in urban areas.

CONSEQUENCES OF ENVIRONMENTAL DETERIORATION

Declining Rainfall and Increasing Aridity

Many of the agricultural development projects, intended to raise productivity and conserve the environment are themselves contributory to the destructive mechanisms significantly accelerating environmental deterioration. This is due primarily to inability of these projects to successfully promote rational integration of improved new technologies into the productive fabric of traditional farming systems. In the absence of compensatory regenerative technologies, the intensity of resource utilisation is having a grave impact on land. According to the World Resources Institute (WRI), deforestation in West Africa is estimated to occur at a rate of 1.5 million hectares a year (Table 2.1) in the eighties out of a total area under forest of over 82 million hectares. Meanwhile reforestation is not matching deforestation. During the eighties, reforestation averaged just 4 percent of the rate of forest destruction. Deforestation is outstripping the rate of new tree planting by 26 to 1.

Deforestation has adverse impacts on the environment, negatively affecting rainfall amounts. Studies by Gorse and Steeds indicate that aridity has been on the increase in the Sahel, beginning in the early seventies (Gorse and Steeds, 1987). Meanwhile Tucker, Dregne and Newcomb documented that the Sahel experienced wide variations in annual rainfall which exhibited a trend consistently and significantly below the long-term annual average by 30 percent since the seventies (Tucker, Dregne and Newcomb, 1991: 299 - 301). The entire Sahel has a significant chance of experiencing drought each year. In years of adequate rainfall it may not be in keeping with established cropping patterns. Even coastal countries of West Africa have been experiencing significant declines in average rainfall. In Cote d'Ivoire, and possibly Ghana, where the rate of deforestation has been highest (Table 2.1), there has been a strong and accelerating decline in mean annual rainfall at all monitoring stations. Consequently, the dry savanna is steadily expanding southward. According to Uma Lele there was a sharp decrease in rainfall in northern Nigeria (Uma Lele (ed.) 1989). Meanwhile in Senegal annual rainfall had been decreasing by 2.2 percent per annum beginning in the early eighties. In general, therefore, there is a *prima facie* evidence that the forest and vegetation destruction is responsible for the long-term decline in rainfall and the increasing aridity. Even humid West Africa is beginning to experience erratic rainfall and random occurrence of drought. According to UNEP, some 85 percent (or 473 million hectares) of land in the Sahel (Table 2.2) had already been degraded by

TABLE 2.1:   Forest Area and Deforestation in West Africa

			Deforestation 1980's		
	Forest and Woodland 1980 (000 ha) (a)	Percent per year (b)	000 ha per year (c)	Reforestation 1980's 000 ha per year (d)	d as % of c
Sahel					
Burkina Faso	4,735	1.7	80	3	4
The Gambia	215	2.4	5	0	0
Mali	7,230	0.5	36	1	3
Mauritania	554	2.4	13	0	0
Niger	2,530	2.5	67	3	4
Senegal	11,045	0.5	50	4	8
Average	26,349	0.5	251	11	4
Coastal West					
Guinea	10,630	0.3	86	0	0
Guinea-Bissau	2,105	2.7	57	0	0
Liberia	2,040	2.3	46	3	6
Sierra Leone	2,055	0.3	6	0	0
Average	16,850	2.3	195	3	1
Coastal East					
Benin	3,867	1.7	67	0	0
Ghana	8,693	0.3	72	3	4
Cote d'Ivoire	9,834	5.2	510	8	2
Nigeria	14,750	2.7	400	32	8
Togo	1,684	0.7	12	1	8
Average	32,828	3.1	1,061	44	4
West Africa	82,027	2.3	1,507	58	4

Source: African Development Indicators, 1992, World Bank/UNDP.

TABLE 2.2:   Extent of Soil Degradation in Sub-Saharan Africa

	Productive Drylands					
	Million ha		% Degraded			
	Sudano-Sahelian	South Africa	Sudano Sahelian	South Africa	Total SS Africa	% Degraded
Total Productive drylands						
Area	473	304	—	—	777	85
Degraded	416	243	88	80	659	—
Productive dryland types	473	304	88	80	777	85
Rangelands	380	250	—	—	630	86
Degraded rangelars	342	200	90	80	542	—
Rainfed croplands	90	52	—	—	142	80
Degraded cropland	72	12	80	80	114	—
Irrigated lands	3	2	—	—	5	32
Degraded irrigate lands	1	0.6	30	30	1.6	—

Source: United Nations Environment Programme - Assessment of dessertification, 1977.
            The World Bank has substituted degradation for desertification, a term considered
            to be more accurate.

the eighties. In West Africa in general, losses of 10 to 20 tons of soil per hectare have been reported on even very gentle slopes.

Implication for Food Security

The key feature for food security is availability of adequate food supplies at all times, stability of food supplies and access to food by the poor. The West Africa sub-region has experienced to varying degrees all the three distinguishable elements of food insecurity: (i) seasonally, when food becomes scarce each year during pre-harvest periods, (ii) transitory when a temporary decline in access to food arising from instability in food production and food prices occurs; and (iii) chronically, when food availability is persistently insufficient to supply an adequate diet or when pervasive poverty precludes access to sufficient amounts of food. Essentially, therefore, food security can be gauged from the point of view of physical (supply) and economic access (income purchasing power) to food.

Physical Access to Food

Physical access to food is greatly influenced by the adequacy of domestic production and internal distribution. Domestic production, as discussed earlier, is a result of dynamic interactions and inter-relationships of biological, socio-cultural, economic and ecological factors. For West Africa, food supply is conditioned not only by access to productive resources; but to a great extent by the progressive destruction of the basic life-resource-land. Because of the extractive and destructive agricultural practices, there is now quantitative evidence of net depletion of plant macro-nutrients in all countries of the sub-region (Stoorvogel and Smaling, 1990). Soil fertility depletion is reflected in declining crop yields. Blaikie, for instance, points out that the productivity of many tropical soils is extremely sensitive to even small soil losses. He recalls that, based on studies in Nigeria, there is empirical evidence that crop yields in these tropical soils fall in a negative exponential way with increasing cumulative soil loss (Blaikie, 1982 : 21). Malton documents that in the Sahel, yields of sorghum (the staple crop) declined at about 1.5 percent per annum, despite investment in biological yield-increasing technology (Malton, 1990). Meanwhile a review of average crop yields for cereals, roots and tubers in countries in the sub-region, based on FAO data (Table 2.3), indicate that declines in yields intensified in the late eighties, despite heavy investment in agricultural research to improve the technical base for agricultural production. The declines are attributable to reductions in rainfall, soil degradation and the encroachment on marginal land by the expanding agricultural population.

These declines in yields are reflected in the index of per capita food production (Table 2.4) which, on a sub-regional basis, fell by about 17 percent in absolute terms between 1964/66 and 1992. The greatest fall was in Mauritania (-40%) and Senegal (-35%) and greatest improvement being achieved by Cote d'Ivoire (+35%). By the sixties, while other sub-regions were recording positive per capita output, in West Africa per capita production declined by 0.3 percent a year, and it increased its deceleration to 1.5 percent per annum in the seventies (FAO, 1978). Because of these declines, the sub-region self-sufficiency ratio (ratio of domestic supply to requirements) has steadily declined. West African countries, therefore, continue to rely heavily on food imports and food aid. During 1980-90 sub-Saharan Africa imported a total of over 94 million tons of cereals. Over 42 million tons, or 45 percent went to West Africa. Out of more than 27 million tons of aid in cereals, almost 6 million tons or 26 percent was received by beneficiaries in West Africa. Despite these imports and aid, caloric intake per person fell by 10 percent in the Sahel, 15 percent in coastal West and remained steady in coastal East Countries. Based on FAO standards of 180 kg of cereals per person a year, in West Africa the supply (domestic production plus imports) averaged about 140 kg, a shortfall of over 20 percent.

TABLE 2.3: Growth Rates of Cereal and Major Export Crop Yields
            in West Africa

| | | Average Annual Percentage Growth | | | | | |
| | | Cereals | | | Major Export Crops | | |
		1975-80	1980-85	1986-90	1975-80	1980-85	1986-90
Sahel							
Burkina Faso	(Maize, cotton)	7.4	-1.9	7.7	5.5	7.3	-5.3
The Gambia	(maize, cotton)	11.8	3.0	-3.1	--	-5.5	0.6
Mali	(maize, cotton)	12.9	2.9	5.2	1.2	2.6	-3.1
Mauritania	(maize, cotton)	11.6	-9.6	-4.2	-2.1	7.0	-1.6
Niger	(sorghum,						
	groundnuts	6.3	-10.2	-0.2	28.7	-11.5	9.2
Senegal	(maize, cotton)	-6.5	9.7	6.4	-3.7	-14	7.6
Coastal West							
Guinea	(maize, coffee)	-3.3	2.4	-0.1	-0.1	-3.7	-2.2
Guinea-Bissau	(maize,	-1.2	3.6	-11.4	-3.5	5.1	2.5
	groundnuts)						
Liberia	(rice, coffee)	0.5	0.0	-1,.0	7.5	3.9	-23.2
Sierra Loene	(rice, coffee)	-3.8	2.5	-4.3	13.7	-11.2	-17.8
Coastal East							
Benin	(maize, cotton)	1.2	3.8	7.0	-2.0	13.6	-0.8
Ghana	(maize, cocoa)	-2.9	-0.5	5.9	-3.6	-1.0	11.2
Cote d'Ivoire	(maize, coffee)	6.3	1.8	-1.2	-6.5	-10.8	-8.4
Nigeria	(maize, coffee)	-1.3	-5.2	1.5	-4.5	-6.7	15.9
Togo	(maize, coffee)	-0.8	—	20.9	-5.7	-13.1	4.9

Source: World Bank/UNDP *African Development Indicators*, World Bank 1992

Economic Access to Food

The incidence of food insecurity can also arise if vulnerable groups are unable to secure sufficient purchasing power to buy food needed for a satisfactory diet. If food supply increases through productivity gains (improved cultural practices), food prices are unlikely to rise and no adverse effects on food consumption need occur. Following the 1974 World Food Conference, in most countries in the Sahel, in particular Senegal, higher food output was achieved largely through increases in producer prices. The on-going macro-economic reforms cover only tradeable commodities and exclude the food sub-sector since food is treated as a non-tradeable commodity. Stabilisation measures involving, amongst other things, currency depreciation actually were accompanied by food price increases.

TABLE 2.4: Food Security Status of West African Countries

	Population Facing Food Insecurity		Per capita daily calories Supply (calories)		Average supply as percentage of minimum[1] requirement '88	Average Annual cereal imports 000t		Index of per capita food production 1979/81 = 100		% increase (+) decrease (-)
	1980/82m/h	Percent	1965	1986/89		1974	1990	1964/66	1988/89	
Sahel										
Burkina Faso	2	33	1,882	2,002	84	99	145	113	114	+1
The Gambia	0.2	19	–	2,339	98	–	–	152	–	–
Mali	3	35	1,938	2,114	90	281	65	100	97	-3
Mauritania	0.2	25	1,903	2,465	107	116	85	143	86	-40
Niger	2	28	1,996	2,321	98	115	86	165	71	-32
Senegal	1	21	2,372	2,162	91	341	534	156	106	-35
Average	9	28	1,987	2,007	88	942	911	116	90	-22
Coastal West										
Guinea	–	–	2,187	2,007	87	63	210	106	87	-18
Guinea-Bissau	–	–	2,014	2,437	106	–	–	140	–	–
Liberia	1	30	2,158	2,344	101	–	70	95	84	-12
Sierra Leone	1	23	2,014	1,313	79	72	146	99	89	-10
Average	3	26	2,080	2,003	87	177	426	102	85	-17
Coastal East										
Benin	1	18	2,019	2,115	92	8	126	94	112	+19
Ghana	4	36	1,937	2,167	94	177	337	120	97	-19
Cote d'Ivoire	1	8	2,352	2,405	104	172	502	73	101	+38
Nigeria	14	17	2,185	2,454	88	389	502	123	106	-15
Togo	1	29	2,454	2,210	92	6	111	118	88	-25
Average	21	17	2,109	2,052	87	752	1,578	116	100	-14
West Africa	33	19	2,065	2,033	86	1,881	2,915	114	95	-17

Source: African Development Indicators, 1992 World Bank UNDP p.322 for per capita calorie supply; World Development Indicators, 1992 for indices of food production, cereal imports, per capita calorie supply 1965.

Note: Average weighted by population size.   1/ Average per capita daily calorie supply for 1986/89 divided by requirement established by WHO for each country

In a separate study (Okai, 1994) the impact of currency depreciation/money supply and income was regressed on the rate of domestic inflation using cross-country data for 33 sub-Saharan African countries. Using Ghana data in the same study, the impact of changes in the real effective exchange rate (REER) and changes in domestic terms of trade were regressed on changes in per capita agricultural income. The regression outputs indicate that every one percent depreciation of domestic currency results in an increase in the domestic price by 0.75 percent, whereas a one percent decline in the domestic terms of trade for agriculture results in a fall in real agricultural income per capita of 0.38 percent. Average annual percentage inflation rates in the sub-region escalated from about 5 percent in 1965/73 to about 16 percent during 1973/80 and to about 17 percent during 1980/89 (Table 2.5). While high domestic inflation reduces income purchasing power, the income of the agricultural population does not rise sufficiently to compensate for higher food prices. Because the domestic terms of trade continue to be to the disadvantage of the agricultural sector, it leads to a shift of resources away from the sector as a result of a decline in real agricultural income. Domestic food supply is unlikely to increase because of the relatively low return on investment in agriculture, since the stimulative effects of domestic currency devaluation are limited to the tradeable commodities (Okai, 1994). Pinstrup-Anderson in 1985 documented that in Gasau in Nigeria, a 10 percent increase in the price of food decreased the income of the poorest decile by 9 percent as compared with only 5.9 percent for the highest decile (Pinstrup-Anderson, 1988).

Accordingly, the progressive loss of the productive base and the bias against the development of the food sub-sector, combine to undermine the sustainability of food security. The income of the agricultural population has been declining (Table 2.6). During 1965-80, the sub-region's per capita agricultural value-added declined by 0.8 percent per year but decelerated to a decline of 0.4 percent per annum during 1980-90. Based on the World Development Report (WDR) ranking, there were in 1976, seven countries in West Africa (Togo, Mauritania, Nigeria, Senegal, Liberia, Ghana and Cote d' Ivoire) that ranked as middle-income. Ten years later, in 1986, only three countries (Liberia, Nigeria and Cote d'Ivoire) remained in that category and none of the low-income countries moved to the middle-income status during the period. This is largely due to the retrogression of the agricultural sector whose percentage share of GDP (Table 2.6) declined from 51 percent in 1965 to 36 percent in 1990.

Implication for Rural Poverty Alleviation

The retrogression of agriculture has also been intensifying rural poverty. During 1965-88, countries in the Sahel recorded the highest growth of the number of absolute poor in rural areas. In the Gambia, while the average annual percentage growth rate of the rural population was 2.6 percent, that of the rural absolute poor was 4.6 percent, a differential of 2 percent. Of the 4.6 percent growth, 2.6 percent was due to the growth of rural population who were unable to climb above the poverty line and the differential of 2 percent was accounted for by the rural population who were originally above the poverty line but fell below it. In Mali, during the same period the respective average annual growth rates of the rural population were 3.3 and 2.3 percent for the two groups of people, yielding a differential of 1 percent, a measure of rural population originally above the poverty line but falling below it at a rate of 1 percent a year. In Ghana the respective growth rates were 2.5 and 2.2 percent, giving a differential of 0.3 percent as the growth rate of the rural population originally above the poverty line but falling below it.

TABLE 2.5: West African Countries: Selected Socio-Economic Indicators

	Population in mid-1989 in millions	Area thousands of square kilometres	US Dollars 1989	GNP Per Capita Average Annual Growth Rate Percent			Average Annual Inflation Rate Percent			Life Expectancy Birth (Years) 1989
				1965/73	1973/80	1980/89	1965/73	1973/80	1980/89	
Sahel										
Burkina Faso	8.8	274	320	1.2	2.5	2.2	1.9	11.2	4.6	48
The Gambia	0.8	11	240	1.7	0.2	-0.6	3.0	13.5	14.6	44
Mali	8.2	1,240	270	–	4.3	1.3	–	10.8	3.6	48
Mauritania	2.0	1,026	490	1.2	-0.6	-1.8	4.1	8.5	9.4	46
Niger	7.5	1,267	290	-3.7	2.6	-5.0	4.1	8.5	3.4	45
Senegal	7.2	197	650	-0.8	-0.5	0.0	3.0	8.8	7.3	48
Average	34.5	4,015	362	-0.7	2.0	-0.3	2.9	9.5	4.9	45
Coastal West										
Guinea	5.5	246	430	1.2	1.3	–	2.9	2.4	–	43
Guinea-Bissau	1.0	36	180	–	-4.2	1.5	–	5.6	53.2	40
Liberia	2.5	111	–	2.4	-0.7	-5.3	1.6	9.1	1.6	50
Sierra Leone	4.0	72	220	2.3	-0.8	-1.1	2.1	14.5	54.1	42
Average	13.0	465	323	0.0	-1.2	-2.1	2.3	7.5	36.1	43
Coastal East										
Benin	4.6	113	380	0.0	-0.3	-1.8	2.9	11.6	7.5	51
Ghana	14.4	239	390	1.0	-2.1	-0.8	8.1	45.4	43.6	54
Cote d'Ivoire	11.7	320	790	4.5	1.2	-3.0	8.7	16.0	3.1	53
Nigeria	113.6	924	300	1.7	-0.6	-1.8	5.9	17.3	20.1	51
Togo	3.5	57	390	2.0	1.5	-2.3	2.9	8.2	5.1	54
Average	147.8	1,653	338	1.7	-0.5	-1.7	5.5	18.8	19.7	50
West Africa	195.3	6,133	338	1.4	-0.1	-1.4	4.9	16.3	17.1	48

Source: World Bank, World Bank Tables

Low-income are those countries with per capita of US$ 580 or less while middle-income countries with per capita of more than US$ 580 but less than US$ 6000.

TABLE 2.6: Agricultural Performance in West Africa

	Agricultural GDP Average Annual Growth (percent)		Average Annual Growth of Population (Percent)		Average Annual per Capita Agric GDP Growth[1] (Percent)		Agriculture's Percentage Share in GDP	
	1965-80	1980-90	1965-80	1980-90	1965-80	1980-90	1965	1990
Sahel								
Burkina Faso	–	3.3	2.1	2.6	–	0.7	37	32
The Gambia	–	7.1	3.0	3.3	–	3.8	–	–
Mali	2.8	2.3	2.1	2.4	0.7	-0.1	65	46
Mauritania	-2.0	0.7	2.3	2.6	-4.3	-1.9	32	26
Niger	-3.4	1.0	2.6	3.5	-6.0	-2.5	68	36
Senegal	1.4	3.1	2.9	3.0	-1.5	0.1	25	21
Average	1.1	2.3	2.3	2.7	-1.2	-0.4	45	32
Coastal West								
Guinea	–	–	1.5	2.4	–	–	–	28
Guinea-Bissau	–	5.7	2.9	1.9	–	3.8	–	–
Liberia	–	–	3.0	3.2	–	–	27	–
Sierra Leone	3.9	2.6	2.0	2.4	1.9	0.2	–	–
Average	3.9	3.2	2.0	2.5	1.9	0.9	–	–
Coastal East								
Benin	–	3.6	2.7	3.2	–	0.4	59	37
Ghana	1.6	1.0	2.2	3.4	-0.6	-2.4	44	48
Cote d'Ivoire	3.3	1.0	4.1	4.0	-0.8	-3.0	47	47
Nigeria	1.7	3.3	2.5	3.3	-0.8	0.0	55	36
Togo	1.6	4.1	3.0	3.5	-1.4	0.6	46	59
Average	1.7	2.8	2.5	3.2	-0.8	-0.4	52	37
West Africa	1.6	2.7	2.4	3.0	-0.8	-0.4	51	36

Source: World Development Indicators, World Bank Washington, D.C. Various years

1) Difference between agricultural GDP growth rate and population rate.

The high average annual growth rate of the rural poor in the Sahel in the sixties was due to the severe drought during the late sixties, seventies, and to some extent the mid-eighties. During 1988-92 the growth rates decelerated and the rural population joining the absolute poor grew at an average annual rate of 0.9 percent in the Gambia and 0.1 percent in Mali but accelerated moderately to 0.4 percent in Ghana. The proportion of the rural absolute poor grew from about 53 percent in 1965 to about 59 percent of the rural population in 1988 — slightly less than the regional average of 61 percent.

The intensification of rural poverty had devastating social consequences especially in the Sahel. As part of adjustment to environmental stress many Sahelian pastoralists had to abandon their traditional resource-use techniques and migrate into other areas even up to coastal areas. Such migration resulted in the breakdown of supportive social arrangements and increased social disparities rather than solidarity. Having suffered physical debilitation and demoralisation, the migrants' adaptive abilities to the new social and political systems were usually impaired. Others fleeing the advance of the desert were forced to resort to cropping in limited well-watered areas while others were forced to seek wage-labour outside their ancestral areas. Malnutrition was widespread but the gravity was ameliorated because the Sahel became more integrated into global economic system that permitted rapid and effective compassionate response provision of food aid on a massive scale. Where mortality occurred, it was due to nutrition-related diseases which largely affected not the active population but the young, aged, infirm and pregnant women. Consequently, life expectancy at birth in the Sahel is only 45 years as compared to 50 years in coastal countries (Table 2.5). Because of continued food aid to the sub-region, the countries facing food insecurity in the sub-region is low (Table 2.7). However, there is a large number of countries facing severe poverty (10) while those classified as being very needy (15) out of 17 countries is relatively big.

Recommendations

As the farmers, over the years, have reached optimum combinations of resource-use through generations of trial and error, only with positive technological intervention do opportunities arise to improve the performance of the agricultural sector through the recombinations of resources. These possibilities arise from the introduction of new technologies, or from changes in factor and product prices. The manner by which technological, institutional and organisational innovation is introduced exerts a profound influence on the direction of technical change in the agricultural economy. Technological intervention should be guided by relative scarcity of factors of production, its resource conserving property, technical feasibility, economic viability and social acceptability.

A number of recent field investigations into technology adoption (Barret *et al.*, 1982, Collinson, 1982; Crawford, 1982; Eicher, 1987; Newman *et al.*, 1987; Tiffen and Mortimore, 1992) have documented that factors which facilitate diffusion of technological innovation include (i) increases in population pressure on land resources, which spur the motivation to invest in land conservation and improvement in order to sustain and/or increase agricultural output per unit area (productivity criterion); (ii) agricultural intensification requires availability of improved technologies, accretionary growth of biological capital investment in upgrading animal breeds (including fisheries); (iii) institutional and organisational innovation (institutional changes such as the individualisation of landed property to motivate on-farm capital development, improvement in marketing services, credit, agricultural research to make available improved technologies, advisory services to facilitate the diffusion of innovation, and development of farmers' own institutions and organisations); (iv) expansion in the availability of credit needed for the acquisition of improved technologies, services, etc.; (v) upgrading and

TABLE 2.7: 1988 Rural Poverty Profile in Sub-Saharan Africa - Number of Countries

Indices	Sub-Saharan Africa					Asia	Lan America	NE & N Africa	Regional Total
	Eastern	Southern	Central	Western	Sub-Saharan				
The Food Security									
Low Food Security	9	5	2	2	18	6	7	4	34
Medium Food Security	3	2	4	11	20	5	16	2	44
High Food Security	—	1	1	4	6	12	9	7	34
	12	8	7	17	44	23	32	13	112
The Integrated Poverty									
Severe Poverty	10	8	5	10	33	14	14	2	63
Moderate Poverty	—	1	2	7	10	6	11	3	20
Relatively Little Poverty	—	1	—	1	1	4	7	8	30
	10	10	7	17	44	24	32	13	113
Basic Needs									
Very Needy	5	5	3	15	28				
Moderately Needy	5	4	4	2	15		Not given		
Relatively Less Needy	—	1	—	—	1				
	10	10	7	17	44				

Source: *The State of World Rural Poverty: A Profile of Africa,* International Fund for Agricultural Development, Rome, Italy, 1994.

Note: *The Food Security Index* combines food production and consumption variables including those reelecting growth and variability; while the integrated poverty index is calculated by combining head-count measures of the poverty with the income ratio, income distribution below the poverty line, and the annual growth rate of per capita GNP. The income gap measure is a national measure, the difference between the highest GNP per capita from among the countries in the study and the individual country per capita GNP expressed as a percentage of the former. *The basic needs index* reflex social status by including education and health indices and access to services.

strengthening human capital development and managerial skills by investing in educational and training institutes and on-the-job training; (vi) availability of markets for agricultural products; (vii) commercial and economic infrastructure (processing, trade, marketing, roads, communication, etc.) to facilitate horizontal and vertical integration of primary production with processing; and (viii) a generally favourable economic environment including the harmonisation of regional macro-economic policies to expand available markets through the expansion of intra-regional trade and development. Such innovations will help in integrating and mutually re-enforcing positive interactions, and forge closer inter-relationships between natural resources on the one hand and biological, socio-cultural, economic and political systems on the other to help regain equilibrium production relations and ecological stability, for enhancing food security and reducing rural poverty.

REFERENCES

Alan, W. *1965. The African Husbandman*, Oliver and Boyd, Edinburgh.

Anderson, D. and Fishwick, R., 1984. "Fuelwood Consumption and Deforestation in African Countries". *World Bank Staff Working Paper* No. 704, World Bank, Washington D.C.

Arecchi Alberto "Rural Settlement Experiences in Post-Colonial Africa". *Ekistics*, 304 pp 47-56.

Baldus, R.D.,1978. "The Introduction of Cooperative Livestock Husbandry in Tanzania". *Land Reform* pp. 37-47.

Barret V. Lassifer, G. Wilcock, D. Baker D. and Crawford, E., 1982. "Animal Traction in Eastern Upper Volta: A Technical, Economic and Institutional Analysis". *MSU International Development Paper* No.4.

Behnke Roy, 1994. "Natural Resource Management in Pastoral Areas". *Development Policy Review* Vol.12, 1994, pp 5-27.

Blaikie P. 1982. "Environment and Access to Resources in Africa". Accra 59 (1).

Collinson, M.P. 1982. "Farming System Research in East Africa: the Experience of VYMMYT and some Agricultural Research Systems, 1976-81". *MUS International Development Paper* No. 3.

Crawford, E.W. 1982. "A Simulation Study of Constraints on Traditional Farming Systems in Northern Nigeria". *MSU International Development Papers* No.2.

Dahl, G. and Hjort, 1976. A. *Haying Herds,* Studies in Social Anthropology; 2 Stockholm, Sweden.

Eicher, C.K. 1987. "Famine prevention in Africa: The Long Views". *MSU International Development Paper* No.3.

FAO, "Ecological Management of Arid and Semi-Arid Rangelands Programme". FAO Rome various years.

Fielder, R.J. 1973. "The Role of Cattle in the Economy". *African Social Reviews* pp. 327 - 361.

Feder G. and Noronha, 1987. "Land Rights Systems and Agricultural Development in Sub-Saharan Africa". *Research Observer* No. 2, World Bank.

Gorse, J.E. and Steeds D.R., 1987. *Desertification in the Sahelian and Sudanian Zones of West Africa* World Bank Technical Paper No. 61, Washington.

Kohen, Peter, 1983. "State Land Allocation and Class Formation in Nigeria". *Journal of Modern African Studies* 21, No.3 461 - 81.

Lo, H.M. and Sene, 1983. A Human Action and Desertification of the Sahel *International Social Science*. Journal ISSJ 121.

Madely J., 1994. "Dangers of Rice Production in West Africa". *Development and Cooperation*. D & C 1994, Berlin, Germany pp 27 - 28.

Mensching, H.G. "Nomads and Farmers in the West African Sahel - Problem of Competing Land Use". Translated reprint from *Die Durre in Sahel* (Drought in the Sahel) Ver lag Justus Perthes, Darmstadt, Germany, undated.

Newman, M. Sow, A. and Ndoye, O., 1987. "Private and Public Sectors in Developing Country Grain Markets: Organisational Issues and option in Senegal". MSU *International Development Papers* No.12.

Okai M., 1994. "The Gravity of Rural Poverty in Sub-Saharan Africa". A Paper Presented at the PANAFEST Collequim, University of Cape Coast, Ghana.

Pinstrup-Anderson, 1965. Per "Food Prices and the Poor in Developing Countries". *European Review of Agricultural Economics* 12, Nos 1-2, pp 69 - 85.

Picardi, A.C. and Seifert, W.W., 1977. "A Tragedy of the Commons in the Sahel". *Ekitics* 258 pp 297 - 3304.

Stoorvogel, J.J. and Smaling, 1990. A.M.A. "Assessment of Soil Nutrient Depletion in Sub-Saharan Africa: 1983-2000". *Main Report* the Winard Staring Centre, Wageningen the Netherlands.

Tiffen M. Mortimore, M. and Giohuki, F., 1993. *More People, Loss Erosion, Environmental Recovery in Kenya*. Chichester, John Wiley.

Tucker, C.J., Dregne, H.W. and Newcomb, W.W., 1991. "Expansion and Contraception of the Sahara Desert from 1980-1990". *Science* Vol. 253, pp 299 - 301.

Uma Lele (ed.), 1989. *Managing Agricultural Development in Africa: Lessons from Experience*. MADIA Discussion Paper 2, World Bank.

Section 2
POVERTY AND HOUSEHOLD FOOD SECURITY

CHAPTER 3

CHARACTERISING THE POOR IN GHANA: A LOGIT APPROACH

W. K. Asenso-Okyere, N.N.N. Nsowah-Nuamah and *Peter Albersen*

SECTION 1

Introduction

The World Bank estimates that 1.1 billion people were poor in 1990; more than one-fifth of the world's population (World Bank, 1992). Half of these people were in South Asia and almost 20 percent in Sub-saharan Africa. However, the Bank estimates that by the year 2000, Sub-saharan Africa's share of the world's poor would increase dramatically to 27 percent; the only region in the world where the poverty situation is expected to deteriorate. In Sub-saharan Africa and South Asia, about three-quarters of the total population live in rural areas where poverty is concentrated. Evidence shows that, in 1988, more than 85 percent of the world's poor lived in the rural areas (Jazairy *et al.*, 1992). Of these rural poor, data suggest that 60 percent are women.

Findings by Boateng *et al.* (1990), Ewusi (1976, 1984), Awusabo-Asare (1981/82), and Brown (1984) have confirmed that poverty is widespread in Ghana, especially in the rural areas. Setting a poverty line at two-thirds of the mean per capita household expenditure per annum or ¢32,981 (in 1988 cedis), Boateng *et al.* (1990) concluded that around 80 percent of the national incidence of poverty is rural (Table 3.1). According to their definition, 36 percent of Ghana's total population and 44 percent of the rural population were below the poverty line in 1987/88. It is also documented that apart from urban-rural differences in poverty, there are disparities in the distribution of poverty between regions and agro-ecological zones (Table 3. 2).

TABLE 3.1: Incidence of Poverty in Ghana, 1987-88

Area	Share of Population (%)	Proportion of population in Poverty (%)	Contribution to National Poverty (%)	Proportion of Population in Hard Core Poverty (%)	Contribution to Hard Core Poverty
Rural	64.95	43.88	79.30	9.54	83.80
Urban excl. Accra	26.76	26.54	19.80	4.48	16.20
Accra	8.29	3.97	0.90	0.00	—
Ghana	100.00	35.93	100.00	7.39	100.00

Source: Boateng *et al.* (1990)

36

TABLE 3.2: Poverty by Location in Ghana, 1987/88

Locality	Population Share (%)	Population below poverty line (%)	Contribution to National Poverty
Accra Metropolis	8.2	5.5	1.3
Mid Coast	8.8	51.2	12.5
West Coast	9.9	16.2	4.5
East Coast	9.1	20.6	5.2
East Forest	10.6	19.9	5.9
Mid Forest	9.1	47.3	11.9
West Forest	11.8	40.8	13.4
Upper Forest	9.0	39.8	10.0
Volta Basin	11.6	52.3	16.8
Savanna	11.8	55.9	18.4
All Ghana	100.0	35.9	100.0

Source: Boateng *et al.* (1990)

When human deprivation is considered, out of a population of 15 million Ghanaians in 1990, there were about 5.9 million without access to health services, 6.5 million without access to safe water and 10.4 million without access to sanitation. Some 10,000 children die before the age of 5 years, there are 2.0 million children not in primary or secondary school and nearly 4 million adults are illiterate (UNDP, 1991).

Taking into consideration the size of the poverty problem effective development cannot take place without tackling the issue of poverty. Attempts to eliminate poverty are made in most developing countries. Direct measures through food supplements, employment guarantees and the like are found to be costly and can entail distortions in the economy. Structural and longer-term measures through education, health care or redistribution of assets appear more promising but do little to change prevailing poverty. The World Bank emphasises the role of structural adjustment in creating options for the poor in an environment with lesser distortions. It attaches considerable importance to improving the functioning of labour markets. A major problem in the case of direct approaches has been the inability to target the poor and so the programmes have been spread too thinly among all sections of the population with limited impact. For instance, the ineffectiveness of the Programme of Actions to Mitigate the Social Costs of Adjustment (PAMSCAD) in Ghana was largely due to the inability of the programme to target the poorest of the poor.

Whether or not targeting is more appropriate in policies to reduce poverty than more general non-discriminating policies, the need for identifying poor households remains to be felt by many policy makers. There are two central questions related to targeting: who are poor and how can their situation be improved? The answer to the first question is crucial for the proper execution of a targeting programme. The second question is also posed in the framework of more general policy analyses and debates. It requires a model of the economy which takes the behaviour of the households into account. A first step in answering the first question is to define a yardstick by which poverty is measured. Usually this is done by establishing an (arbitrary) poverty line. This poverty line is then used as an indicator of poverty (head-count index, poverty gap index). Theoretical objections against the commonly used head-count index are many. Still, due to its simplicity and clear (though limited) interpretation the head-count is used by many policy makers. The obtained measures may give the defined extent of the problem, it certainly does not provide a tool for the actual identification of households at risk. It may well be that the discriminating characteristics of households are not very sensitive to the level of the poverty line. So with respect to identification and targeting the issue of setting a poverty line may be of limited importance.

Objectives

This chapter is aimed at providing guidelines for identifying poverty groups in Ghana. Specifically, the chapter will:

(i) Present the indicators which can be used to classify individuals into poverty groups;

(ii) Ascertain whether the classification yardsticks are sensitive to the level at which the poverty line is fixed; and

(iii) Discuss changes in classification yardsticks in a two-period regime.

This chapter is organised as follow: In Section 2 the description of poverty measures will be described based on the per capita expenditure and applied to survey information from Ghana. Section 3 starts with the description of the method of analysis, followed by a description of the data used. Estimation results and their interpretation are presented in Section 4. The policy implications of the findings are stated as part of the summary and conclusion in Section 5.

SECTION 2

Poverty Measures Applied to Ghana

Whether or not a person or a household is poor is often assessed by the level of the per capita income[1]. Discussion on how to set the appropriate level of income has been extensive. Methods like cost-of-basic-needs-of-food-energy-intake have been proposed and applied (see e.g. Greer and Thorbecke (1986), Paul (1989), and Ravallion & Bidani (1994)). Since poverty does not have 'natural' origin none of these methods produce an absolute poverty line. Where the level is set may influence the poverty measures. Several measures and desirable properties of measures have been discussed in the literature (see Foster (1984) for a survey). Foster, Greer and Thorbecke (1984) proposed a class of decomposable poverty measures P_α:

$$P_\alpha = (1 / N) \, \Sigma_{i=1..q} \, \{(z - y_i)/z\}^\alpha \qquad (3.1)$$

where N is total population, z is the poverty line, q is the number of persons below the poverty line and y_i is the total income of an individual or household, i. The two cases for which α is 0 and 1 are well-known. The headcount index is P_0, i.e. the percentage of persons below the poverty line. P_1 is known as the poverty gap index, i.e. the average income shortfall in the population as a proportion of the poverty line. $N*Z*P_\alpha$ is the total amount needed to get the poor to the level of the poverty line. The ratio P_1/P_0 gives the average poverty gap of the poor population. The less common used index of depth of poverty P_2 is sensitive to the distribution of poverty. Households far below the poverty line have a higher weight in this index.

Before presenting these measures as applied to the first and second round of the Ghana Living Standards Surveys (GLSS) the basic variable is discussed, i.e. the income(expenditure)

---

[1] Given the budget equation of an economic agent total income and total expenditure are equal. In budget surveys total expenditure often exceeds total income. However, total expenditure is assumed to give a better approximation of long-term income instead of the current income. In this chapter we will use the terms income and expenditure interchangeably to denote total expenditure.

per capita. In Figures 3.1a (rural) and 3.1b (urban) the distributions of income[2] of both rounds are shown. The ¢30,000 line is used as reference and is approximately half of the mean. It is the line used by Twum-Baah (1994) as poverty line. In an earlier study by Boateng *et al.* (1990) a distinction was made between poor (two-thirds of the mean) and hard-core poor (one-third of the mean). The shape of the distribution across the years is the same for both localities. There was a small shift to the left in the second year, meaning an increase in poverty. For the rural populations the reference line coincides by and large with the mode of the distribution. For both years a clear difference exists between the distribution of the rural and urban populations. The rural distribution is much more skewed to the left and the kurtosis is higher. A joint distribution at national level would reveal, by and large, a horizontal line in the range of 30 to 50 thousand cedis.

Table 3.3 presents the average per capita income and the index of income distribution, the Gini-coefficient, by locality. In real terms the average income decreased in the second round of the GLSS[3]. The urban population was relatively better off than the rural. The increase in 1988/1989 in the skewness of the distribution is reflected in the increase in the Gini-coefficient. The poverty measures will reveal many of these characteristics.

TABLE 3.3:   Per capita Expenditure and Gini-coefficient

	1987/1988		1988/1989	
	Per capita Expenditure	Gini-coefficient	Per capita Expenditure	Gini-coefficient
National	61.5	.395	56.0	.389
Rural	52.8	.376	47.1	.364
Urban	78.1	.392	74.8	.382

Source: Computed from the Ghana Living Standards Surveys 1987/1988 and 1988/1989.

Note: Per capita expenditure is deflated and in '000 cedi.

Four levels of the poverty line will be used to present the values of the various poverty measures: one-third, half, and two-thirds of the mean and the mean. The value of the lines is kept fixed for both years. The information in Table 3.4 confirms the distributions presented in Figures 3.1a and 3.1b. The head-count index at ¢30,000 increased by some 19 percent for rural areas. The rural poverty gap index increased by the same amount. This means that the poverty gap index of the poor, reflected in P_1/P_0 was constant. Also depth of poverty increased since P_2 increased. More or less the same applies to the other levels of the poverty line (one-third and two-thirds especially).

---

[2] The figures 1a and 1b are based on the deflated per capita expenditure. The range of the income classes is 5000 cedis. So coordinates are smoothly connected to ease interpretation. Deflation is based on a monthly national deflator, taken from the Statistical Newsletter of Ghana Statistical Service. See Seini *et al.* [1995] for a discussion on income and expenditure profile, based on the GLSS.

[3] This is not in line with the national per capita national income as given in the Quarterly Digest (GSS, [1992]). In that period a (moderate) real growth of 3 percent occurred.

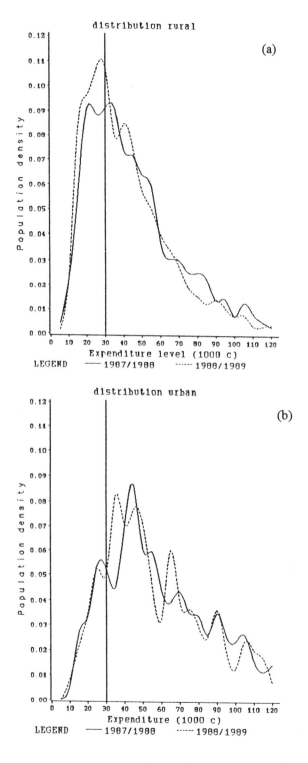

Figure 3.1a and 3.1b: Per Capita Expenditure by Locality

40

TABLE 3.4: Poverty Measures at Different Levels of the Poverty Line

Poverty line	1987/1988				1988/1989			
	¢20,000	¢30,000	¢40,000	¢60,000	¢20,000	¢30,000	¢40,000	¢60,000
Headcount index ($\alpha=0$)								
national	.099	.256	.404	.657	.120	.299	.459	.704
rural	.127	.311	.482	.750	.158	.370	.544	.786
urban	.046	.150	.255	.539	.042	.148	.278	.531
Poverty gap index ($\alpha=1$)								
national	.023	.074	.138	.273	.027	.087	160	.306
rural	.030	.093	.169	.319	.035	.110	.197	.358
urban	.010	.039	.079	.185	.010	.037	.081	.195
Poverty gap index of the Poor (P_1/P_0)								
national	.231	.291	.343	.416	.224	.290	.349	.434
rural	.234	.299	.352	.435	.223	.297	.363	.456
urban	.219	.260	.309	.362	.233	.252	.290	.366
Distribution sensitive index ($\alpha=2$)								
national	.008	.031	.064	.145	.009	.036	.075	.166
rural	.011	.039	.080	.175	.012	.046	.094	.201
urban	.003	.015	.034	.089	.003	.014	.034	.093

Source: Computed from the Ghana Living Standards Surveys 1987/1988 and 1988/1989.

To summarise the results: (i) the income distribution of the populations by locality is different; the rural distribution is more skewed to the left; (ii) the shape of both the rural and urban distributions is the same in both years; and (iii) as a result of the first conclusions the poverty indexes by locality P_0, P_1 and P_2 measure the same changes in time.

SECTION 3

Method of Analysis and Data

In the previous section, households have been categorised into two groups, poor and non-poor based on a single variable, the level of expenditure as a proxy for long-term income. In addition, the two groups differ with respect to several variables each giving some indication as to the group in which the household should be placed. The problem of utilising two or more variables is obviously not simple unless one is found to be sufficient in itself for putting each household in its appropriate group, in which case it will be superfluous to consider more than one variable. Though more complex it may be preferable to use measurements regarding a number of variables for an individual household in combination as the appropriate way to assign a household to a particular group in the poverty profile.

Let the two poverty profiles, non-poor and poor be denoted by Π_1 and Π_2 respectively and the group density function on R^2 associated with each group Π_j be $f_j(\mathbf{x})$ ($j=1,2$), where $\mathbf{x}$ is the p-dimensional vector of characteristics. Suppose that associated with each group there is a probability density function (p.d.f.) on R^2, so that if the household comes from group Π_j, it has

41

p.d.f. $f_j(x)$. Then the objective of the analysis is to allocate an individual household to one of these groups on the basis of a set of common measurements on x. The classification rule d is to allocate households to Π_j if $x \in \Pi_j$.

Discriminant analysis and the logit (probit) model are considered appropriate analytical techniques for this purpose, especially since the dependent variable, namely the poverty profile is categorical. The discriminant function depends strongly on the multivariate normality assumption. If the independent variables are normally distributed the discriminant-analysis estimator is the true maximum-likelihood estimator (MLE) and therefore is asymptotically more efficient than the logit MLE. However, if the independent variables are not normally distributed there is scattered evidence to show that the linear discriminant function does not perform well with discrete data of various types, especially if the components of x are dummy variables. The discriminant analysis estimator is not consistent but the logit MLE is consistent and therefore more robust (Maddala, 1983; Greene, 1991).

Using the logit model, the posterior probabilities P that individuals with observed vector of independent variables x come from population Π_j are:

$$P(\Pi_1|x) = P_i$$
$$= \exp(\beta'x) / (1+\exp(\beta'x)) \quad (3.2)$$

$$P(\Pi_2|x) = 1-P_i$$
$$= 1 / (1+\exp(\beta'x)) \quad (3.3)$$

The parameters of the nonlinear logit model are estimated using the maximum likelihood method. Instead of maximising the likelihood function itself, it is common to maximise instead the logarithm of the likelihood function. Let $O_i = P_i / (1-P_i)$, the estimated conditional odds that an individual household i will be in a specified group, be defined as:

$$L_i = \ln (P_i/(1 - P_i)) = \beta'x. \quad (3.4)$$

Anderson (1972) has pointed out that the log odds-ratio is linear in x for a range of different assumptions about $f_i(x)$. The regression coefficients β_j estimate the change in the log odds of being poor for a one-unit increase in the jth independent variable, controlling for all other independent variables in the model. The $100[\exp(\beta_j) -1]$ is the estimated percentage change in the odds for a one-unit increase in the jth independent variable.

The predicted probability of a household being in a particular group, based on the model is obtained as

$$P_i = O_i / (1+O_i). \quad (3.5)$$

Data

The data for the analysis are derived from the Ghana Living Standards Survey, 1987/1988 and 1988/1989. The first year sample has information on 3,172 households and 15,648 persons; the second year has 3,434 households and 15,882 persons. The GLSS provides information about the characteristics of households and their members, various aspects of household economic and social activities and the interaction between these activities. The data include comprehensive information on demographic characteristics, education, health, employment, migration and anthropometric measurements. Half of the first year set is used as a panel. Those households were revisited the second year. However, it was not possible to trace all previously interviewed households. This resulted in a panel data set of 1,200 households. Given the current analysis where the characteristics of the heads of households are used, a one-to-one match at the level of the head of household between the years is necessary. This decreases the panel set to 1,174 households.

The characteristics of the household are divided into two groups. One set of variables is related to characteristics of the head of household. The second set is related to characteristics of the household as a whole. In Tables 3.5 and 3.6 information on the variables are given for the two years. Since part of the analysis uses the panel of the GLSS, the information at that level is given also.

The head of household responds to the majority of the questions regarding the household. He or she also has an important role on the decisions concerning the welfare of the household and its members. His or her characteristics influence the income generating capacity of the household. Table 3.5 shows that there are small differences for some variables between the years and the panel versus non-panel sets.

TABLE 3.5: Characteristics of the Head of Household

| | Year 1987/1988 | | | Year 1988/1989 | | |
	Panel	Non-Panel	Total	Panel	Non-Panel	Total
Sex (% male)	69.8	70.5	70.2	69.6	70.7	70.3
Age in years (average)	43.9	43.6	43.7	44.7	42.6	43.3
Education in years (average)	5.4	4.9	5.1	5.5	5.6	5.6
Migrant (% yes)	54.1	53.6	53.8	54.6	50.5	51.9
Main occupation (% farmer)	55.9	55.8	55.8	50.8	52.2	51.7
Average Body Mass Index	21.2	21.2	21.2	1.2	21.1	21.1

Source: Computed from the Ghana Living Standards Surveys 1987/1988 and 1988/1989.

The age of the panel group is of course higher in the second year. The age distribution of the head of household is skewed to the left for the males as can be seen from figure 3.2a. The age distribution of the female heads (figure 3.2b) is less skewed. The drop around 40 can be caused by misreporting of the age. There is no difference for the second round of the GLSS. Young men and women become the head of household when they are in their late twenties. Many households headed by young men or women start from scratch with respect to assets. Over the years capital accrues to the household, i.e. the position of the household will become less vulnerable. On the other hand the responsibilities of the head of household also grow with the years. For the elderly it can be said that the labour potential decreases and so is the resilience to changes. The variable 'years of education' has two important frequencies. Around 45 percent of the head of households did not have any formal education, while 30 percent had finished elementary education (10 years). With respect to the poverty question one may hypothesise that education decreases the risk of falling below the poverty line. The migration figures refer to the number of heads of households not living in the village or city where they are socio-culturally regarded as being indigenous. Of the native inhabitants three quarters have lived for some time in a different location. Many migrants start usually with only their relatives or friends in their new environment.[4] They were looking for better opportunities and are perhaps more eager to achieve their objective since they have no other endowments apart from their labour and relations.

---

[4] Over half of the male migrants report a work related reason for moving, against twenty percent for females. Marriage or joining family is the reason for a large number of women (50 percent compared with 20 percent for men).

43

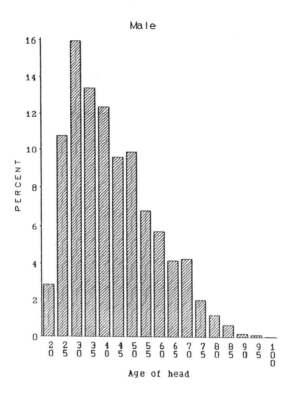

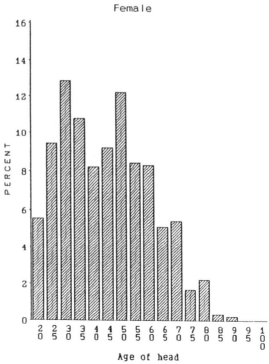

Figure 3.2a and 3.2b: Age Distribution of the Head of Household by Sex, 1987/1988

44

The reported main occupation of the head of household, i.e. farmer or not, shows that in the panel some persons went out of agriculture in the course of the year. Given the prevalence of poverty in rural areas farming can be associated with poverty. The Body Mass Index (BMI) is an indicator of nutritional status of an adult measured as the weight in kilograms divided by the squared height in metres. The majority of observations lie within the range of 15 to 27. It is common to use 17.5 as a cut-off point below which the person is said to be in bad shape. In this chapter the BMI is used as a continuous variable. The BMI has a positive relationship with the labour productivity of a person (Boom *et al.*, 1997), thus a positive relation with the income of the household can be expected.

Table 3.6 gives information at the household level. These variables give important characteristics of the households. Again as in Table 3.5 there are on the average only small differences between the years for the panel vs. non-panel figures. The variables 'Dependency ratio' and 'Size' relate to the household composition. The economic dependency ratio[5] ranges from 0 to 8, but is skewed to the left as can be expected. Ninety percent of the observations has a ratio below 2.5. A similar distribution can be found for the size of the household, with 90 percent of the household size equal to or smaller than 8 persons. The variables 'Land in use', 'Income from remittances' and 'Total value of assets' refer to three different sources of income. The fourth source of income (earnings from labour) is already attributed to the head of household. The three variables are given in per capita terms. The latter two variables are also deflated to adjust for inflation during the survey periods (September 1987=100)[6]. The last two variables 'Treated water' and 'Electricity' refer to conditions of the locality. Treated water may contribute to better sanitation and thus lead to relatively better living conditions. Those conditions may lead to higher labour productivity. The availability of electricity (light) may among other things increase productivity through more working hours.

TABLE 3.6:   Characteristics of the Household

| | Year 1987/1988 | | | Year 1988/1989 | | |
	Panel	Non-Panel	Total	Panel	Non-Panel	Total
Size of household	4.7	5.0	4.9	4.7	4.5	4.6
Dependency ratio (%)	111.2	107.5	108.9	109.2	103.4	105.3
Land use (acres)	1.8	1.9	1.9	1.6	1.8	1.7
Income from remittances	3.9	3.7	3.8	4.4	5.7	5.3
Value of assets	36.7	37.1	37.0	94.5	31.2	52.8
Availability of treated water (%)	35.2	29.8	31.8	37.0	33.2	34.5
Availability of electricity (%)	29.8	24.0	26.2	30.5	26.5	27.8

Source: Computed from the Ghana Living Standards Surveys 1987/1988 and 1988/1989.

Notes: Income from remittances and value of assets are deflated per capita figures in '000 cedis. Land use also given in per capita figures.

---

[5] The economic dependency ratio of the household is defined as the ratio of economically inactive persons (age of 0-14, 65 and above) over the economically active persons (age 15-64). For households with only economically inactive persons (e.g elderly, still taking care of themselves) the denominator is set to 1.

[6] Seini *et al.* (Chapter 4) give a description and discussion on the income pattern of Ghanaian households based on the GLSS 1 and GLSS 2.

Estimation Results and Policy Implications

The logit model specified in the last section for the odds of being poor under the poverty line was estimated using data from GLSS 1 and GLSS 2. Estimations were also done separately for the panel data comprising the 1,200 households which were interviewed in both GLSS 1 and GLSS 2. The logit model is applied for 5 levels of the poverty line. In this section the results of the 5 levels of the first year and the comparison between the first, second, and the panel will be presented.

The estimated results indicate that most of the characteristics/attributes for the five levels were significant at the 0.01 level (Table 3.7). The parameter estimates represent the effects of the characteristics/attributes on the log odds of being poor. The age, sex and Body-Mass-Index of the head of household do not significantly affect the log odds of being poor in the lower levels of the poverty line. The household characteristics and those about the locality are significant at all levels.

To investigate the predictive ability of the model it was validated by predicting the number of households correctly classified by it at different levels of the classification rule d (Table 3.8). The model predicts always better than a simple rule of classifying all households in the largest group. i.e. non-poor (bottom line of Table 3.8).

The percent contribution of the various characteristics/attributes in the model for the odds of being poor are presented in Table 3.9. Discussion at this point will be limited to the change in the contribution of the attributes. The main occupation (farmer or not) stays high on all levels. More or less the same holds for the sex of the head of household and whether he or she is a migrant. The land availability contributes only substantially at the lower levels of the poverty line. The local conditions, water and electricity are high, but the contribution decreases somewhat as the poverty line is set at a higher level. The same pattern was observed when data for the second year of the GLSS were used. The comparison of the two rounds of the GLSS and the panel household data will be limited to the poverty line at 50 percent of the mean (following Twum-Baah (1994)). This implies that any person whose total annual expenditure was less than ¢30,750 in 1987/88 and less than ¢28,000 in 1988/89 is considered to be poor. The results of the four estimations are presented in Table 3.10. The age, sex and occupation of the head of household do not significantly affect the log odds of being poor. Whereas remittances and migrant status of the individual were significant for GLSS 1, they turned out to be insignificant for GLSS 2. The size of the household, land availability, total value of assets and availability of electricity were found to be significant in all four estimations.

The predictive probability of the model is higher than the simple rule of assigning each household to the group of households in poverty (see Table 3.11).

The percent contribution of the various characteristics/attributes in the model to the odds of being poor are presented in Table 3.12. The body mass index (BMI) negatively affects the odds of being poor. The body mass index has been found to be directly related to labour productivity (Boom et al., 1997). Thus, an increase in the BMI would tend to raise household income and therefore reduce poverty. However, significant results were only found in the second round.

Education of the head of household is inversely related to the odds of the household members being poor but its contribution is less than 8 percent in either 1987/88 or 1988/89. Apart from the prospects of good income, education of the head of household positively affects the care household members obtain.

TABLE 3.7: Parameter Estimates for Logit Models for Different
Poverty Lines-GLSS 1

Characteristics/Attributes	Poverty line				
	PL_2	PL_3	PL_4	PL_5	PL_6
Intercept	-3.8832	-4.4545	-2.9500	-2.1312	-1.4548
	(17.3***)	(58.4***)	(40.7***)	(26.2***)	(14.1***)
Age in years	0.0035	0.0091	0.0092	0.0018	0.0034
	(0.3)	(4.1)	(6.1**)	(0.3)	(2.1)
Body Mass Index	-0.0467	-0.0152	-0.0469	-0.0554	-0.0621
	(2.16)	(0.6)	(8.4***)	(14.1***)	(20.0***)
Main occupation	-0.7057	-0.1992	-0.3268	-0.2299	-0.1859
	(11.1***)	(2.4)	(9.5***)	(5.5**)	(3.8*)
Education	-0.1134	-0.0065	-0.0362	-0.0481	-0.0409
	(20.2***)	(19.7***)	(10.0***)	(21.1***)	(17.2***)
Sex	-0.2327	-0.0306	0.1708	0.2454	0.3593
	(1.2)	(0.0)	(2.1)	(4.8**)	(10.5***)
Migrant	0.5798	0.3957	0.3450	0.2307	0.2360
	(9.7***)	(10.7***)	(12.2***)	(6.2**)	(6.9***)
Size of household	0.1459	0.1774	0.1815	0.2203	0.2405
	(30.0***)	(80.6***)	(100.4***)	(136.9***)	(146.4***)
Economic dependency	0.2281	0.2097	0.2621	0.3046	0.3104
	(9.2***)	(16.3***)	(33.4***)	(44.6***)	(41.4***)
Land availability	-0.6100	-0.2418	-0.1999	-0.1825	-0.1375
	(34.4***)	(30.4***)	(39.5***)	(45.3***)	(37.1***)
Remittances income	-0.0436	-0.0441	-0.0353	-0.0222	-0.0344
	(2.4*)	(6.4***)	(9.0***)	(6.3**)	(16.4***)
Value of assets	-0.0116	-0.0088	-0.0055	-0.0033	-0.0037
	(8.2***)	(16.9***)	(16.4***)	(12.2***)	(18.9***)
Electricity	1.183	1.3369	0.9963	0.9708	0.6876
	(7.1***)	(25.3***)	(31.1***)	(41.3***)	(26.1***)
Treated water	0.6640	0.6805	0.6598	0.6731	0.6059
	(4.4**)	(11.8***)	(18.9***)	(25.1***)	(23.2***)

Note PL_2 = One-third of mean per capita expenditure per annum

PL_3 = Half of the mean per capita expenditure per annum

PL_4 = Two-thirds of the mean per capita expenditure per annum

PL_5 = Five-sixths of the mean per capita expenditure per annum

PL_6 = Mean per capita expenditure per annum

Figures in parenthesis are $\chi 2$ values

*   significant at $p<=0.1$

**   significant at $p<=0.05$

***   significant at $p<=0.01$

TABLE 3.8: Percent of Households Correctly
Classified by Model

Probability	Poverty Line				
Level	PL_2	PL_3	PL_4	PL_5	PL_6
0.5	94.7	86.9	78.0	75.5	75.1
0.6	94.8	86.9	75.6	73.7	72.9
0.7	94.9	86.6	76.7	71.7	68.5
0.8	94.8	86.5	76.1	68.8	63.3
0.9	94.9	86.4	75.9	66.3	57.7
Percentage in largest group	93.0	81.2	69.4	56.8	54.6 (poor)

Note: See note of Table 3.7 for the description of the 5 lines.

The economic dependency ratio increases the odds of being poor. This expected result has a lot of consequence for Ghana which has a large young population and an increasing aged population.[7] Large household size directly affected poverty and contributed almost 20 percent to the odds of being poor. The rate of growth of Ghana's population which is over 3 percent per annum is high and so any poverty reduction programme will have to address the demographic trends to be successful.

TABLE 3.9: Percent contribution of characteristics/attributes to poverty,
GLSS 1

Characteristics/	Poverty line				
Attributes	PL_2	PL_3	PL_4	PL_5	PL_6
Age in years	0.4	0.9	0.9	0.2	0.5
Body Mass Index	-4.6	-1.5	-4.6	-5.4	-6.0
Main occupation	-99.0	-99.0	-96.2	-90.6	-80.6
Education	-10.7	-6.3	-3.6	-4.7	-4.0
Sex	-98.4	-98.9	-93.8	-84.8	-66.6
Migrant	-96.3	-98.3	-92.6	-85.1	-70.4
Size of household	15.7	19.4	19.9	24.6	27.2
Economic dependency	25.6	23.3	30.0	35.6	36.4
Land availability	-45.7	-21.5	-18.1	-16.7	-12.8
Remittances income	-4.3	-4.3	-3.5	-2.2	-3.4
Value of assets	-1.2	-0.9	-0.6	-0.3	-0.4
Electricity	-93.3	-95.6	-85.8	-68.7	-53.6
Treated water	-96.0	-97.7	-89.9	-76.7	-57.2

Note: See note of Table 3.7 for the description of the 5 lines.

---

[7] The age dependency ratio for Ghana in 1984 was 0.96 (Nuamah, 1994). For the GLSS 1987/1988 and 1988/1989 these ratios were 1.09 and 1.05 respectively.

Assets and remittances marginally contributed to the odds of not being poor. Remittances have been part of Ghana's social safety net. However, these tend to provide short-term support. A long-term solution to poverty reduction lies in policies and programmes that would increase the earned incomes of people so as to minimise the dependence on uncertain remittances which cannot be used effectively in planning at the household level.

Availability of light and water which are community characteristics reduces the odds of being poor. Absence of these amenities contribute over 95 percent to the odds of an individual being poor. Thus, the provision of light and water can be seen clearly as a component of poverty reduction programmes. Electricity raises the options for diversification in rural areas and may increase the working hours of people and so may help reduce the incidence of poverty. The use of treated water reduces the incidence of many water-borne diseases and so raises the health status of people which in turn increases their incomes through productivity increases. Costs of health care are also reduced, both for the individual and the government.

TABLE 3.10: Parameter estimates for LOGIT models for poverty

Characteristics/ attributes	Poverty line[*]			
	GLSS 1	GLSS 2	panel yr 1[+]	panel yr 2[+]
Intercept	-4.4545	-2.3734	-3.4221	-3.4355
	(58.4***)	(9.9***)	(12.4***)	(7.6***)
Age in years	0.0091	0.0046	-0.0040	0.0036
	(4.1)	(0.8)	(0.3)	(0.2)
Body Mass Index	-0.0152	-0.1193	-0.0508	-0.0668
	(0.6)	(18.5***)	(2.1)	(3.2)
Main occupation	-0.1992	-0.1033	-0.2479	-0.4618
	(2.4)	(0.1)	(1.3)	(3.5*)
Education	-0.0065	-0.0767	-0.0395	-0.0290
	(19.7***)	(19.7***)	(2.6*)	(0.9)
Sex	-0.0306	0.1987	0.3541	0.5958
	(0.0)	(1.4)	(2.3)	(4.1**)
Migrant	0.3957	0.0199	0.1992	0.1312
	(10.7***)	(0.0)	(1.0)	(0.3)
Size of household	0.1774	0.1729	0.1996	0.2139
	(80.6***)	(61.3***)	(33.2***)	(28.8***)
Economic dependency	0.2097	0.1415	0.2437	0.0571
	(16.3***)	(6.4***)	(8.9***)	(0.3)
Land availability	-0.2418	-0.3683	-0.1758	-0.6019
	(30.4***)	(30.8***)	(5.9*)	(17.9***)
Remittances income	-0.0441	-0.0101	-0.0380	-0.0723
	(6.4***)	(0.8)	(2.5)	(3.9**)
Value of assets	-0.0088	-0.0079	-0.0052	-0.0080
	(16.9***)	(7.6***)	(3.2*)	(2.6*)
Electricity	1.3369	1.1286	1.3806	1.4082
	(25.3***)	(11.5***)	(12.6**)	(7.0***)
Treated water	0.6805	1.1036	0.4202	0.9210
	(11.8***)	(18.4***)	(1.9)	(5.9**)

Note: [*] The poverty line was set at half of the mean per capita expenditure per annum
[+] The panel represents 1200 households who were interviewed in both GLSS 1 and GLSS 2

Figures in parenthesis are χ^2 values
* Significant at $p \leq 0.1$
** Significant at $p \leq 0.05$
*** Significant at $p \leq 0.01$

TABLE 3.11. Percent of Households Correctly
Classified by Model

Classification rule d	Poverty line*			
	GLSS 1	GLSS 2	panel yr 1[+]	panel yr 2[+]
0.5	86.9	90.8	83.4	84.8
0.6	86.9	90.7	83.4	84.9
0.7	86.6	90.8	83.6	84.9
0.8	86.5	90.8	83.7	84.8
0.9	86.4	90.8	83.7	84.8
Percentage in largest group	81.2	82.9	76.0	73.6

Note: * The poverty line was set at half of the mean per capita
expenditure per annum

[+] The panel represents 1200 households who were interviewed in
both GLSS 1 and GLSS 2

TABLE 3.12: Percent Contribution of Characteristics/Attributes to Poverty*

Characteristics/ Attributes	Percent			Percent		
	GLSS1	GLSS2	Change	Panel yr1[+]	Panel yr2[+]	Change
Age	0.9	0.5	44	-0.4	0.4	198
Body Mass Index	-1.5	-11.2	-646	-5.0	-6.5	-31
Migrant	-98.3	-90.5	8	-96.0	-96.3	-0
Main occupation	-99.0	-91.6	8	-97.5	-97.9	-0
Education	-6.3	-7.4	-18	-3.9	-2.9	26
Sex	-98.9	-88.6	10	-95.3	-94.2	1
Size of household	19.4	18.9	-3	22.1	23.9	8
Economic dependency	23.3	15.2	-35	27.6	5.9	-79
Land availability	-21.5	-30.8	-43	-16.1	-45.2	-181
Remittances income	-4.3	-1.0	77	-3.7	-7.0	-87
Value of assets	-0.9	-0.8	10	-0.5	-0.8	-52
Electricity	-95.6	-71.2	26	-87.0	-86.8	-0
Treated water	-97.7	-71.9	26	-95.0	-91.9	-3

Note:

* The poverty line was set at half the mean per capita expenditure per annum
+ The panel represents 1200 households who were interviewed in both GLSS 1 and
  GLSS 2.

Availability of land to the household contributed 21.5 percent and 30.8 percent in 1987/88 and 1988/89 respectively to the odds of not being poor. Substantial changes occurred in the contribution of characteristics/attributes to poverty for the households in the panel from 1987/88 to 1988/89. Body mass index, land availability, remittances and total assets increased their contribution to the odds of being non-poor from GLSS 1 to GLSS 2 whereas the contribution by education was reduced. The contribution of the dependency ratio to the odds of being poor declined between GLSS 1 and GLSS 2 in the panel data.

SECTION 5

Summary and Conclusion

Irrespective of where the poverty line is fixed it is possible to classify households and find out the characteristics associated with the classes. The distinction between poverty groups improves as the poverty line is lowered considerably and only a few households can be mis-classified under it.

The impact of the availability of social amenities on poverty reduction has often been emphasised. In this chapter water and electricity contribute significantly to the odds of non-poverty. The provision of electricity has been an independent policy of the government, fostered by supply considerations: rural electrification programmes are a sequel to the investment in hydropower generation in the Volta River. Providing the population with electric power can be a major stimulus to rural diversification and growth, leading to poverty reduction. But the findings of the present study should not be construed to suggest putting all one's eggs into the basket of rural electrification. Rural development must be tackled in its entirety to be effective. Electricity and treated water are (important) conditions which give incentives to economic development in that broader context.

Land availability is found to be associated with lesser poverty. The livelihood of many Ghanaians depends upon farming. The major input used in Ghanaian agriculture is land and due to population pressure and reduced fertility of the soils as a result of intensive cultivation, it is increasingly difficult to obtain land for farming. One solution is increased intensification of land use, but this would only be feasible if the use of fertilizer or organic matter increases. Where land is in relatively abundance, communities and landowners must be encouraged to simplify and reduce the tenancy requirements to encourage landless people to gain access to land.

Migration status is found to contribute positively to poverty reduction. If the acquisition of land is made easier many more people might migrate to new areas to farm and to reduce the high incidence of poverty. Such a move would also reduce rural-urban migration which is a growing problem in many urban centres.

The contributing effect of large-sized households and high dependency ratios on poverty is demonstrated by this study. Public education on the adverse effects of large household size and the adoption of family planning practices should continue especially in the rural areas.

There was an increase of poverty from 1987/88 to 1988/89 as it was revealed by distribution of the expenditure data. The rural situation deteriorated more than that of the urban areas.

ACKNOWLEDGEMENT

The data for the research was based on the Ghana Living Standards Survey which was obtained with the kind permission of the Ghana Statistical Service.

REFERENCES

Boom, G.J.M. van den, M. Nubé and W.K. Asenso-Okyere, 1997. Nutrition, Labour Productivity and Labour Supply of Men and Women in Ghana, In: Asenso-Okyere *et al.* (eds.) Sustainable Food Security in West Africa. Kluwer Academic Publishers, Dordrecht.

Awusabo-Asare, K. 1981/82. Toward an Integrated Program of Rural development in Ghana. Ghana Journal of Sociology, Vol. XIV, 2.

Boateng, E.O., Ewusi, K., Kanbur, R., Mckay, R, 1990. A Poverty Profile for Ghana, 1987-1988. Social Dimensions of Adjustment in Sub-saharan Africa. Working Paper No. 5 World Bank, Washington, DC.

Brown, C.K., 1984. Social Structure and Poverty in Selected Rural Communities in Ghana. ISSER Technical Publication No. 54, University of Ghana, Legon.

Ewusi, K., 1976. Disparities in Levels of Regional Development in Ghana. Social Indicators, Vol. 3.

Ewusi, K., 1984. The Dimensions and Characteristics of Rural Poverty in Ghana. ISSER Technical Publication No. 43, University of Ghana, Legon.

Foster, James E., J. Greer and E. Thorbecke, 1984. A Class of Decomposable Poverty Measures, Econometrica 52: 761-766

Ghana Statistical Service, 1992. Quarterly Digest of Statistics, Vol. X. No.2., Accra.

Greene, W.H. 1991. Econometric analysis, MacMillan, New York.

Greer, J., and E. Thorbecke, 1986. A Methodology for Measuring Food Poverty Applied to Kenya, Journal of Development Economics 24: 59-74.

ISSER and SOW-VU, 1993. Background to Food Security in Ghana. Institute of Statistical, Social and Economic Research, University of Ghana, Legon, and Centre for World Food Studies (SOW-VU), Free University, Amsterdam, Netherlands.

Jazairy, Z.M. Alamgir, and Panuccio. 1992. The State of World Poverty - An Inquiry into its Causes and Consequences. International Fund for Agricultural Development (IFAD), Intermediate Technology Publications, London.

Maddala, G.S., 1983. Limited Dependent and Qualitative Variables in Econometrics, Cambridge University Press, Cambridge.

Nuamah, N.N.N.N., 1994. Statistical and Demographic Measures for Population Studies, RIPS, University of Ghana, Legon.

Paul, Satya, 1989. A Model for Constructing the Poverty Line, Journal of Development Economics 30: 129-144.

Seini, A. Wayo, V.K. Nyanteng and G.J.M. van den Boom, 1997. Income and Expenditure Profiles of Households in Ghana, ISSER, University of Ghana, In: Asenso-Okyere *et al.* (eds.) Sustainable Food Security in West Africa, Kluwer Academic Publishers, Dordrecht.

Ravallion, M., and Bidani, B., 1994. How Robust is a Poverty Profile? The World Bank Economic Review, Vol. 8, No. 1, pp. 75-102.

World Bank. 1992. Development and the Environment. World Bank Report. Oxford University Press, New York.

CHAPTER 4

INCOME AND EXPENDITURE PROFILES AND POVERTY IN GHANA

A. Wayo Seini, V.K. Nyanteng and *G.J.M. van den Boom*

515

I32

Introduction

The Ghana government, in 1983, launched a four-year Economic Recovery Programme to redress the economic malaise and foster growth through liberalisation. A second phase of economic reforms, the structural adjustment programme covered the period 1987 to 1989 and sought to tackle the economy's deep-rooted structural imbalances and to build a productive base for the economy.

To provide data on a regular basis for measurement of living standards of the population and the progress made in raising them, the government instituted the Ghana Living Standards Survey (GLSS) in 1987. This chapter uses data from two rounds of the GLSS, covering the period September 1987 to September 1989, to explore income and expenditure profiles of households in Ghana, recognising that such profiles are among the most important indicators of the standard of living.

This chapter draws attention to some specific groups of households in which poverty is concentrated through a comparison of income and expenditure profiles of selected income classes. Those whose expenditure levels are considered inadequate are identified by their distribution spatially and by key household characteristics.

We start our discussion with a summary on aggregate national income and expenditure estimates and then turn to the definition and classification of poverty. This is then followed by a discussion of income distribution within socio-economic groups as well as the socio-economic characteristics of income classes. A presentation of the composition of income and expenditure by income classes precedes the summary and conclusions from this chapter.

The identification of the main sources of income and types of expenditure of Ghanaian households follows conventional methods for household surveys and is in line with earlier work on the Ghana Living Standards Survey (Ghana Statistical Service, 1989; Johnson *et al.*, 1990; Boateng *et al.*, 1990; Coulombe *et al.*, 1993; Ghana Statistical Service, 1995).

Table 4.1 gives aggregate income and expenditure estimates for the two survey years 1987/88 and 1988/89.[1]Based on current prices, average per capita income is estimated to be 48,600 cedis per year in the first year and 49,600 in the second year of the survey, while per capita expenditure averaged 71,500 and 81,600 cedis per year in the respective years. After

---

[1] The representativeness of the GLSS sampling scheme is supported by the tables in Appendix 1 which give the spatial distribution of the sample as well as the distribution by selected household characteristics. Details on the household income and expenditure estimates can be found in Appendix 2, together with some explanatory notes on the way by which the income and expenditure figures have been derived from several sections of the questionnaire.

correction for increasing cost of living[2] the income figure converts into 41,700 and 35,300 cedis per capita per year, while per capita expenditure at constant September 1987 prices averaged 61,500 and 56,000 cedis in the two years, respectively.

TABLE 4.1: Income and Expenditure Estimates, 1987/88 and 1988/89
('000 Cedis Per Capita Per Year, Constant September 1987 Prices)

	Average Income	Average Expenditure	Income as % of Expenditure
GLSS year 1 (1987/88)	41.7	61.5	68%
GLSS year 2 (1988/89)	35.3	56.0	63%
Year 2 as % of year 1	85%	91%	

Source: Estimated from Ghana Living Standards Survey (GLSS)

The estimates indicate a considerable shortfall of household income as compared to household expenditure. On the average estimates of income fall short of expenditure estimates by as much as 32 and 37 percent in the respective years. With such a large shortfall of income as compared to expenditure it is difficult, if not impossible, to reconcile income and expenditure levels as estimated from the GLSS survey. Also the shortfall differs considerably among households. In fact, for some households estimated income actually exceeded expenditure, whilst for others estimated income was only a very small proportion of estimated expenditure.

To see to what extent variation of income is correlated with variation of expenditure, household expenditure has been regressed on household income, while controlling for time, for spatial aspects and finally for household characteristics. It appeared that almost half of the variation of the expenditure estimates may be explained by variation of income, although the income-elasticity of expenditure is estimated to be only 0.3, whereas one would expect it to be close to one. Also, the correlation between estimated income and expenditure levels is practically the same in the two years.

A second observation on Table 4.1 concerns the difference between the estimates for the two survey years, indicating an average decline of income of 15 percent and a decline of expenditure of 9 percent. Among other things, this might be explained by less favourable weather conditions in the 1988/89 season. Also Ghana's terms of trade deteriorated (due to decreasing world market prices for its main export products), while the benefits of economic recovery and structural adjustment programmes as yet had not matured (Younger, 1989).

Average per capita income and expenditure levels give a general idea about living standards in Ghana. For example note that the equivalence of 60,000 cedis per capita per year at September 1987 prices compares to about 400 US dollars at the prevailing overvalued

---

[2] Income and expenditure figures are deflated with the monthly consumer price index, as reported in Statistical News Letter, Ghana Statistical Service. As price variation in time seems much more pronounced than spatial price variation (see also Glewwe and Twum-Baah, 1991; appendix B) no spatial deflator has been applied.

exchange rate of 150 cedis for a dollar at that time. This level by itself may not suffice to sustain an adequate standard of living, even after allowance for a correction for purchasing power parity (at a parity factor of 2.4, the purchasing power of 60,000 cedis is still less than 1000 dollars, see UNDP, 1994). Besides, an average may be ambiguous if, like in Ghana, two-thirds of the people live in households with per capita expenditure below the average.

Not surprisingly, there is a wide variation of both income and expenditure levels in Ghanaian households. For per capita expenditure the standard deviation in both years is close to the average, so that the coefficient of variation is about one. Income estimates exhibit even larger variations and the standard deviation is almost 50 percent bigger than the average. In the two surveys, about two thirds of Ghanaians lived in households with per capita expenditure below the national average of 61,500 and 56,000 cedis per year, respectively. This is also reflected in the median, which equals 46,600 and 42,700 cedis in the two years, indicating that half of Ghanaians spent three-quarters or less of the national average. For slightly more than 25 percent it was the case that expenditure levels were less than half the national average, while expenditure levels of almost 5 percent of all Ghanaians were equal to or less than as little as one quarter of the national average.

The most frequently occurring expenditure level (or the mode of the distribution) was around 30,000 cedis per capita per year, which is only half the average expenditure level and gives further evidence that the expenditure distribution is skewed to the right. This inequality is also reflected in the Lorenz curves in Figure 4.1a and 4.1b.

The diagonal represents perfect equality in incomes or expenditures and the closer the curve is to the diagonal the more equality. In both surveys the lowest quintile of the population spent less than eight percent of total national expenditure, while the highest quintile spent about 45 percent.

The income distribution shows a similar picture, but the income estimates actually exhibit larger disparities, witness a Gini-coefficient of the expenditure distribution of less than 0.4 in both years as compared to a coefficient near 0.5 for the income data. In both years the lowest quintile of the population earned less than five percent of the national income, whereas the quintile with the highest incomes earned as much as 55 percent. This accentuates and reinforces the picture of a distribution of living standards around a low average and skewed to the right.[3] The implications of this for the extent of poverty in Ghana are discussed in the succeeding sections.

Definition and Classification of Poverty

Poverty alleviation is the ultimate objective of development policies. To define what is meant by poverty, a common starting point is a poverty line. Such a line may be defined as the budget necessary to acquire the goods and services necessary to achieve some minimum standard of living.[4]

---

[3] Note however that inequalities in terms of expenditure are less pronounced than inequalities in terms of income. This reflects a well-known phenomenon in Ghana that income increases inside the household have consumption effects both inside and outside the household (extended families etc.).

[4] Similar definitions of a poverty line include "... the cost of a bundle of goods *deemed to assure* that basic consumption needs are met" and "... the local cost of a *normative* food and non-food consumption bundle" (Lipton and Ravallion, 1995:2576; italics added). For Ghana it would seem that the minimum wage level used in the paper forms a reasonable base for estimating such a normative budget.

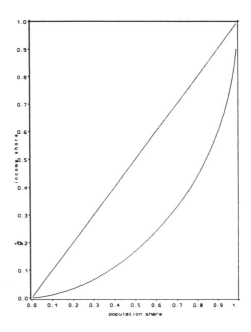

Income 1987/88, GINI=0.515

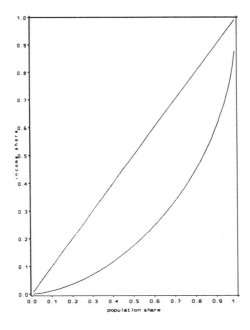

Income 1988/89, GINI=0.507

Figure 4.1a: Lorenz Curves of Income Distribution in Ghana, 1987/88 and 1988/89

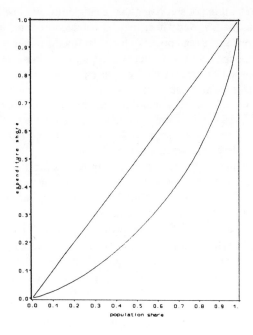

Expenditure 1987/88, GINI=0.395

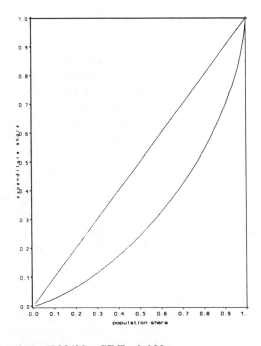

Expenditure 1988/89, GINI=0.389

Figure 4.1b: Lorenz Curves of Expenditure Distribution in Ghana, 1987/88 and 1988/89

59

From this definition it is not surprising that a single poverty line has not been adopted universally. In fact, as many poverty lines may be obtained from the definition, as there are different views on the quantities (and prices) which identify the minimum. As a result, poverty lines may differ both among countries and between regions within a country, depending on the specific context. Ghana is a case in this respect as a consensus on poverty lines seems nonexistent. Here, we take the minimum wage as a benchmark for poverty and consider persons whose consumption entirely depends upon this minimum to be at the edge of poverty. By implication and in the absence of individual consumption data in the GLSS, the living standard of those Ghanaians living in households with a per capita expenditure[5] smaller than the minimum wage is considered inadequate.

Although the minimum wage offered by government is proposed as the benchmark for poverty, it is recognised that the adequacy of a minimum wage income has been challenged by, for example, the Trade Unions Congress. Still, for a single individual without dependents the minimum wage at the time of the survey might be considered to be just sufficient to escape from poverty, i.e. to cover the expenses for his or her own basic needs. If however, the person has to share the minimum wage with others and his or her household does not have any other income, then income may no longer suffice to cover the basic needs of the household. Against that background, a household of two persons depending on one minimum wage is considered to be at the edge of extreme poverty and similarly a household of four persons who depend on an income of twice the minimum.

Before presenting these poverty lines and the implied classification of households, we mention a few other widely used poverty lines, usually based on cut-off points of per capita expenditure. For example, the World Bank employs a poverty line of one US dollar per capita per day at 1985 US purchasing power parity and this line was developed to enable international comparison of poverty (World Bank, 1993). In a low income country like Ghana, the first decile of the per capita expenditure at a certain point in time (the poorest 10 percent) might be another reasonable cut-off point for poverty, while the second quintile (the poorest 40 percent) is sometimes used to define more moderate poverty (Glewwe and Twum-Baah, 1991; Kakwani, 1993). Alternatively, one could choose some fixed fraction of mean per capita expenditure, for example, one-third and two-thirds as proposed by Boateng et al., 1990), to define poverty and 'hard core' poverty in Ghana, respectively. Still another way used to define the poverty line is to take some fixed fraction of the median per capita expenditure.

As any poverty line based on a cut-off point of per capita expenditure is somewhat arbitrary, it is impossible to come up with an unambiguous poverty line for Ghana. However, the GLSS expenditure estimates suggest that adoption of either of the above rules gives poverty lines at rather similar levels of per capita expenditure, which are also comparable to the minimum wage level. This is explicit in Table 4.2, suggesting poverty lines somewhere near 40,000 and 20,000 cedis per capita per year (September 1987 prices) to define respectively moderate poverty and extreme poverty. Thus, different cut-off points to define a poverty line for Ghana yield comparable results when tested with the GLSS expenditure data and are within ranges employed elsewhere (Sarris and Shams, 1991; World Bank, 1991; UNDP, 1994).

---

[5] It should be noted that *per capita* expenditure is only an unbiased estimate of individual consumption if an equal intra-household distribution is presumed and economies of scale in consumption are not significant. For poor households in Ghana such assumptions seem reasonable. For example, the scale assumption has been relaxed (using adult equivalence scales based on nutritional requirements), but appeared to have no significant effect on the characterisation of the poor. Differential impacts of poverty within households have not been assessed; however, despite some evidence of intra-household inequalities, the inadequacy of *average* consumption levels in a household will also be detrimental to those whose consumption is least affected.

TABLE 4.2: Comparison of Poverty Lines for Ghana, (September 1987
Per Capita Expenditure) ('000 Cedis)

	Moderate Poverty		Extreme Poverty	
	GLSS 1	GLSS 2	GLSS 1	GLSS 2
Minimum wage and half of minimum		43.8	21.9	
Amount in dollar equivalents				
(300/150 US dollars)		45.0	22.5	
Position w.r.t. average				
(two thirds/one third of average)	41.0	37.7	20.5	18.9
Position w.r.t. middle				
(median / half of median)	46.6	42.6	23.3	21.3
Percentiles (second quintile, first decile)	39.8	36.2	20.1	18.7

Notes: Minimum daily wage in the survey period was 146 cedis; assuming 300 working days
gives the figure of 43,800 cedis per year.

— The amount in dollar equivalents is based on World Bank (1993); using Ghana's Purchasing
Power Parity factor of 2.4, the cut-off point of 150 US dollars is equivalent to 360 US PPP
dollars per capita per year, i.e. 1 dollar per capita per day for extreme poverty.

Source: Estimated from Ghana Living Standards Survey (GLSS).

In what follows, we shall employ the minimum wage and one-half of that minimum to
define moderate and extreme poverty, respectively, i.e. cut-off points of 43,800 and 21,900
cedis per capita per year. These poverty lines shall be used to locate the poor and the non-poor
both spatially and in groups of households with certain characteristics. Also an assessment shall
be given of some significant differences between income and expenditure profiles of poor and
non-poor Ghanaian households.

With respect to the measurement of poverty at a given poverty line, the head-count index
is taken, that is, the number of poor persons as a percentage of the total number of persons in
the sub-sample. The head-count provides a straightforward measure of the incidence of poverty
and its spread in the country. Yet it is recognised that the head-count is not measuring the depth
of poverty, i.e. it provides no information on the disparities in consumption amongst the poor.
Other measures take account of this by considering the poverty gap or its square (the P_α-
measures with $\alpha=1$ and $\alpha=2$, respectively) or by inclusion of the poverty gap and the Gini-
coefficient of the poor (the Sen-measure), see for example Sen, 1976 and Foster *et al.*, 1984[6].
However, for purposes of presentation and because the shape of the distribution of per capita
income and expenditure of the poor is not very sensitive in time, nor in space or over the

---

[6] Most poverty measures built on axioms delineated in Sen (1976). The axioms state that a poverty measure should
only depend on the incomes of the poor (focus axiom; hence the head-count), that it increases as the incomes of the
poor decrease (monotonicity axiom; hence the poverty gap which is defined as the money needed to bring
consumption levels of the poor to the poverty line level) and as inequality amongst the poor increases (weak transfer
axiom; hence the Gini-coefficient of the poor in the Sen-measure and the poverty gaps squared in the P_2- measure).
For an overview, see Blackwood and Lynch (1994) and Ravallion and Lipton (1995) for example.

Table 4.3: Income Classes in Ghana, 1987/1988 and 1988/1989
(Shares in Total Population)

	Extreme poor	Moderate poor	Non-poor	Rich	ALL
1987/1988	12.7	33.0	47.6	6.7	100.0
1988/1989	14.6	37.2	42.6	5.6	100.0

Note: Cut-off points relative to the minimum wage of 43,800 cedis per year:
per capita expenditures of half, once and thrice the minimum wage are
used to demarcate the four income classes.

Source: Estimated from Ghana Living Standards Survey (GLSS)

socio-economic groups (see Appendix 3), the discussion is confined to the head-count index[7].

In order to add an extra dimension to the comparison of households considered poor with those considered non-poor, the latter are also classified into two income classes. The group of households with per capita expenditures between once and thrice the minimum wage shall be referred to as the non-poor, while the highest income bracket with per capita expenditure above thrice the minimum wage shall be referred to as the rich. Thus, households classified as rich can afford an expenditure of 131,400 cedis per capita per year, which is more than twice the average per capita expenditure in the country and represents the purchasing power of about 2000 US dollars comparable to average per capita income in a middle income country like Brazil (World Bank, 1991). Table 4.3 summarises the occurrence of the four income classes in the country.

Living standards of some 10 to 15 percent of Ghanaians are way below adequate levels (the extreme poor with per capita expenditure below half the minimum wage), while adequacy of living standards of another 35 percent of the population may be questioned (the moderate poor with per capita expenditure between the minimum wage and half the minimum wage). At the upper tail of the distribution, consumption of only about 6 percent are at or above average levels in middle income countries (the rich with per capita expenditure above thrice the minimum wage).

In 1987/1988 about 46 percent of Ghanaians were estimated to be poor of which about 13 percent can be classified as extremely poor while the remaining 33 percent were designated as moderately poor. The level of poverty, however, increased at all levels in the 1988/1989 survey to 37 percent moderately poor and almost 15 percent extremely poor.

Poverty Distribution Within Socio-Economic Groups

From the preceding sections it appears that there is a considerable variation in the levels and composition of per capita income and expenditure across the country and these disparities reflect themselves when the national picture of Table 4.3 is dis-aggregated. This is done in Table 4.4, which shows among other things where and in which types of households poverty is

---

[7] Other papers also find that poverty comparisons on the basis of head-counts are reasonably robust (although not necessarily); for example, Quisumbing et al., 1995 compare poverty with three P_α measures (head-count, $\alpha=0$; poverty gap, $\alpha=1$; poverty gap squared, $\alpha=2$) and find only minor differences.

relatively severest. The table provides information on the depth of poverty in selected parts of the population as well as on the effects of the general decrease of expenditure levels between the two survey years.

Table 4.4 also shows the relatively high living standards in Greater Accra and Western region, where extreme poverty is rare, although moderate poverty at or below the minimum wage level (but above half that minimum) still applies to some 15 percent of the population of Greater Accra in both years and to 25 and 40 percent of the people in Western region in the respective years. On the other hand, poverty is most severe in the three northern regions and in the Volta region, where 25 to 35 percent of the population can be considered extreme poor. It is further noteworthy that despite decreasing average per capita expenditure levels, the extent of extreme poverty decreased for most regions in the 1988/89 sample. However, it increased substantially for the Eastern region and worsened in the Northern region. The intensification of poverty in these regions can be ascribed to the decline on major cash crops produced there, cocoa in the Eastern region and rice in the Northern region.

For the agro-ecological zones the population shares of extreme poor is about half the national average in the coastal zone, and twice the national average in the savannah zone, while the income distribution of the forest zone is close to the national average, with some 11 to 14 percent extremely poor and another 35 to 40 percent moderately poor. The distribution of the poor by agro-ecological zone and locality further confirms that there was a general increase in the incidence of poverty between the two survey periods. However, it is noteworthy that the proportion of the urban extreme poor decreased between the two survey years, whereas extreme poverty increased appreciably in rural localities.

The level of poverty also differs according to socio-economic characteristics. For example, poverty is highly correlated with household size. With about 20 percent prevalence, extreme poverty occurs twice as frequently in large households (more than 6 members) than in households of average size (4 to 6 members), while the spread of extreme poverty among small households (1 to 3 members) is only limited to 3 percent of that of population. Table 4.4 also corroborates the general observation that poverty is directly correlated to the level of education (Ewusi, 1984). The proportion of both extreme and the moderate poor reduces steadily as the level of education increases. Of those living in households headed by persons without education 20 percent are classified in the lowest income class, while another 40 percent of them may be considered moderately poor. Comparison with households headed by someone who received middle school education shows that extreme poverty in the latter group decreases by about 60 percent, whereas moderate poverty reduces by some 30 percent. It is significant to note that there is no extreme poverty in households headed by those educated beyond middle school.

Other household characteristics of interest are gender and migration. In gender terms, there is little difference in the prevalence of extreme poverty among male and female headed households, although female headed households are somewhat overrepresented in moderate poverty.[8] Table 4.4 also shows that a household headed by an indigenous person suffers more from extreme poverty than one headed by a migrant, although moderate poverty is comparable between the two groups.

---

[8] Employing the first round income data and a 33 percentile poverty line, Quisumbing et al. (1995) conclude that female headed households in Ghana are significantly poorer (viz. 38 percent of those in female headed households are below the poverty line, as compared to only 31 percent of those residing in male headed households). Although for *moderate* levels of poverty this may be true, it appears that *extreme* poverty is equally severe among male and female headed households.

TABLE 4.4: Poverty Distribution Across Ghana, 1987/1988 and 1988/1989
(Contribution to Total Population or Region or Socio-Economic Group)

SPATIAL	Extreme Poor		Moderate Poor		Non-poor		Rich		ALL
Region									
Ashanti	10.5	(12.9)	44.3	(37.4)	40.8	(43.6)	4.4	(6.1)	100
Brong-Ahafo	7.9	(6.5)	38.3	(38.1)	51.0	(50.8)	2.8	(4.6)	100
Central	18.0	(12.6)	34.8	(36.5)	43.2	(46.5)	3.9	(4.5)	100
Eastern	8.4	(18.6)	32.7	(44.4)	53.5	(34.7)	5.4	(2.3)	100
Greater Accra	0.4	(2.7)	15.0	(16.3)	64.7	(64.3)	19.9	(16.7)	100
Northern	24.8	(35.9)	33.0	(38.8)	38.2	(23.7)	4.0	(1.7)	100
Upper East	31.8	(24.8)	37.1	(36.9)	29.7	(35.4)	1.4	(2.9)	100
Upper West	31.8	(34.9)	33.5	(41.3)	29.7	(23.6)	5.0	(0.2)	100
Volta	23.6	(19.5)	37.5	(49.6)	35.3	(28.8)	3.6	(2.1)	100
Western	2.9	(5.1)	26.2	(38.5)	61.7	(49.5)	9.2	(6.9)	100
Zone									
Coastal	6.9	(7.7)	23.9	(29.2)	57.1	(53.0)	12.1	(10.2)	100
Forest	11.2	(14.1)	38.2	(40.4)	46.2	(41.2)	4.4	(4.3)	100
Savannah	22.8	(24.1)	35.9	(42.0)	37.7	(31.9)	3.6	(2.1)	100
Locality									
Rural	16.0	(18.9)	37.5	(41.6)	42.0	(36.5)	4.5	(3.1)	100
Urban	6.4	(5.7)	24.4	(27.8)	58.3	(55.4)	10.9	(11.1)	100
Ghana	12.7	(14.7)	33.1	(36.8)	47.5	(42.8)	6.7	(5.7)	100

HOUSEHOLD CHARACTERISTICS	Extreme Poor		Moderate Poor		Non-Poor		Rich		All
Size of household									
Small (1-3)	3.0	(4.3)	15.2	(17.6)	56.5	(58.0)	25.4	(20.1)	100
Average (4-6)	9.3	(10.4)	29.9	(36.7)	55.6	(50.3)	5.2	(2.6)	100
Large (over 6)	18.3	(22.2)	40.7	(44.8)	38.6	(30.1)	2.4	(2.9)	100
Gender of head									
Male	13.2	(14.2)	31.1	(36.5)	48.8	(43.1)	6.9	(6.1)	100
Female	11.3	(15.9)	38.6	(39.0)	44.0	(40.9)	6.0	(4.2)	100
Years of education									
None	17.8	(21.9)	38.6	(43.1)	39.7	(32.5)	3.9	(2.5)	100
Primary (1-6)	11.7	(11.7)	34.9	(37.7)	48.0	(45.4)	5.3	(5.2)	100
Middle (7-10)	7.5	(8.5)	26.2	(32.4)	56.9	(51.4)	9.3	(7.7)	100
Secondary (11-17)	0.4	(3.5)	24.4	(23.5)	61.3	(56.6)	13.9	(16.4)	100
Tertiary (18-25)	0.0	(0.0)	13.1	(17.2)	62.4	(68.9)	24.5	(13.9)	100
Migratory status									
Indigenous	15.8	(16.7)	33.8	(38.1)	44.8	(40.8)	5.6	(4.4)	100
From elsewhere	10.2	(12.7)	32.5	(36.2)	49.7	(44.3)	7.6	(6.8)	100
Sector of work									
Agriculture	17.0	(20.1)	38.2	(42.7)	40.6	(34.6)	4.2	(2.6)	100
Industry	4.2	(8.5)	26.0	(28.8)	58.3	(53.6)	11.5	(9.2)	100
Services	5.5	(5.2)	23.2	(27.1)	60.2	(56.5)	11.1	(11.2)	100
Ghana	12.7	(14.7)	33.1	(36.8)	47.5	(42.8)	6.7	(5.7)	100

Note: 1988/1989 figures in parenthesis
Source: Estimated from Ghana Living Standards Survey

The above picture is relative to the population in the groups, some of which are small as compared to others. Thus, in order to provide further insight into the types of households contributing most to national poverty, the size of the diverse population groups has to be taken into account and to this we now turn.

Socio-Economic Characteristics of Income Classes

Table 4.5 gives the spatial and household characteristics of the four income classes. Between the two survey periods, the contribution to national poverty increased for one-half of the regions while it decreased for the other one-half. The table shows that the Ashanti, Volta, Central, Eastern, Brong Ahafo and the three northern regions harbour most of the poor in Ghana, each of these regions accounting for 10 to 20 percent of the extreme poor. The distribution of the poor over the agro-ecological zones shows that most of the extreme poor live in the forest and savannah zones, harbouring about 32 and 44 percent of all extreme poor, respectively. The two zones exchange positions with respect to moderate poverty, with the forest zone having the highest share of moderate poor, viz. 45 to 50 percent, as compared to less than 30 percent of them living in the savannah zone. With the inclusion of the relatively wealthier regions, it is not surprising that the coastal zone harbours the smallest proportion of the extreme poor (17 percent) and highest proportion of the rich (some 60 percent).

Table 4.5 further confirms the widely supported view that poverty in Ghana is mainly a rural phenomenon (Ewusi, 1976; Boateng *et al.*, 1990). About two thirds of Ghanaians live in rural localities and account for three-quarters of the moderate poor as opposed to one-quarter of them living in urban areas. Looking at extreme poverty it appears that rural poverty goes even deeper as 83 to 88 percent of the extreme poor of Ghana lived in the rural areas in the respective years.

Extreme poverty is concentrated in large households and households headed by persons with no formal education contain about two-thirds of the extreme poor and more than half of the moderate poor, even though the proportions are slightly smaller in the second year.

Comparison of the poverty picture between the two survey years suggests that moderate poverty increased over time but remained reasonably stable among zones and localities and socio-economic groups. However, with respect to extreme poverty the estimates suggest some redistributions. Extreme poverty increased in the country as a whole, but improved slightly in urban areas as compared to a considerable worsening of rural poverty. The distribution of extreme poverty over groups of households with different socio-economic characteristics has remained largely the same, although in the second year more of the extreme poor lived in female headed households and industrial households.

The general worsening of the poverty situation in Ghana seems to reflect a lack of opportunities for increasing agricultural income, which is the major source of income for the poor, in the late eighties. However, the 13 percent increase of the incidence of poverty between 1987/88 and 1988/89 is to be contrasted with a 15 percent reduction between 1987/88 and 1991/92 (World Bank, 1995; Ghana Statistical Service, 1995); apparently, there has been a drastic reduction in the level of aggregate poverty between 1988/89 and 1991/92 in the magnitude of some 25 percent.

TABLE 4.5: Characterisation of Income Classes in Ghana, 1987/1988 and 1988/1989
(Percentage Contribution to Total Population of Income Class)

SPATIAL	Extreme Poor		Moderate Poor		Non-poor		Rich		ALL	
Region										
Ashanti	13.2	(12.8)	21.4	(14.6)	13.7	(14.9)	10.5	(15.9)	16.0	(14.5)
Brong Ahafo	6.5	(4.4)	12.0	(10.2)	11.1	(11.8)	4.4	(8.2)	10.4	(9.9)
Central	12.9	(9.1)	9.6	(10.4)	8.2	(11.5)	5.4	(8.4)	9.1	(10.6)
Eastern	9.3	(17.3)	14.0	(16.2)	15.9	(11.0)	11.4	(5.5)	14.1	(13.6)
Greater Accra	0.4	(2.5)	6.0	(5.9)	17.9	(20.3)	39.2	(39.8)	13.2	(13.4)
Northern	14.0	(19.2)	7.2	(8.2)	5.8	(4.4)	4.3	(2.3)	7.2	(7.8)
Upper East	10.6	(10.0)	4.8	(5.9)	2.6	(4.9)	0.9	(3.0)	4.2	(5.9)
Upper West	11.3	(6.2)	4.6	(2.9)	2.8	(1.4)	3.4	(0.1)	4.5	(2.6)
Volta	19.4	(15.0)	11.9	(15.1)	7.8	(7.6)	5.7	(4.1)	10.5	(11.3)
Western	2.5	(3.6)	8.6	(10.7)	14.1	(12.1)	14.9	(12.7)	10.8	(10.4)
Zone										
Coastal	17.5	(17.2)	23.4	(25.7)	38.7	(40.7)	58.4	(59.1)	32.2	(32.8)
Forest	37.2	(39.5)	49.1	(44.6)	41.2	(39.6)	28.0	(31.2)	42.4	(41.0)
Savannah	45.4	(43.2)	27.6	(29.7)	20.1	(19.7)	13.6	(9.7)	25.3	(26.3)
Locality										
Rural	82.9	(87.5)	74.7	(75.9)	58.1	(58.1)	44.1	(36.9)	65.8	(67.8)
Urban	17.1	(12.5)	25.3	(24.1)	41.9	(41.9)	55.9	(63.1)	34.2	(32.2)
Ghana	100		100		100		100		100	

HOUSEHOLD CHARACTERISTICS	Extreme poor		Moderate poor		Non-poor		Rich		ALL	
Size of household										
Small (1-3)	3.3	(4.9)	6.4	(7.8)	16.6	(22.4)	53.3	(58.5)	14.0	(16.4)
Average (4-6)	27.9	(27.6)	34.5	(38.5)	44.5	(46.1)	29.6	(18.3)	38.1	(39.0)
Large (over 6)	68.8	(67.5)	59.1	(53.7)	38.9	(31.5)	17.1	(23.1)	47.9	(44.5)
Gender										
Male	77.3	(72.3)	70.3	(73.3)	76.5	(75.5)	77.0	(81.2)	74.6	(74.5)
Female	22.7	(27.7)	29.7	(26.7)	23.5	(24.5)	23.0	(18.8)	25.4	(25.5)
Years of education										
None	72.5	(70.0)	59.8	(54.2)	42.6	(35.6)	29.9	(20.0)	51.2	(46.6)
Primary (1-6)	8.4	(8.4)	9.5	(10.7)	9.1	(11.2)	7.1	(9.7)	9.0	(10.6)
Middle (7-10)	18.9	(19.9)	25.0	(29.8)	37.6	(41.2)	43.5	(46.0)	31.5	(34.1)
Secondary (11-17)	0.2	(1.7)	5.0	(4.6)	8.7	(9.6)	13.9	(20.7)	6.8	(7.2)
Tertiary (18-25)	0.0	(0.0)	0.6	(0.7)	2.0	(2.5)	5.6	(3.7)	1.5	(1.5)
Migratory status										
Indigenous	55.8	(55.4)	45.9	(49.8)	42.4	(46.4)	37.7	(37.7)	44.9	(48.5)
From elsewhere	44.2	(44.6)	54.1	(50.2)	57.6	(53.6)	62.3	(62.3)	55.1	(51.5)
Sector of work										
Agriculture	85.3	(83.6)	73.7	(70.8)	54.5	(49.4)	39.6	(28.1)	63.8	(61.1)
Industry	3.4	(6.8)	8.2	(9.2)	12.8	(14.8)	17.9	(18.8)	10.4	(11.8)
Services	11.3	(9.6)	18.1	(20.0)	32.6	(35.8)	42.6	(53.1)	25.8	(27.1)
Ghana	100		100		100		100		100	

Note: 1988/1989 figures in parenthesis; figures in column 'ALL' represent population shares
Source: Estimated from Ghana Living Standards Survey.

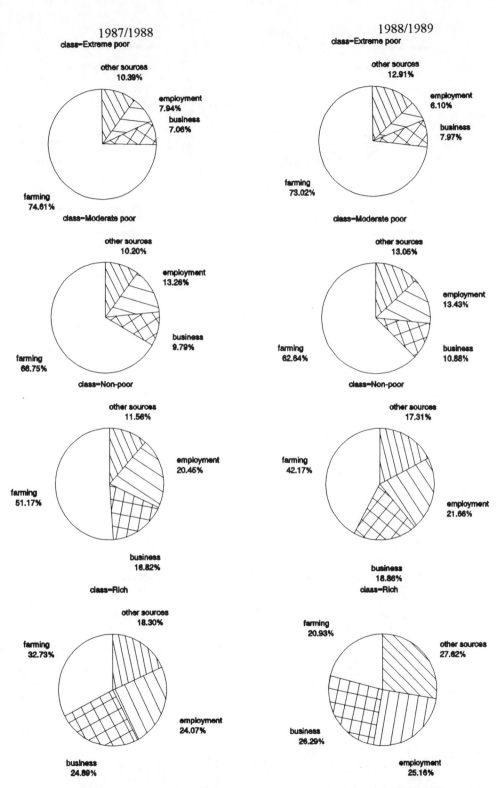

Figure 4.2: Sources of Income by Income Class, 1987/1988 and 1988/1989

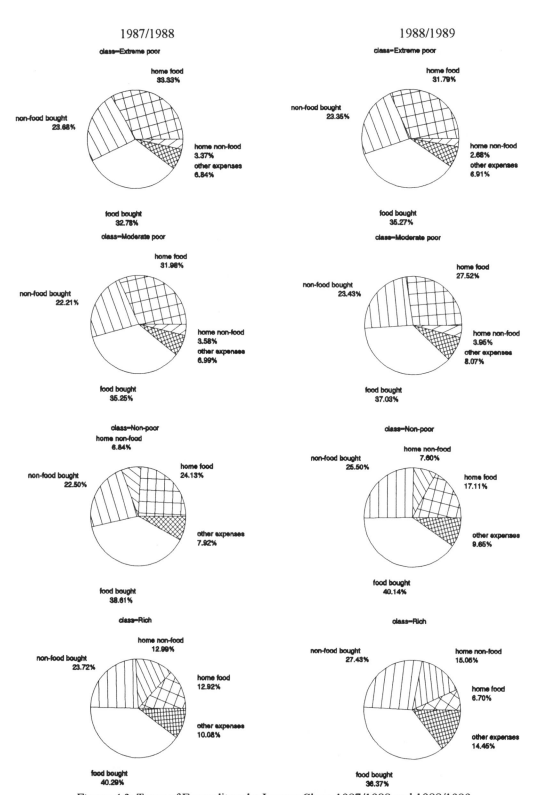

Figure 4.3: Types of Expenditure by Income Class, 1987/1988 and 1988/1989

Composition of Income and Expenditure by Income Classes

Figures 4.2 and 4.3 compare the composition of income and expenditure among income classes. With a share of almost three quarters, the importance of agricultural self-employment income is even more apparent for extreme poor households than on average in Ghana (with a share of 50 to 55 percent). Poverty in Ghana is not only rural but it is also largely agricultural in phenomenon. More than 60 percent of Ghanaians live in agricultural households contributing as much as 85 percent of extreme poverty in the country. A majority of 70 to 75 percent of the moderate poor is also mainly dependent on agriculture. This share decreases steadily over the income classes, the rich however still depending on farm income for more than 20 percent. It is further significant to note that the dependency on agriculture decreased between the two years for moderate poor and non-poor households, but the extreme poor remained dependent on agriculture. About 10 percent of the extreme poor depend on the services sector while only about 3.5 to 7 percent are to be found in industry. Employment income and business income are much less important for the poor and contribute only slightly more than half of the share they have on average.

The picture on the expenditure side remains rather stable, with food covering close to two-thirds of total expenditure in all income classes except the rich, which spend 40 to 50 percent of their income on food. However, the share of home produced food in total food is substantially higher for the poor: the poor get about half of their food from home production as compared to less than one third for the non-poor.

Summary and Conclusion

On the basis of two rounds of the Ghana Living Standards Survey, covering a two-year period from September 1987 onwards, the chapter presents and discusses income and expenditure profiles of households in Ghana and uses the minimum wage as a benchmark for poverty. The profiles reflect low levels and wide disparities of living standards across Ghana, both spatially and over time and according to socio-economic characteristics, and point to problems of pervasive poverty and food insecurity.

About two-thirds of the population live in households with per capita expenditure below the mean of 61,500 and 56,500 cedis in the respective years, which is the purchasing power parity of slightly less than 1,000 US dollar. Furthermore, close to half of the population had per capita expenditure below the minimum wage of 43,800 cedis prevailing at the time of the survey, while about one out of seven had less than one-half of that minimum to spend. The latter are demarcated as extreme poor and have to survive from the purchasing power of less than one US dollar per capita per day. At the other end of the income distribution some 6 to 7 percent of the Ghanaians could afford to spend thrice the minimum wage or more and those have been classified as rich.

The three northern regions and the Volta region had average per capita income and expenditure levels below three-quarters of the national average and, although harbouring only one-quarter of the population, these four regions accounted for one-half of the extreme poor. On the contrary, expenditure levels of the 13 percent of the population living in Greater Accra exceeded the national average by more than 50 percent; extreme poverty is practically absent, but about 40 percent of the rich live there. In general, income and expenditure levels are found to be markedly below the average in the savannah zones and rural localities of the country and in relatively large agricultural households, whose members are hardly educated and produce a major part of their own food.

The lowering of 9 percent of average per capita expenditure suggests a decline of living

69

standards during the survey period, which is also confirmed by a lowering of 15 percent of income estimates. The decrease manifested itself in increased rural poverty, whereas urban poverty declined somewhat. However, note that the level of aggregate poverty has decreased considerably between 1988/89 and 1991/92.

With an average share close to 60 percent in total expenditure and 55 percent in total income, food and agriculture assume primary importance for the economy of Ghana and this importance amplifies in poor population groups. In both years, incomes of the extreme poor depended for three-quarters on self-employment in agriculture, while food took almost two-thirds of total expenditure, half of which originated from home production. This dependency on agriculture and home produced food decreases steadily over the income classes and is also less persistent over time at higher income and expenditure levels. The latter suggests that, for reasons of food security, the extreme poor are more reluctant to engage in more remunerative crops and more remunerative employment.

Using the minimum wage as a benchmark for poverty, poverty in Ghana remains largely a rural phenomenon, rural localities accounting for about 85 percent of the extreme poor with per capita expenditure below half the minimum wage. The savannah zone stands out as the poorest, harbouring close to 45 percent of the extreme poor, corresponding to almost one quarter of its population (as compared to one seventh of the total population). From the profiles by household characteristic it appears that poverty is strongly correlated with household size, with a lack of education and with a dependency on agriculture and home produced food. The relationship with female headship is less pronounced, despite women's apparent arrears in education. Still, female headed households are somewhat overrepresented in moderate poverty, not however in extreme poverty.

The demarcation of poverty stricken groups may help to identify policies which can promote their engagement in more remunerative (self-) employment. Poverty alleviation policies in Ghana should enable large rural agricultural households to move towards higher and more stable incomes and away from the production of their own food. Although a quantitative evaluation of certain policies to achieve the growth of productivity inside agriculture and the growth of employment outside agriculture needed for that purpose has not been pursued, our observations suggest that sector policies alone cannot establish this growth. In addition and complementary to sectoral programmes, it seems that educational reforms that speed up literacy rates and middle school enrolment rates, along with programmes that turn women's time into better account and encourage family planning are among the most effective policies to improve living standards of the poor.

ACKNOWLEDGEMENT

The data for the research was based on the Ghana Living Standards Survey which was obtained with the kind permission of the Ghana Statistical Service.

REFERENCES

Blackwood, D. L. and R. G. Lynch, 1994. The Measurement of Inequality and Poverty: a Policy Maker's Guide to the Literature, *World Development* 22, 567-78.

Boateng E., K. Ewusi, R. Kanbur and A. Mckay, 1990. A Poverty Profile for Ghana, 1987-88. Social Dimensions of Adjustment in Sub-saharan Africa. Working Paper No. 5, World Bank, Washington.

Coulombe H., A. McKay and J. Round, 1993. The Estimation of Household Incomes and Expenditures from the First Two Rounds of the Ghana Living Standards Surveys, 1987/1988 and 1988/1989. Ghana Statistical Service, Accra.

Ewusi, K., 1984. The Dimensions and Characteristics of Rural Poverty in Ghana. ISSER Technical Publication No. 43. University of Ghana, Legon.

Foster, J.E., J. Greer and E. Thorbecke, 1984. "A Class of Decomposable Poverty Measures". *Econometrica* 52, 761-66.

Glewwe P. and K. Twum-Baah, 1991. The Distribution of Welfare in Ghana, 1987-88. Working Paper 75, Living Standards Measurements Study, World Bank, Washington.

Ghana Statistical Service, 1987. *Ghana Population Census 1984.* Accra.

Ghana Statistical Service, 1989. *Ghana Living Standards Survey: First Year Report,* Sept. 1987 - Sept. 1988. Accra.

Ghana Statistical Service, 1993. Sample Design for the first two rounds of the Ghana Living Standards Survey. Accra.

Ghana Statistical Service, 1995. *Ghana Living Standards Survey: Report on the Third Round ,* Sept. 1991 - Sept. 1992. Accra.

Kakwani N., 1993. Measuring poverty: Definitions and significance tests with application to Cote d'Ivoire. In: M. Lipton and J. van der Gaag (eds.), *Including the Poor.* Proceedings of a symposium organised by the World Bank and the International Food Policy Institute.

Lipton, M. and M. Ravallion, 1995. Poverty and Policy. In: J.R. Behrman and T.N. Srinivasan (eds.), *Handbook of Development Economics.* Volume 3B, Chapter 41, Amsterdam: North-Holland.

Quisumbing A.R., Haddad, L. and C. Peña, 1995. Gender and Poverty: New Evidence from 10 Developing Countries. Discussion Paper 9, Food Consumption and Nutrition Division, International Food Policy Research Institute, Washington DC.

Sarris, A. and H. Shams, 1991. *Ghana under Structural Adjustment: The Impact on Agriculture and the Rural Poor.* New York University Press for the International Fund for Agricultural Development.

Sen, A.K., 1976. "Poverty: An Ordinal Approach to Measurement". *Econometrica* 44, 219-31.

UNDP, 1994. *Human Development Report 1994*. UNDP/Oxford University Press, New York.

World Bank, 1989. Ghana: Structural Adjustment and Growth. Report No. 7515-GH.

World Bank, 1991. Ghana: Progress on Adjustment. Report No. 9475-GH.

World Bank, 1995. Ghana: Growth, Private Sector, and Poverty Reduction. A Country Memorandum. Report No. 14111-GH, World Bank, Washington, D.C..

Younger, S.D., 1989. Ghana: Economic Recovery Programme - A Case Study of Stabilisation and Structural Adjustment in Sub-saharan Africa. Analytical Case Studies No. 1, Development Policy Case Series, Successful Development in Africa, EDI, World Bank, Washington, D.C.

APPENDIX 1: Sample Characteristics of the GLSS

Table 1A gives the distribution of the GLSS sample for the year 1987/88 and compares the population shares derived from the sample with those of the 1984 census. The figures support the representativeness of the sampling scheme and similar figures apply for the second year.

TABLE 1A: Distribution of the GLSS Sample by Space
(1987/88, 3172 Households)

	Share households	Average household size	Population share	
Region				
Ashanti	16.7	4.7	16.0	(17.0)
Brong-Ahafo	9.1	5.6	10.4	(9.8)
Central	10.4	4.3	9.1	(9.3)
Eastern	14.1	5.0	14.2	(13.7)
Greater Accra	15.6	4.2	13.2	(11.6)
Northern	5.5	6.4	7.2	(9.5)
Upper East	4.0	5.2	4.2	(6.3)
Upper West	3.5	6.3	4.5	(3.6)
Volta	10.0	5.2	10.5	(9.9)
Western	11.1	4.8	10.8	(9.4)
Zone				
Coastal	36.1	4.4	32.2	
Forest	42.8	4.9	42.4	
Savannah	21.1	5.9	25.4	
Locality				
Rural	62.6	5.2	65.8	(68.0)
Urban	37.4	4.5	34.2	(32.0)
GHANA	100.0	4.9	100.0	

Note: Shares in 1984 population census in parentheses (Ghana Statistical Service, 1987).

TABLE 1B: Distribution of the GLSS Sample by Household
Characteristics, 1987/88 (3172 Households)

	Share households	Average household size	Population share	
Size				
Small (1-3)	35.5	1.9	13.9	(16.2)
Average (4-6)	38.9	4.9	38.6	(30.5)
Large (over 6)	25.6	9.1	47.4	(53.2)
Gender				
Male	71.3	5.2	74.6	(71.2)
Female	28.7	4.4	25.4	(28.8)
Education				
None	46.2	5.4	50.6	
Primary (1-6)	9.6	4.7	9.1	
Middle (7-10)	35.2	4.5	31.9	
Secondary (11-17)	7.5	4.5	6.8	
Tertiary (18-25)	1.5	4.8	1.5	
Migratory status				
Indigenous	45.7	4.8	44.5	
From elsewhere	54.3	5.0	55.5	
Type of work				
Agriculture	60.2	5.2	63.9	(61.1)
Industry	11.3	4.6	10.6	(12.8)
Services	28.6	4.4	25.5	(26.1)
GHANA	100.0	4.9	100.0	

Note: Shares in 1984 population census in parentheses (Ghana
Statistical Service, 1987).

APPENDIX 2: DETAILS ON THE ESTIMATION OF INCOME
AND EXPENDITURE

Main Sources of Income

Employment income

Hours worked and payments received by individual household members are reported, both by type of payment (normal, bonuses, in-kind) and by job (main, secondary). Also, information on the type of work (agriculture, trade, other types) and on the relation of the employer to the household (family worker, employee) shall be considered. The latter distinguishes between employment provided through an enterprise owned by the household and other enterprises. Employment income of all household members is added to obtain household employment income.

Farm income (agricultural self-employment income)

Monetary revenue is composed of sales of 32 crops, 14 processed crop products and 6 animal products. With respect to crop production, there are 7 types of crop-specific expenses (seed, fertilizer, manure, insecticides and herbicides, transport, sacks and containers, storage), while in addition crop-specific expenses are reported regarding 12 types of labour (male, female, child; clearing, planting, harvesting, other) and 6 types of capital (animal rental, equipment rental, maintenance, irrigation, fuel oil, other). For processed crop products total inputs expenses by product are reported, while the 8 types of expenses for raising livestock (herding labour, pen maintenance, feed, veterinary products, transport, commissions, packaging, etc.) are accounted for as costs for the production of the animal products. Subtracting expenses from sales, we arrive at gross farm income as the first part of agricultural income.

The second main part of farm income is imputed as the value of the consumption of home produced food. This part is derived from the expenditure side of the GLSS.

Finally, an allowance can be made for depreciation of farm assets. Following Coulombe *et al.* (1993) we assume a geometric depreciation and estimate the use value of farm assets as a share $\delta / (1 - \delta)$ of the current value (where δ is the depreciation rate, taken to be 0.10).

Total gross agricultural self-employment income is estimated as the sum of the gross farm income (sales less expenses) and the value of home produced food consumption. Net agricultural income equals gross agricultural income less depreciation. In several cases the estimated agricultural income thus obtained is negative and as this is considered to be unrealistic. The value of home produced food consumption is taken as the lower bound for agricultural self-employment income.

Business income (non-agriculture self-employment income)

Sales and expenses are reported for up to three non-agricultural businesses. As in the case of agricultural income, the value of the consumption of home produced non-agricultural products can be added to obtain gross business income. Depreciation is again estimated as a fixed portion of the current value of business assets. Net income equals gross income less depreciation. Many estimates thus derived are negative and as in the case of agricultural self-employment income the value of home produced consumption is considered to be a reasonable lower bound for non-agricultural self-employment income.

Remittances and other income

This source of income consists of remittances received, rental income from land, buildings, equipment and durables, receipts from welfare, retirement and insurance funds, dividend and interest receipts, scholarships, gifts, lottery prizes and a miscellaneous category.

Main types of expenditure

Food bought

For a list of 61 food products households are asked to report the amounts spent, both on two-week recall basis (amount spent since last visit) and on an annual basis (number of months product bought, number of times product bought per month, amount spent each time). Multiplying the two-week recall figure with 26 produces an estimate which is likely to be subject to considerable bias due to seasonality, while also outliers are more likely. The alternative estimate based on the annual recall is also prone to a significant bias, mainly as a result of lapse of memory of the respondents. Still the latter is preferred as the figures are consistent with those for home produced food consumption and as seasonality differences are considered more serious than measurement errors.

Home produced food consumption

The consumption of home produced food of the 60 food products is reported on an annual basis only. Households are asked to give the number of months in which the product was consumed, the number of times the home produced food was eaten per month, and how much it would cost to buy the amount eaten each time. This information is processed into an estimate for the value of home produced food.

Non-food items bought

Expenses are reported for a list of 9 daily goods and 31 annual goods.

Home produced non-food consumption

This item consists of the value reported for household consumption from non-farm business activities.

Other expenditures

This aggregate includes rent and utilities, school and medical expenses, use value of durables, remittances paid and wage payments in kind.

APPENDIX 3:Distribution of per Capita Income and Expenditure in Ghana,
1987/88–1988/89

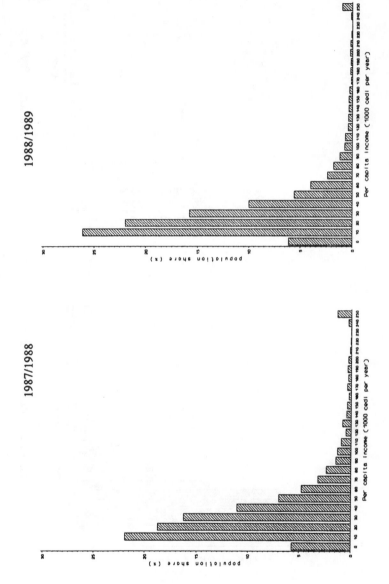

Figure 3.1: Distribution of *Per Capita* Income in Ghana, 1987/1988 and 1988/1989

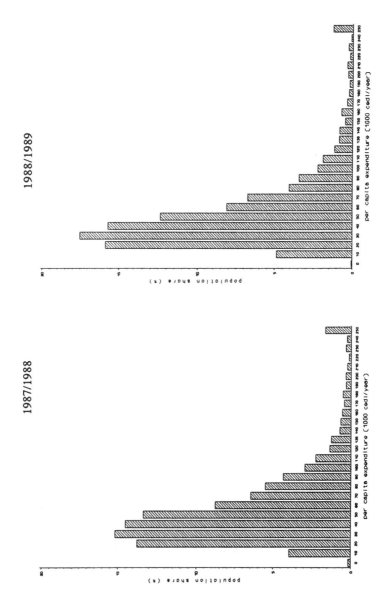

Figure 3.2 Distribution of *per capita* expenditure in Ghana, 1987/1988 and 1988/1989

Section 3
FARMERS' STRATEGIES FOR FOOD SECURITY

CHAPTER 5

ANALYSIS OF FARMERS' STRATEGIES IN THE CENTRAL AND NORTH-WEST REGIONS OF BURKINA FASO; APPROACH AND SOME RESULTS

Kaboré, P.D., Kaboré, T.S., Maatman, A. Ouédraogo, A.A., Ruijs, A., Sawadogo, H., and *Schweigman, C.*

SECTION 1

Introduction

The population of the Central and North-West regions in Burkina Faso faces a gloomy prospect. As a matter of fact, the situation is critical: the prevailing systems of production and distribution do not prevent serious food shortages for the majority of people, and, through the force of circumstances, natural resources are overexploited.

Despite, or rather owing to this critical situation, farmers have taken important initiatives. On farm and village level, they have progressively adapted their strategies to the new conditions. With regard to agricultural methods, we can mention for instance, the intensification of water and soil conservation methods, the reinforcement of the link between cattle keeping and cultivation and the intensification of tree and shrub planting on the fields. Moreover, through reforms in their organisation, by cooperation and in particular, by creating cooperatives, the people endeavour to implement new forms of social life that would allow them to work out new strategies by using the scarce available resources. More and more supported by the Government, credit banks, Non-Governmental Organisations, etc., those initiatives of the people also deserve to be supported by scientific research.

The objective of this study was to better understand the determinants on which farmers' strategies are based, and to know the most influential factors for those strategies. To what extent do these strategies contribute, both in the short and long term, to increase food security? It is hoped that the findings would help in working out recommendations that would be useful for decision makers at several levels (farmers, farmers associations, agricultural information services, credit banks, Non-Governmental Organisations, Ministry of Agriculture and Livestock, etc.): which strategies could be adopted in order to achieve, in the short and long term, an increased food security; what role should Government and Non-Governmental Organisations play in this process?

The farmers' strategies and the way they act are greatly influenced by external factors from various levels: the national or international level (e.g. prices on the world market, monetary policy of the Government, national policy about prices, tax system, etc.), the regional or provincial level (e.g. environmental conditions, role of merchants, presence of credit banks, etc.), and the level of the village (e.g. access to local markets, presence of village associations, etc.). Other hierarchical levels, related for instance to ethnic and clan systems, interfere with the regional and socio-political hierarchical systems. The kinship system plays an important role, since farmers' strategies are considerably influenced by factors deriving from this system (e.g. the right of land use, traditional processes of decision making, etc.). It should be noted that the farm may comprise several socio-economic units — and several decision-makers — each with their own responsibilities, objectives and activities.

The target region of the study consists of the Central and North-West regions of Burkina Faso (Figure 5.1). The analysis is carried out at two levels: farm level and regional level.

There exists a strong relation between the two levels of analysis. The resources and tools of analysis to be used consist of the following four elements:

(i) Secondary sources, in particular all village level studies previously effected in Burkina Faso: the studies of ICRISAT (e.g. Matlon and Fafchamps, 1988; McIntire 1981, 1983; Kristjanson, 1987), ICRISAT and IFPRI (Reardon and Matlon, 1989); of the programme FSU/SAFGRAD (e.g. Lang, Roth and Preckel, 1984; Nagy, Ohm and Sanders, 1986; Roth et al., 1986; Roth, 1986; Singh et al., 1984; Singh, 1988); of CEDRES of the University of Ouagadougou (Thiombiano, Soulama and Wetta, 1988), of the University of Wisconsin (e.g. Sherman, Shapiro and Gilbert, 1987); Delgado (1978), Broekhuyse (1982, 1983), M.J. Dugué (1987), P. Dugué (1989), Kohler (1971), Marchal (1983), Imbs (1986) and Prudencio (1983, 1987); the results of farming systems research published e.g. in Matlon et al. (1984) and Ohm and Nagy (1985) have been consulted as well.

(ii) Primary sources: results of studies and interviews in three villages in the North-West region: Baszaïdo, Kalamtogo and Lankoé. These studies focus on the study of the factors which determine farmers' behaviour and strategies and on possibilities of changes in view of sustainable food security. For the analysis of *zaï* use has been made of data obtained in village studies in Donsin, in the Central region. The primary data have been collected by regional research teams of the national Farming Systems Research Programme (RSP) of the Institute of Agricultural Studies and Research (INERA) of Burkina Faso.

(iii) A mathematical model to carry out analysis of farmers' strategies at the household level. A distinction is made between different production systems. An important element is distinguishing a *representative household* for each production system. A *systems approach and linear programming* are used as tools to analyse farmers' strategies in each representative farm household.

(iv) A mathematical model to analyse strategies at the regional level. This regional sector model combines the models for the representative households and elements which describe the interaction between various actors: producers, merchants and consumers. Price formation is the central issue in the model.

The level of analysis (household or regional) depends on the questions to be answered, e.g. the importance of the application of organic manure can very well be studied at the household level. However, the influence of the introduction of Structural Adjustment Programmes and in particular the influence of price liberalisation on farmers' strategies has to be studied at the regional level. In section 2 we start with an analysis of one representative household. This representative household is called `Exploitation Centrale'. The systems approach and the linear programming model are briefly presented anticipating a discussion of results.

In section 3 we concentrate on a technique to improve agricultural methods. It is the technique *zaï* which is applied in order to improve the availability of water for cultivation, to optimise the use of organic manure and even to regain uncultivated land. Section 4 deals with zoning, classification of categories of production systems and the choice of representative households. In section 5 some elements of the analysis of the regional level are discussed. The last part of the chapter are the conclusions that are drawn from the study.

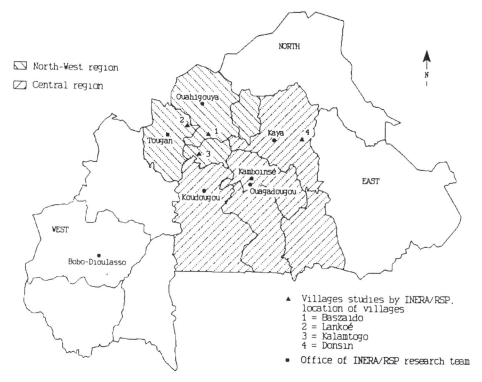

Figure 5.1: The Central and North-West Regions of Burkina Faso

SECTION 2

'EXPLOITATION CENTRALE': ANALYSIS OF FARMERS' STRATEGIES

In this section we present an analysis of farmers' strategies at the household level. The typical household is taken to be representative of a large group of Mossi households on the Central Plateau[1] in Burkina Faso. This representative household, the `Exploitation Centrale', is involved in a production system that occurs most frequently in the study area. It concerns a production system without any 'modern' inputs: Animal traction is not used and chemical fertilizers are not applied either.

---

[1] For this analysis we have limited our target region to the Central Plateau, which covers the largest part of the Central Region (except for a small area in the South-West) and the entire North-West Region, with the exception of the province of Sourou in the western part of this region.

82

Systems Approach

In order to structure the analysis of farmers' strategies for the 'Exploitation Centrale' we start by applying a systems approach. This also serves as a basis for the formulation of the linear programming models. A systems approach requires an accurate description of important concepts. A distinction is made between a descriptive analysis and a normative analysis of farmers' strategies. A *descriptive analysis* supplies a description of strategies and of factors influencing them. Questions to be studied in such an analysis are called *empirical questions*. They refer to actual situations in the present and past; examples are: what is the social organisation, what was the reason of migration, what is the method of storage, what were the deficits etc.? In a *normative analysis* of farmers' strategies, a study is made of the question, which strategies do well under various conditions and how strategies can be improved? A normative analysis refers to required changes and to measures to effectuate these changes. A descriptive analysis focuses particularly on present and past and the normative analysis on present and future. Results of descriptive analyses are important inputs of the normative analysis. A normative analysis can be an important tool for drawing conclusions from descriptive analyses. In this chapter, we restrict ourselves to a normative analysis of farmers' strategies.

The normative analysis focuses on some principal activities of a household and their most important determinants. It is set up with the help of a discussion on the following questions:

— Which decision questions should be taken into account and who take(s) these decisions?

— Which factors influence the decisions and which of these factors should be taken into account?

— How do the factors influence decision-making?

— Which criteria play a role in the decision-making process?

Which decision questions should be taken into account?

The decisions taken on the farm level are of a remarkably diverse nature and refer to various domains. In this chapter we concentrate on decisions relating to:

— crop production,

— commercialisation,

— storage, and

— consumption.

Crop production decisions are decisions which have a *direct* influence on the quantity of harvested produce. For instance, decisions on the areas of land to be sown are examples of production decisions. However, a decision to send some members of the family abroad to earn some money is not considered to be a production decision, although the decision can have a considerable influence on the production capacity of the family. But this influence is *indirect* via the availability of labour, the decisions of areas to be cultivated and other production decisions.

Agricultural production decisions involve choice of crops, varieties, fields, surface areas, dates for sowing and weeding, fallow periods, cultivation and soil preservation methods. Where storage is concerned, it is mainly a matter of determining the volume of the safety stock and the quantities that should be reserved as seed for the following season. Decisions on commercialisation concern the choice of quantities and the periods of buying and selling as well as sources and allocations of revenues. Decisions on consumption relate to the part of the harvest to be used for own consumption and on daily consumed quantities per product.

Who take(s) the decisions?

The socio-economic organisation of the Mossi, which is founded on the social family organisation, comprises several levels. Each level plays an important part in the social and economic life of the Mossi. Ancey (1975) distinguishes between the following levels: (1) the individual level, with a distinction between elder men, younger men and women; (2) the level of the 'restricted' production unit; (3) the level of the 'consumption group'; (4) the level of the 'farming' group; (5) the level of the compound (the 'residential unit'); (6) the level of the family extended to the clan or the hamshed; (7) the village level; (8) the supra-village level. In this chapter we concentrate on two decision-making levels: the level of the farming group and the level of the restricted production unit. The *Mossi farm* (level 4) can be defined as 'the human community putting together its efforts on one or several (large) common fields whose harvest serves the collective consumption of the members participating in the work and their inactive dependents' (adapted from Ancey, 1975).[2] The Mossi call these large fields *pukasinga*. It is the head of the farm who organizes the work on the common fields, who decides which fields are to be cultivated, which crops will be grown and what methods will be used. He manages the harvested crops of the common fields, and decides for instance on the allocation of cereals stored in the common granaries for the preparation of meals. He also manages the livestock of the farm, and organises the work concerning feeding and watering the cattle; he decides about financial transactions and is responsible for social exchanges between the different farms. He is the principal decision maker within the farm, and even the seemingly independent decisions which are taken by other members belonging to his farm, are directly or indirectly under his control. It should be noted that in this study the terms *household* and *farm* are interchangeable, i.e. they refer to the same socio-economic unit. A 'restricted' production unit is formed by a person (or group of persons) within the farm who cultivate one or several 'individual' fields (*beolga*). In particular, the wives of the farm head cultivate individual fields.

Which factors influence the decisions?

Decisions to be taken at a certain time can be influenced by decisions taken at earlier times and by various exogenous factors. The interrelationship between decisions taken at different times will be discussed elsewhere. Here, we discuss only the influence of exogenous factors, that is those which the farmers' decisions have no influence. Exogenous factors can be rainfall and market prices, but also various conditions to be reckoned with in the analysis, for instance: the availability of fertilizers, the possibilities of obtaining credit for the purchase of cereals, the loss

---

[2] The definition given by Ancey (1975) was more general in order to cover several ethnic groups in West Africa (Lobi, Mossi, Senoufo, Haussa, Wolof etc.). See also Tallet (1985) for a comprehensive discussion on the concept of a farm in rural areas of West Africa (especially based on field studies in Burkina Faso).

of nutritional values of the agricultural products during the preparation of meals, the revenues from off-farm activities, the money transferred by migrants etc.

Some of the exogenous factors, such as the time of the first rainfall and the total rainfall in the growing season, are of a *stochastic nature*. Other factors are of a *deterministic nature*. The outcomes or realisations of the deterministic factors are certain, even if they are unknown and have to be estimated. An important concept in our systems approach is the concept of *scenario*. A scenario is a realisation of exogenous factors, for instance:

Composition of the household = 2 men, 3 women and 5 children; available surface area that can be cultivated = 6.5 ha; time of the first rainfall = 1st June; total rainfall in the growing season = 850 mm; the level of the grain reserve of the previous year = 0; ...

At this stage of our systems approach, we will say more about the concept of *strategy*. Before we give a description of this concept, we will first illustrate an example: when the first rains fall, the head of the farm will have to decide how much seed he will sow. His decisions might be different in situations when the first rains fall early or late. With early rains he will reserve a part of seeds to anticipate a situation when the rains do not persist and after some time, he would be obliged to resow. Even the choice of the crops (and sometimes of varieties) can change, if the rains appear later or if he has to resow. In fact, the head of the farm will take into account various scenarios (early rainfall, late rainfall) of exogenous factors. He will choose a strategy where he can react adequately on various scenarios. He will more or less have a plan for different scenarios. This example shows the dynamic nature of strategies. A strategy does not refer so much to a fixed sequence of activities to be carried out, but rather to the decisions from which the activities follow. A strategy is defined here as the set of the decisions to be taken, it means the set of answers to the decision questions.

How do factors influence decisions?

First we will discuss the question of when decisions are taken. We look at the agricultural production in one single year. The planning period includes on the one hand the *growing season* and on the other hand the period in which the harvested products are consumed, commercialised and stored. The latter period we call the *target consumption period*. The questions regarding production decisions relate to the first period. The questions regarding consumption, commercialisation and storage relate to the target consumption period, when the levels of the harvests are known. Decisions are taken at different times of the year. We split up the planning horizon in which are numbered $t = 1, 2, ...$ The beginning of a period t is also called the moment t. If a decision at the moment t ($t \geq 1$) is determined by all decisions at the moments $1, 2, ..., t-1$ and by the exogenous factors in the periods $1, 2, ..., t-1$, we call that a *conditional* decision. An example of a conditional decision: let us suppose that at period $t = 1$ it is necessary to respond to the following decision question:

What is the area of land where white sorghum is to be sown, and at the period $t = 2$?
How much labour should be used in period 2 for the cultivation of white sorghum?

It will be assumed here that the amount of labour necessary to cultivate one hectare of white sorghum in period 2 is an exogenous factor. Such an assumption is quite common in farming systems studies. It is, however, not evident. There can be situations when farmers have to decide whether they will put much or little effort per hectare in the cultivation of a certain crop, depending on rainfall conditions, the appearance of herbs, diseases, etc. In studies of farmers' strategies in Burkina Faso great importance must be attached to these issues. However, the

85

conditional decisions, where, for instance, the phenomena of resowing and intensive or superficial weeding depending on the rains at the beginning of the growing season are taken into account, are not considered in this chapter. We refer to Maatman and Schweigman (1995b), chapters 7 and 8. Here we start from the hypothesis that at the beginning of the growing season, that is at the period $t = 1$, all exogenous factors that play a role in the planning period are known.

We will first start with a *given scenario*. This scenario can, for example, be an 'expected scenario' (corresponding to 'average rainfall', 'average yield', 'average market prices'), to an 'optimistic scenario' ('high yields', 'high producer prices') or a 'pessimistic scenario' ('low yields', 'low producer prices'). We will consider a certain decision question, for example, on the size of the area to be sown with a certain crop on a certain piece of land. It is once more assumed that all conditions of climate and soil on the piece of land are everywhere the same and that one specified method of cultivation is applied. For a given scenario the yield is known. Therefore, the decision on the size of the area is equivalent to the decision on the amount of produce to be harvested. This decision depends on two questions: how much can we produce and how much do we want to produce? *How much we can* produce is especially determined by the availability of the production factors, land, labour and capital. The restrictive conditions are evident: no more land be cultivated than there is available and during each period $t = 1, 2, ...$ no more labour can be mobilised than there is available. These restrictive conditions can easily be formulated in quantitative terms. Restrictive conditions can also result from requirements of crop succession, crop rotation or land to be left fallow. These types of conditions are usually more difficult to formulate. There can be many more restrictive conditions, for example there is only a limited amount of cash available to buy fertilizers, the amount of manure produced by the farmers' cattle and to be used on the land is limited, etc. *How much we want to produce* can be formulated as a requirement, for instance enough should be produced to satisfy the consumptive demand. These types of requirements have to be handled with care. It is not certain that a given requirement can be satisfied. Often a requirement has to be reduced to the objective to produce 'as much as possible' and deficits may occur. A strategy which satisfies all restrictive conditions and requirements is called a *feasible strategy*.

What are the criteria for decision-making?

What is a good decision and what is the best decision? First we note that all decisions depend on the restrictive conditions and requirements discussed above. It is possible that the decisions are completely determined by these restrictions. In such a situation the objective of determining 'better' strategies lacks appropriateness. What is important is that we find out whether there exists a feasible strategy at all. But if there is still some room for different strategies, the question of good and 'best' strategies is valid. The farmers aim at realising several objectives, as is shown by many village studies: sufficient production which allows them to satisfy their own consumption, revenues from sales, a safety stock and minimisation of risks. Some of these objectives are conflicting, e.g. production for own consumption and keeping a safety stock to reduce the sales. How should the three objectives be weighed: (1) maximum production for own consumption, (2) maximum revenues by sales, (3) sufficient safety stock? Obviously, there does not exist a blueprint for such a weighting. For the various heads of households on the Central Plateau the order of importance of these objectives depends on the situation in which the household finds itself. Broekhuyse (1988) for example pictures the development of the farms on the Central Plateau going from a surplus economy through a subsistence economy to a survival economy. According to him the majority of the households on the Central Plateau are in the last phase: *the survival economy*, a phase in which long-term investments are no longer feasible for the households. Consequently, the farmers' strategies are determined by short-term considerations. The only objective is to get a maximal harvest which permits survival for as many months as possible. All means are used to achieve this objective, even if the long term consequences would be disastrous.

Linear Programming Model Under the `Exploitation Centrale'

This section is devoted to the formulation of a linear programming model. Several researchers have used linear programming models: Roth *et al.* (1986) to find out how far certain technical possibilities (especially the mechanical ridge tier) may be successfully applied within the existing farming systems on the Central Plateau; Delgado (1978) and later Jaeger (1984, 1987) to study the possibilities of using animal traction in agriculture. Whereas in these studies the linear programming models aimed at answering specific questions, the present study is of a much wider scope and directed to farmers' strategies in general, including the interaction between production strategies and the strategies of commercialisation, storage and consumption during the target consumption period.

There is an unequivocal relation between the systems approach and the construction of the linear programming model. Decision questions in the systems approach correspond to *decision variables* in the model, the exogenous factors to the *parameters*, the scenarios to a choice of values of the parameters. How the various factors influence the decisions and the decision criteria is dealt with through the *constraints* and the *objective function* of the model.

When discussing the key elements of the linear programming model, Appendix 1 is frequently referred to. There we present the definitions of all decision variables and parameters and the entire model. All equations in the model are numbered. In the text below, we refer to these numbers between brackets.

Figure 5.2 contains a schematic representation of the planning period. In Appendix 1 the division in periods t = 1, 2, ... is described in detail.

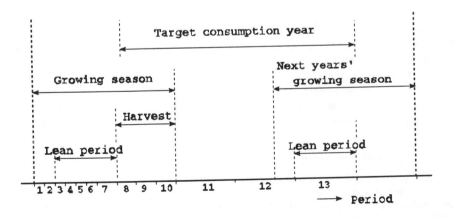

Figure 5.2: Schematic Representation of the Growing Season and Target Consumption Year

For the 'Exploitation Centrale' we assume the following composition: 10 persons, of whom 5 are over 15 years of age, including the head of the household, and who are supposed to make an important contribution to the work on the common fields of the household. These are the 'active members'. Of these active members three are women, two are men. The persons under 15 are not active members, although it happens that they have to lend a hand, for instance by chasing away birds or by keeping small ruminants at a distance etc.[3] This composition is based on data furnished by village studies collected by the Ministry of Agriculture in the Central-Northern region (MAE, 1988), by the National Institute of Statistics and Demography, INSD (figures of the general censuses held in 1975 and 1985); see also Konaté (1988), and by the village studies referred to in the introduction.

Crops, Soil, Methods of Cultivation

A distinction is made between five main crops: maize, red sorghum, white sorghum, millet and groundnuts and a secondary crop, cowpea which is only cultivated intercropped with sorghum and millet. Crops that do not occupy at least an average of 3% of the total area per household are not taken into account. This refers to, for instance, voandzou, sesame, sorrel and gombo. The practice of leaving land fallow will be dealt with explicitly. The fields will be distinguished in accordance with three different criteria. According to the '*toposequence*' fields situated in the higher areas are distinguished from those situated in the lower areas. And according to the distance between fields and compounds we distinguish three '*rings*'.

The first ring includes fields situated less than 100 metres from the compound, the second those situated 100 to 1000 metres from the compound and the third those situated more than 1000 metres away. The distances chosen are based on data collected by Prudencio (1987). Relations between the rings and the agricultural strategies concern the choice of crops, the intensity of fertilisation, the area of fallow land and the sequence of crops as well as anti-erosion measures. The time spent on travelling between the compound and the field also plays a role in the choice of strategies. Finally, a difference is made between *common fields* and *individual fields* belonging to the women. The various categories of land, also referred to as 'soil types', are shown in Appendix 1.

A distinction is made between the use and the non-use of organic manure. The quantities of organic manure applied may vary. For the purpose of the model five dosages of organic manure are considered: 0, 800, 2000, 4000 and 8000 kg/ha, although from time to time in practice even bigger doses are applied. We recall that the possible application of any chemical fertilizers is not taken into account.

A distinction is made between the various sowing periods for monocrops or intercropping with red sorghum, white sorghum and millet; these are the periods t = 1, 2, 3, 4 and 5. For all cereals in this model two levels of weeding intensity are distinguished. The first level, called 'intensive', is based on a very good field management, which means weeding in time, before the

---

[3] The distinction between active and non-active members of a household does not reflect the often important participation in field work by children between the ages of 10 and 15; on the other hand it probably overestimates the activities of old people in the household who do not do a great deal of work on the field and who often only cultivate a small individual field. For the types of analyses intended in this chapter these slight distinctions are not explicitly taken into account; all the same, the effects of the variations in availability of labour can be studied with sensitivity analyses.

weeds cause too many problems and the employment of sufficient labour to get rid of all weeds. The second weeding level, called 'extensive', is based on a less thorough field management; weeding starts a little later and is carried out quicker, with less man power.

We distinguish among different *plots*. A plot is a piece of land where the same crop or a combination of crops is grown, on one soil type; one dose of organic manure is applied; the plot is sown in one of the periods 1, 2, 3, 4 or 5, and weeded in an 'average' way (in the case of groundnut plots), intensively or extensively (for the other plots).

We restrict ourselves to plots, i.e. to a combination of crops, of soil types, of organic manure applied, of sowing dates and of weeding levels, which are regularly found on the Central Plateau and for that reason are '*representative*'. Moreover '*alternative*' combinations are considered in order to study their feasibility and possible advantages.

Availability of land and fallowing systems

The surface area available per soil type depends on the total surface area the 'Exploitation Centrale' has at its disposal and on the distribution of this land over the various soil types. On the Central Plateau, especially in its northern part, the extent of the available surface area also depends on the possibility of reclaiming eroded land, for instance, by applying the '*zaï*' technique. The estimated values of the available areas per soil type are based on data obtained from village studies, e.g. data on cultivated area per household, intensity of land use, location of fields on the toposequence and the distance of the field from the compound, and on the difference between common fields and individual fields. Data from agricultural statistics (cultivated areas, arable land, population density) have also been taken into consideration. Land-use intensity has explicitly been taken into account by means of the practice of fallowing. We assume that each plot j corresponds to an area of fallow land (see equation (3) of Appendix 1). The parameter $\lambda(j)$ in the expression (3) represents the ratio between the surface area of the fallow plot and the surface area of the cultivated plot. We choose $\lambda(j)$ in such a way that it is equal to the ratio between the duration of the fallow period and the duration of the period of cultivation of plot j. This choice of $\lambda(j)$ allows us to follow a rotation schedule over a succession of years, where the cultivated area remains the same in all those years and where the fallow plots and those cultivated may alternate. In order to determine the fractions $\lambda(j)$, we have first of all used data on the differences in land-use intensity between varieties grown and rings, and especially the levels of land-use intensity as are shown in a study of ICRISAT of the different crops in the region of Yako (Matlon and Fafchamps, 1988). With the help of these data we determined the 'average' fractions $\lambda(j)$, which vary from 0, for the cultivation of maize with 8000 kg of organic manure, to 1.50, for the cultivation of sorghum without any organic manure being applied. In the absence of information we have assumed that the soil type, in particular the high or low situation of the fields, does not have any influence on the 'desired' fractions $\lambda(j)$. We will also study another scenario for the agricultural planning problem of the households, which we will call scenario 2. In scenario 2 the fallow periods are longer with a view to a sustainable exploitation of the soil, i.e. the values of $\lambda(j)$ exceed the 'average' values.[4] For each soil type the 'land constraints' are given by equation (2) of Appendix 1.

---

[4] In fact, the values of $\lambda(j)$ correspond to the land-use intensity factors (R of Ruthenberg, 1980) at plot level (cultivated and fallow: R (j) = $100/(1+\lambda(j))$).

Availability of organic manure

There are not many data available on the use of organic manure on the individual fields of the women. However, it seems that the women have little access to organic manure. In this model we assume that the women do not have any organic manure at their disposal to apply to their individual fields. The condition that the required quantity of manure must at most be equal to the available quantity is formulated in equation (4) of Appendix 1. The available quantity of manure per household has been calculated on the basis of the estimated quantity of manure per head. For the different village studies this estimated quantity varies from 160 to 250 kg/hbt. When we assume that the available quantity of organic manure is 200 kg/inhabitant, the entire household has therefore 2000 kg at its disposal. The quantity of manure used per hectare on a certain plot is integrated in the definition of a plot.

Availability of labour

For all the periods t of the growing season the labour constraints for work on the common fields and the individual fields are shown in equations (5) and (6) of Appendix 1. Data on labour required per hectare in each period t are based on results of the village studies taken into consideration. These data have allowed us to deduce the working hours for each activity and each crop under the most common conditions, as well as the working hours under different conditions (toposequence, application of manure), see Appendix 2. Next an average agricultural calendar has been set up on the basis of the agricultural calendars observed in the different village studies. The data on working hours in each period t have been obtained by combining the estimates of working hours per activity with the agricultural calendar. Finally the time needed to walk to the field and back has been estimated for each plot and each period in relation to the distance covered (which depends on the ring where the plot is located) and the working hours needed per month to cultivate the plot (in order to calculate the number of trips to the fields).

As regards the available quantity of labour per household in each period, we have assumed that an active member of the 'Exploitation Centrale' can work in the fields an average of either 7.5 hours per day (men) or 6.5 hours per day (women) for 26 days per month. We presume that the quantity of available labour does neither increase in the peak periods, nor diminish in the months when work is less intensive. It is the amount of spent labour which will vary. Extension of available labour to the organisation of invited working parties or by employing paid labourers is not introduced in the equations (5) and (6) of Appendix 1. If we assume that there are no differences in effectiveness between the various active members and that all tasks can be carried out by all of them, it is easy to calculate the total labour available to the 'Exploitation Centrale'. The distribution of work between the common fields and the individual fields is subject to strict rules on the Central Plateau. There is only little information available on the social codes and they apparently differ considerably between the various regions. On the basis of data produced by the village studies (and in particular those of ORSTOM) we assume that the women can work on the individual fields 30% of their time during the peak periods 3, 4, 5 and 6, and 40% of their time during the periods 7, 8, 9 and 10. In the first two periods the work is concentrated on the common fields. The division is strict, which means that it is not possible to transfer labour from the individual fields to the common fields or the other way around.

An important aspect which has not been taken into account in the constraints (5) and (6) (Appendix 1) is the restricted number of days available for sowing (and ploughing) at the beginning of the growing season. For sowing and ploughing, the soil needs to be humid which depends on the rainfall (they can sow approximately 2 or 3 days after a major rainfall). However, the first showers are rare and very irregular. The constraints of labour for sowing at the beginning of the growing season (t = 1, 2, 3, 4) are given in equations (7) and (8) of Appendix 1. The values of the number of days available for sowing in each period are based on rainfall data on the Central Plateau over the last few years and on data produced by village studies on the number of days spent on sowing.

Stocks of agricultural produce

We now arrive at the equations describing stock levels of the agricultural produce. These stock balances contain production, sales, purchases, consumption, volumes reserved for sowing in the following year and stock losses (see equations (9) to (12) of Appendix 1). Stock losses include loss of weight during storage of products, because of damage by insects, rats and mice, etc.. Annual losses are estimated, making use of the few data which are available (for instance, Nagy, Ohm and Sawadogo, 1989; Eicher and Baker, 1982; Yonli, 1988). The following annual fractions are assumed: for red sorghum 6% per year; for white sorghum 8% per year; for millet and maize 10%; for groundnuts 15% and for cowpea 30% per year. Maize is never stored for a long time.

The values for the yields are based on results of village studies that were pursued and have allowed us to deduce the yields of each crop under the most common conditions (see Appendix 2), and moreover the yields of the same crops grown under different conditions. In this model the production of the individual fields is added to the production of the common fields; so the place of origin of the agricultural products does not play a role in the stock management of the agricultural produce in this example. The quantity of seeds to be reserved is estimated as the product of an average quantity per hectare to be reserved for each crop, taking into account a certain number of resowings, and the area to be sown in the following season, which we take to be the same as the area sown in the present season. Data are furnished by Matlon and Fafchamps (1988), Dugué (1989) and Yonli (1988). The stock at the beginning of the harvesting period is taken as a parameter in this model. Later we will revert to the safety stock.

Financial balances

The financial balances of the 'Exploitation Centrale' given in equation (16) of Appendix 1 contain sales revenues of the harvested produce, cost of purchasing cereals, extra-agricultural revenues and extra-agricultural expenses. The interest rate has also been included. The expected prices deserve an explanation. At first, in the course of the year following the harvest, the market prices rise: They are low just after the harvest, when supply is high, they gradually increase and reach a maximum level during the lean period, when supply is low. The 'Exploitation Centrale' is situated close to a local market, it only sells or buys cereals at this market. This market is representative of the rural markets on the Central Plateau: during and just after harvest the producers (in general the women) offer their produce on the market to sell them to merchants/retailers; during the lean period the merchants/retailers usually are the sellers and the producers are the buyers (see for instance, Yonli, 1988). Here we assume that the 'Exploitation Centrale' knows that it cannot sell its cereals on the local market except just after the harvest, as later the merchants will not be very much interested in buying them. The selling price of the cereals in the financial balance is therefore the producer price, realised just after the harvest. Similarly we assume that the household knows that it cannot buy cereals from the merchants except just before or during the lean period. The purchasing price of the cereals is therefore the consumer price paid during these periods. The assumptions result in the equations (14) and (15) in the model presented in Appendix 1.

If the above-mentioned assumptions look reasonable from the point of view of the actual state of affairs, they do not correspond to the desired selling and buying strategies. In fact, the producers should be able to profit from the low level of purchasing prices at the time of harvesting and from the higher level of selling prices at the end of the year following the harvest. The above-mentioned assumptions refer to cereal prices. For the selling prices of groundnuts and cowpea, we take the average market prices per year. The purchasing prices of groundnuts and cowpea do not play any role in this chapter. The estimated selling prices are 64, 56, 61, 66, 108 and 78 FCFA/kg respectively for maize, red sorghum, white sorghum, millet, groundnuts and cowpea. The estimated purchasing prices are 120, 96, 100 and 107 FCFA respectively for maize, red sorghum, white sorghum and millet. These estimates are based on selling prices observed during and after the harvest (October-December) in several rural markets on the Central Plateau and on the purchasing prices observed during the lean period (July-September), during the years 1981-91.

In anticipation of the discussion of the results of our calculations at the end of this chapter, the importance of the values of extra-agricultural revenues and expenses should be stressed. There are only few data available on these revenues and expenses for the households on the Central Plateau. Basing ourselves on data provided by Sherman *et al.* (1987), Broekhuyse (1983) and Thiombiano *et al.* (1988), we have assumed the following values: 7500, 7500, 0, -25000, 20000, 20000 FCFA for the periods t = 8, 9, 10, ..., 13 (see Appendix 1) of the target consumption year. Note that in period 11 (the period after harvest) the expenses are greater than the off-farm revenues, which forces the farmer to sell part of his harvest in order to meet his financial needs.

Nutritional value of consumption and food balances

In the model we concentrate on the needs for energy, measured in 1000 kilocalories, and in proteins, measured in 1000 grams. The demand for food is based on the composition of the 'Exploitation Centrale' and the nutritional needs of each member (man, woman, child) per year. These figures are provided by Bakker and Konaté (1988) and by Agbessi Dos Santos and Damon (1987). The energy requirement equally depends on the activities of the members of the household. Bleiberg *et al.* (1980) and Brun *et al.* (1981) have conducted studies to determine the variations in the energy spent during the year, depending on the various activities of the population (men and women) on the Central Plateau. For the calculation of energy requirements per period we have taken these variations into consideration (for instance, increased needs in the peak periods in the growing season). For the calorie and protein contents of the different agricultural products we have taken into consideration the most important consumption patterns. The *tô*, made of cereal flour, is the traditional dish on the Central Plateau. Millet and white sorghum are used to prepare the *tô*, millet is also used to prepare porridge and a number of other dishes. Maize is usually consumed fresh (grilled) during the lean period. Red sorghum is especially used for the preparation of *dolo*, the local beer. Cowpea and groundnuts are the ingredients for sauces served with the *tô*. The different phases in preparation of foodstuffs may result in the loss of nutritional values in the agricultural products used. This is particularly the case when the *tô* is prepared on the basis of white sorghum. Since the grains of white sorghum are very hard and the grinding of whole grains with a stone is very difficult, the white sorghum losses[5] are estimated at 25% of the total weight of the grains (Bakker and Konaté, 1988). The values used for the nutritional needs of the 'Exploitation Centrale' and the nutritional values of the products are given in Appendix 2.

The consumption of nutritional elements is expressed in equation (18) of Appendix 1 and the nutritional deficits are defined in equations (19) and (36) of Appendix 1. We assume that 80% of the demand for energy and 70% of the demand for protein (the values of the parameters $\Theta_1(n)$ in equation (19) have to be covered by the consumption of cereals, groundnuts and cowpea. These values have been adapted from the study by Bakker and Konaté (1988), who assume that a satisfactory nutrition is obtained when roughly 75% of the nutritional needs is covered by the basic cereals. The cereals form the greatest part of the products of the 'Exploitation Centrale' consumed. The percentage taken for the protein needs to be covered is low, since it is known that cereals are in the first place sources of energy.

On the basis of the consumption studies made by the IFPRI (see Reardon and Matlon, 1989), carried out in a very bad year after the great drought of 1984, we have fixed a minimum limit below which consumption should never fall (in any of our periods), see constraint (23) in Appendix 1. The values of the parameters $\Theta_2(n)$ in the constraint are 0.60 and 0.50 respectively.

The consumption of red sorghum which is consumed in the form of *dolo*, should not surpass a certain maximum quantity, (see equation (21) of Appendix 1). The maximum quantities, expressed in kg of red sorghum, are 8, 8, 9, 50, 50, 25 respectively in the periods 8, 9, 10, ..., 13. These values are based on the quantities of *dolo* consumed on the Central Plateau for different periods of the year, which is highest for the period after the harvest.

Finally we introduce the constraint (25) in Appendix 1 to ensure that the daily meals, with the exception of the consumption of *dolo*, are based on cereals and are not dominated by the ingredients, groundnuts or cowpea. The values of the parameters $\Theta_3(n)$ in constraint (25) are 0.85 and 0, respectively.

'Optimal' strategy

We recall that the objectives of the 'Exploitation Centrale' are the following:

— production of a maximum quantity of cereals for the provision of its own food;

— to realise a maximum of revenues from the sale of agricultural products;

— to have a safety stock which allows them to compensate for bad harvests in one or more seasons;

— reduction of risks.

The first objective is considered a priority. This objective is reflected in the model by the introduction of a constraint of 'subsistence' (equation (26) of Appendix 1). It is presumed that in any case the 'Exploitation Centrale' produces part of its cereals consumption. The constraint (26) plays a crucial role. Omitting constraint (26) from the model could lead to a different strategy: selling its own production and purchasing cereals for its own consumption. This strategy would be fully acceptable, if one could be sure in advance that one could find buyers

---

[5] In the opinion of certain farmers reasons of preservation and quality of the *tô* would explain this practice. When the bran of the sorghum is not removed, the *tô* is of a lower quality and can only be preserved for a shorter period.

for the production and get prices that would be more or less fixed and, moreover, find sellers of cereals for prices that would be more or less fixed. In reality that is not the case. In actual fact, one runs the risk that prices may develop contrary to one's expectations; the selling prices may be lower and the purchasing prices higher. In fact, the purpose of constraint (26) is to avoid these risks. A value of $\alpha = 1$ would imply a safe strategy that could, however, result in a loss of financial gains. Moreover, such a strategy will often not be feasible. In this exercise $\alpha = 0.6$.

Apart from the constraint (26), the objectives are treated as follows. The 'Exploitation Centrale' tries in the first place to avoid or to minimise food deficits during the target consumption year. If these deficits are 0, they will attempt to maximise their stocks at the end of the target consumption year in order to meet food requirements in the harvest period of the following growing season (see figure 5.2). This objective implies the minimisation of 'deficits' defined in equations (20) and (37) of Appendix 1. If these deficits are also 0, the 'Exploitation Centrale' can also produce for the purpose of selling. In this way they maximise their net revenues. All these objectives are included in the objective function (1) of Appendix 1. The order of importance can easily be represented by selecting the weighting coefficients in equation (1). In a situation where the 'Exploitation Centrale' is able to acquire net revenues, part of the net revenues is invested in a safety stock, (see equation (27) of Appendix 1) and, the coefficient β is chosen to be 0.10.

Discussion of Results of LP Model under 'Exploitation Centrale'

The model that is presented in this chapter has been developed step by step. While developing the models and analysing the results we arrived at an extension of the model which was more in line with the real situation. Some elements in the model are clearly the results of this interactive process. We refer, for instance, to the different sowing periods, the restricted time available for labour and sowing, the distinction between intensive and extensive weeding, etc.

The results of our calculations in the model defined in equations (1) to (37) of Appendix 1, which correspond to an 'average' scenario, have been presented in Table 5.1. In general the results seem to describe the real situation quite well. The levels of agricultural production and of consumption correspond well to the average levels observed in the village studies on the Central Plateau (referred to in the Introduction). A remarkable feature in this study is the heterogeneity of agricultural strategies, i.e. the cultivation of different crops, sometimes 'pure', sometimes intercropped, on different soil types and using a great diversity of growing methods (different sowing periods, with different quantities of organic manure, intensive and extensive weeding). The great diversity in agricultural activities, in response to a complex range of objectives and constraints, is a key element of the farmers' strategies on the Central Plateau. So they grow, for instance, maize (and groundnuts) to meet their consumption needs in the month of September and early October. Other results that conform to observations made in field studies are the small part of the harvest that is sold and the necessity to buy back cereals later in the year. Sales of part of the groundnuts production are necessary to meet urgent needs in the period after harvest. That farmers only engage in the commercialisation of their agricultural products after the harvest corresponds with the findings of Sherman *et al.* (1987) and Thiombiano *et al.* (1988).

In a normative analysis one does not only expect results that conform to reality, on the contrary, diverging results may show some important perspectives. There are some diverging results: for instance, maize cultivation on the lower fields rather than on the fields near to the compound, the absence of white sorghum, i.e. the predominance of millet, the small area where intercropping is applied, the sales of cowpea and the quantities of groundnuts that are consumed.

TABLE 5.1: Some Results of the Linear Programming Model

Crop Fallowed	Soil type	Owner	Manure (kg/ha)	Manure period	Sowing intensity	weeding area (ha)	Cultivated area
Maize	high-1	common	8000	average	intensive	0.127	0.000
Maize	low-2	common	4000	average	intensive	0.011	0.000
R.Sorghum	low-2	common	2000	1	intensive	0.095	0.024
R.Sorghum	low-3	common	800	1	intensive	0.147	0.098
Millet	high-3	common	0	4	intensive	0.391	0.391
Millet	high-3	common	0	5	intensive	0.499	0.499
Millet/Cowpea	high-1	common	800	2	intensive	0.100	0.043
Millet/Cowpea	high-2	common	800	2	intensive	0.688	0.296
Millet/Cowpea	high-3	common	0	2	intensive	0.054	0.054
Millet/Cowpea	high-3	common	0	3	intensive	0.173	0.173
Millet/Cowpea	high-3	common	0	3	extensive	0.174	0.174
Millet/Cowpea	low-3	common	0	1	intensive	0.102	0.102
Groundnut	high-2	common	0	average	average	0.013	0.013
Groundnut	high-3	common	0	average	average	0.499	0.499
Millet	high-1	indiv.	0	4	intensive	0.025	0.025
Millet	high-3	indiv.	0	4	intensive	0.126	0.126
Millet	high-3	indiv.	0	5	intensive	0.173	0.173
Groundnut	high-2	indiv.	0	average	average	0.100	0.100
Groundnut	high-3	indiv.	0	average	average	0.056	0.056
						3.55	2.85

	t=8	t=9	t=10	t=11	t=12	t=13
Production (kg)						
- Maize	129	43	0			
- R.Sorghum	0	110	110			
- W.Sorghum	0	0	0			
- Millet	0	465	465			
- Groundnut	88	179	0			
- Cowpea	21	21	0			
Consumption (kg)						
- Maize	124	41	0	0	0	0
- R.Sorghum	0	8	9	50	50	25
- W.Sorghum	0	0	0	0	0	0
- Millet	0	80	122	325	332	396
- Groundnut	7	14	13	36	37	44
- Cowpea	12	0	0	0	0	0
Sales (kg)						
- R.Sorghum			43			
- Groundnut			49			
- Cowpea			24			
Purchases (kg)						
- Millet	0				0	378
Stocks of products (kg)						
- Maize	0	0	0	0	0	0
- R.Sorghum	0	99	197	102	51	25
- W.Sorghum	0	0	0	0	0	0
- Millet	0	372	698	361	24	6
- Groundnut	62	188	173	83	44	0
- Cowpea	7	25	25	0	0	0
Finances (FCFA)						
- EAR-EAE	7500	7500	0	-25000	20000	20000
- Sales			9570			
- Purchases	0				0	40480
Consumption						
- in 1000 kilocalories	520	520	520	1472	1501	1673
- in proteins (1000gm)	7	8	8	23	24	26
Food shortage						
- in 1000 kilocalories	0	0	0	0	0	0
- in proteins (1000 gm)	0	0	0	0	0	0

Annual net revenues: 0 CFA

Source: Based on calculations in the model defined in equations (1) to (37) of Appendix 1.

We will now discuss the reasons for these divergences. Let us start with the last point. Because of the higher nutritional value of groundnuts (in energy and protein terms) it is apparently more effective to consume groundnuts rather than sell them and buy cereals in return. Bakker and Konaté (1988) and Agbessi Dos Santos and Damon (1987) also stress the importance of the consumption of non-cereal products such as groundnuts in order to get away from ill-balanced nutrition.

The absence of white sorghum is explained by the losses in nutritional value during the preparation of *tô* on the basis of white sorghum. These losses (estimated at 25% of the total weight of the grains, see above) have a strong effect on the results. In fact, a sensitivity analysis (see Maatman and Schweigman, 1995b) shows that without these losses more white sorghum would be grown (substituting millet by white sorghum), and the nutritional value of the total package of agricultural products would improve considerably. Mills that would be able to grind entire grains of white sorghum could make an important contribution to the food security of the households.

The dominant position of millet on the higher fields is therefore the result of the low nutritional value of white sorghum compared with that of millet. Moreover, millet demands less minerals and water, which explains its dominance on the higher fields, in particular in the northern district of the Central Plateau.

Maize cultivation on the lower fields is not common practice on the Central Plateau, in spite of the fact that the soil is quite suitable for the growing of maize,[6] which is probably due to the problems with regard to the transportation of large quantities of organic manure to these fields (they need, for instance, a cart). In addition, the fields close to the concession can more easily be guarded, which protects them against theft and damage (for instance by animals).

The results show that only on 36% of the cultivated area is intercropping practised. But on the Central Plateau in general 70% of the cultivated area is used for intercropping. One can think of several reasons to explain this result. The values taken for the yields of cowpea could be too low and/or the reduction in the yields of the principal crops too high (note that our estimates are based on village studies carried out to a large extent in years of drought[7]). In fact, an increase in the yield of cowpea by 25% is already sufficient to arrive at intercropping on more than 70% of the cultivated area. Besides, some elements that are often mentioned as advantages of intercropping are not included in this model. We refer in particular to residual effects of intercropping (cowpea fixes nitrogen) and increased stability of the yields; and to the drop in labour time spent on weeding (see for instance, Norman, 1974; Steiner, 1984).

The prognoses of the model do not hide the fact that the situation on the Central Plateau is delicate. Investments (purchase of inputs, animal traction) do not seem possible except at the expense of food security (already weak). The production of cereals by the 'Exploitation Centrale' reaches a total of 1322 kg, equal to 132 kg/person. By combining effectively the factors of production, land, labour and capital (and especially organic manure), under average conditions, the crop production system of the 'Exploitation Centrale' meets roughly 99% of the annual energy needs of the household, including the energy values of the production of groundnuts and cowpea. Only 3.5% of the cereal production is sold immediately after the harvest to cover direct expenses. The majority of agricultural revenues come from the sale of groundnuts and cowpea.

---

[6] There is still a risk of inundation in the lower fields. Maize cannot stand inundation.

[7] McIntire (1981) has found that cowpea contributes only modestly to the revenues per hectare of the intercropped crops (data from the village studies of the ICRISAT).

In Table 5.2 the shadow prices for the means of production are given. These figures express the implicit values which these means of production have for the 'Exploitation Centrale', in terms of the objective function, (equation (1) of Appendix 1). It is not difficult to express the values of the shadow prices approximately in FCFA, (see Maatman and Schweigman, 1995b). These values are identical to the marginal prices used in the conventional economics theory. The marginal revenues of labour found in this model are much lower than those found in other studies based on linear programming models on the Central Plateau (see for instance, Jaeger (1987) and Roth *et al.* (1986), who find values between 150 and 620 FCFA/hour in the 'peak' periods). On the other hand, the results correspond with the study by Singh (1988), who used production functions to estimate the marginal revenues of labour.

The disparity of these results can especially be explained by the fact that a large number of technical alternatives were taken into consideration (sowing in different periods, intensive and extensive weeding, several levels of fertilization) for the same crop and the same soil type. Since the results in this model are more detailed and accurate where the agricultural activities are concerned, they are probably more reliable.

The results show, contrary to those of, for instance, Roth (1986), that technology essentially oriented towards the reduction of labour in the peak periods, (no longer?) have great value for the growth in agricultural production on the Central Plateau. This argument gains even more strength when the dynamics of the agricultural systems on the Central Plateau are considered (a growing rural population, saturation of arable land), which will lead to a larger number of active members per hectare in the future (see Matlon, 1987, 1990).

Technology for the development of land resources seems to be more promising. This has, for instance, to do with anti-erosion measures, such as agro-forestry and stone walls constructed during the dry season. Reclaiming of eroded fields, for example by application of *zaï*, is of importance as well. Maintaining the fertility of the soil (physically and chemically) is crucial to a sustainable agriculture. In a cultivation system which becomes more and more intensive (shorter fallow periods), the availability of organic manure and the effectiveness of its application equally gain in importance. This role is even more apparent, if scenario 2, described above is considered. In this scenario the strategies for a more permanent agricultural production are analysed, by imposing longer fallow periods (of course, depending on the crop and the quantity of manure applied). The results of the calculations on the basis of the second scenario show important food deficits. In fact, for the 'Exploitation Centrale' to stay at roughly the same level of food security a double quantity of organic manure is required (4000 kg instead of 2000 kg). These results show the importance of a better integration of agriculture and animal husbandry, even more important if one considers the poor chances the farmers have to buy chemical fertilizers. Extensive studies, which should closely involve the rural population, on possibilities and problems of better integration of animal husbandry into agriculture, are necessary to correct the voids in the knowledge about this subject (management and quality of grasslands, fodder needed for animals, possibilities of fodder cultivation, ...).

On the individual fields millet and groundnuts are grown. The growing of sorghum or maize is less effective because manure is not available. The marginal revenues of labour are not as big on the individual fields as they are on the common fields (Table 5. 2). The area where groundnuts are grown is 0.67 hectare, of which 23% is cultivated on individual fields. This result (of 23%) lies a little below the results of the majority of village studies consulted. Nevertheless, relatively more groundnuts are grown on individual fields than on common fields. Mostly the production of groundnuts on individual fields is particularly explained by the revenues the women can obtain from this production, or by the responsibility the women have to provide the ingredients for sauces. The analyses show yet another reason. In view of the distribution of labour among the members of the household, and especially the low availability

TABLE 5.2: Shadow Prices of Production Resources[1] in the Model

	Shadow price (in FCFA equivalents)[2]
Land:	
- High lands, ring 1, common fields	8674 (in FCFA/ha)
- High lands, ring 2, common fields	7672
- High lands, ring 3, common fields	6319
- Low lands, ring 2, common fields	14021
- Low lands, ring 3, common fields	12156
- High lands, ring 1, individual fields	10572
- High lands, ring 2, individual fields	10693
- High lands, ring 3, individual fields	9638
Organic manure	13 (in FCFA/kg manure)
Labour[1]	
- Common fields: period 4	40 (in FCFA/hr)
- Common fields: period 5	66
- Common fields: period 6	46
- Common fields: period 7	20
Labour for sowing[1]	
- Common fields: period 1	24 (in FCFA/hr)
- Common fields: period 2	26
- Individual fields: period 2	1601
- Individual fields: period 4	97
- Individual fields: period 5	64

Notes:
1) This table contains only the production resources from which the shadow price is higher than 0.
2) Based on the purchase of millet in required quantities.

of labour for the individual fields at the start of the season, the demands on labour for the growing of groundnuts relatively correspond with the availability of labour for the individual fields.

Price variations for the agricultural products, while maintaining the same ratio between selling and purchasing prices, do not seem to have a big influence on the agricultural production strategies and the commercialization of agricultural products (see Table 5.3). This result confirms that the farmers react to the prices only marginally. And yet, when the price of groundnuts increases relatively compared to the price of cereals, the strategies change: the production of groundnuts goes up and sales of groundnuts replace the sales of cereals. The opposite occurs when the price of cereals shows a relative increase compared to the price of groundnuts. However, when the sales of cereals have replaced the sales of groundnuts, an additional increase in the price of cereals will reduce the quantities of cereals sold. These reverse relations between price and supply of cereals are also found in the studies of the University of Ouagadougou (see Thiombiano, 1987). Moreover, a drop in the price of cereals seems to be interesting for the 'Exploitation Centrale'.

At the end of this chapter we will revert to the merits and weaknesses of the linear programming models as they are presented here. We draw attention to the fact that the model represented by equations (1) to (37) in Appendix 1 only deals with one single production system. Models for the other production systems currently used on the Central Plateau, are in the process of being developed and contain modifications of the basic model discussed in this chapter.

TABLE 5.3: Some Results of the Linear Programming Model for Different Prices

Sales and purchase price		Value of the objective function[2]	Production[3]			Sales[3]			Purchases[3]	
CER[1]	AR[1]		CER	GN	CP	CER	GN	CP	CER	GN
-	-	-276,705	1322	267	42	43	49	24	378	-
-	+20%	-264,296	1289	292	38	-	82	-	388	-
+20%	-	-377,334	1428	195	31	131	-	10	316	-
+20%	+20%	-357,728	1295	272	46	-	60	19	316	-
+50%	+50%	-515,174	1361	217	53	35	23	25	251	-
-	-20%	-277,380	1468	140	58	119	-	37	318	44
-20%	-	-213,183	1261	308	53	-	78	38	492	-
-20%	-20%	-234,717	1300	292	40	52	68	23	470	-

Notes:
1.   CER = Cereals (Maize, Red Sorghum, White Sorghum and Millet); GN = Groundnut; CP = Cowpea.
2.   See (1).
3.   Total of the production of one growing season; totals of sales and purchases in one target consumption year.

SECTION 3

WATER AND SOIL CONSERVATION METHODS

The deterioration of land is strong in Burkina Faso: 24% of arable land has deteriorated badly (Kambou *et al.*, 1994). Efforts to preserve water and soil are made by the farmers in order to recover their production capital. Techniques such as that of *zaï*, stone walls, composting and others are developed in order to intensify production. Producers are often supported in their efforts by governmental development bodies or NGO's. In this section a fairly traditional basic technique will be presented, namely *zaï*.

A Traditional Technique: *Zaï* [8]

Zaï is an intensive technique for the management of manure and the preservation of water. In Yatenga in the north of Burkina Faso it is an ancestral practice used for regenerating the poorest parts of the fields. Nowadays it is widespread in all the zones of the country where deterioration is grave. It consists of digging holes in the ground, putting organic manure in them and finally sowing seeds in these holes. The holes have a depth of 10 to 15 cm and a diameter of 15 to 20 cm. The micro-environment thus created allows the plant better resistance against drought and the yields improve substantially. In Yatenga the yields come to 2000 kg/ha against 900 kg/ha on untreated fields (Agripromo, 1993). In a village such as Donsin where much land has deteriorated, *zaï* has just been introduced, average yields of 1200 kg/ha have been found in 1993

---

[8]   *Zaï* is a word in Mooré (language of the Mossi) which means " rushing, running to have to eat, digging".

with maximum levels of more than 4t/ha. (Kaboré *et al.*, 1995). *Zaï* helps to improve the yields of crops, but demands a great deal of effort as regards:

— labour: about 780 hours/ha for the preparation of the *zaï* (digging holes, transport of manure, seeds), for instance to dig between 20,000 and 25,000 holes/ha.

— organic matter: in general manure is usually not available in sufficient quantities and the production of compost requires much water, materials and work. Between 3 and 6 tons of compost has to be produced for each hectare of *zaï* (Agripromo, 1993).

Nevertheless, *zaï* remains a commendable technique for several reasons:

— the digging of holes is done at a time when the farmers are not too busy. The low opportunity costs of labour allows them to invest in this activity.

— *zaï* allows the regeneration of deteriorated soil: it is therefore an investment that has positive effects in the long term. For that reason the farmers in Donsin put it into practice, although in the course of the first year it yields less than a practice like 'mulching' on arable fields.

— the practice of *zaï* allows a change in the production system in that it improves the integration of agriculture/animal husbandry. This increases the productivity of the farm.

Linear Programming Model Under *Zaï*

The practice of *zaï* has been included in the linear programming model of the 'Exploitation Centrale' in three stages: (1) the addition of plots where *zaï* is practised to the total of plots already defined; (2) modification of some constraints in the model; (3) an estimation of the parameters (yields, agricultural calendars and labour hours) for the new plots.

With regard to the definition of the plots we have based ourselves on the studies by teams of INERA/RSP in the Central and North-West regions, which show that *zaï* is practised on the higher fields, and often in the most deteriorated parts (the *zippélé*, in the local language). The set of the soil types taken into consideration for the analysis in this section also includes the *zippélé* (see Appendix 1). In the *zaï* pockets only sorghum and millet are sown, mostly intercropped with cowpea. A particular problem came up with the following phenomenon: the *zaï* holes are used during several successive years, in other words, once the holes have been dug in a specific area, the farmers sow in the same holes for several consecutive years. Therefore the application of organic manure happens only once, just after the holes have been dug. In the following years they do not apply organic manure or just small doses, but they always have the benefit of the organic manure that stays behind in the holes. The number of years the same *zaï* pockets are used varies in the zone we studied between 3 and 5 years. After such a period they start digging holes and applying organic manure again. The cultivation cycles that are created in this way have explicitly been taken into account in this model in the following way:

We take a cycle of four years for the practice of *zaï* on a specific plot on the higher fields and an initial dosage of 16000 kg/ha of organic manure. This cycle can be introduced by dividing the plot where *zaï* is practised into four equal parts (see figure 5.3): one part where they have to dig holes and apply organic manure, and three parts where holes have been in use for one, two or three years; there they use the existing holes and do not add any organic manure. In the following year they use the holes that have been made this year, last year and the holes made two years ago. In the part of the plot that has three-year-old holes, these holes are no longer used. In this part they have to start digging new holes again (this schedule is shown in figure 5.3).

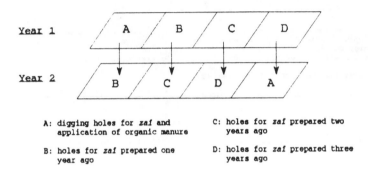

A: digging holes for *zaï* and application of organic manure

B: holes for *zaï* prepared one year ago

C: holes for *zaï* prepared two years ago

D: holes for *zaï* prepared three years ago

Figure 5.3: Schedule of a cultivation cycle for *zaï* on higher fields

For the deteriorated zones such as the *zippélé* the *zaï* also constitutes a method for regenerating land (see subheading `A Traditional Technique: *Zaï*'). We have introduced this concept by first assuming a cycle of three years on deteriorated soil and the same dose of 16,000 kg/ha of organic manure in the first year, and subsequently by assuming that after those three years of cultivation the soil will no longer be a *zippélé*, but will form part of the soil type cultivable high land.[9] In respect of the practice of *zaï* on the *zippélé* it is better to speak of a cultivation sequence, for each year one part of the plot is added to the higher land, thereby including an equal part of the non-cultivated *zippélé* area in the plot (see figure 5.4).

The modifications in the formulation of the constraints of the linear programming model especially refer to the land constraints and labour constraints for the digging of holes. The land constraints described in equation 2 and 3 in Appendix 1 of the linear programming model, also take into consideration the surface area of the *zippélé* that has been regenerated and added to the higher land through the practice of *zaï* (in a single year). The digging of holes can already start before the growing season (March, April) and can be continued until the time when sowing commences. The work is done by only adult men. In general they only do this work during a few hours per day because of the great physical effort demanded. The labour constraints with regard

---

[9] Here the term cultivable is used for the higher 'non-deteriorated' land, defined earlier. Such land should be distinguished from the deteriorated higher land (*zippélé*) that can only be cultivated by applying the zaï technique.

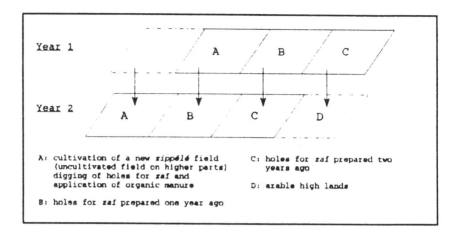

Figure 5.4: Schedule of a cultivation sequence for *zaï* on the *zippélé*

to the digging of holes are given in equation (38) of Appendix 1. The data on yields, the agricultural calendars and the labour hours have been obtained on the basis of the various studies made of *zaï*, especially the studies of INERA/RSP/Central region in Donsin. A lack of accurate information on the *zaï* should be pointed out. Very few data exist on this practice over a period of time, i.e. on the evolution of yields and working hours. Supplementary data are estimated on the basis of a survey in the village Baszaïdo and the experiences of the team of INERA/RSP/North-West (Sawadogo *et al.*, 1995). The following are some important elements: on the higher fields the yields increase enormously in the first year (between 3.75 times for red sorghum and 3 times for millet), because of the organic manure, but also because of the effects on the infiltration of water. The yields on the *zippélé* are lower compared with those on the higher grounds (20% lower). The yields decrease progressively each year (an average of about 15% per year). The working hours spent on digging holes are estimated at 450 hours/ha on the higher grounds and 650 hours/ha on the *zippélé*. The labour needed for fertilization are estimated at 250 hours/ha.

Discussion of the Results of the LP Model under *Zaï*

Table 5.4 shows some results of the linear programming model defined in appendix 1 (equations (1), (2'), (3), ... (38)). The 'Exploitation Centrale' practises *zaï* on approximately a quarter hectare on the higher grounds. They sow red sorghum intercropped with cowpea on these fields.

We recall that the holes are only dug on one quarter of that area (0.07 ha). On the remainder of the area they use holes that are already there. The modest integration of the practice of *zaï* into the agricultural activities of the 'Exploitation Centrale' has hardly changed

TABLE 5.4: Some Results of the Linear Programming Model

Crop	Soil type	Owner	*Zaï*	Manure (kg/ha)	Sowing period	Weeding intensity	Cultivated area (ha)	Fallowed area (ha)
Maize	low-2	common	no	4000	average	intensive	0.130	0.000
Maize	low-3	common	no	4000	average	intensive	0.035	0.000
R.Sorghum/Cowpea	low-3	common	no	800	1	intensive	0.011	0.007
Millet	high-3	common	no	0	4	intensive	0.554	0.554
Millet	high-3	common	no	0	5	intensive	0.581	0.581
Millet/Cowpea	high-1	common	no	800	2	intensive	0.009	0.004
Millet/Cowpea	high-2	common	no	800	2	intensive	0.301	0.129
Millet/Cowpea	high-3	common	no	0	2	intensive	0.350	0.350
Millet/Cowpea	high-3	common	no	0	3	intensive	0.244	0.244
Millet/Cowpea	low-3	common	no	800	1	intensive	0.069	0.030
Millet/Cowpea	low-3	common	no	0	1	intensive	0.150	0.150
Groundnut	high-2	common	no	0	average	average	0.290	0.290
Groundnut	high-3	common	no	0	average	average	0.059	0.059
Millet	high-1	individ.	no	0	4	intensive	0.025	0.025
Millet	high-3	individ.	no	0	4	intensive	0.126	0.126
Millet	high-3	individ.	no	0	5	intensive	0.173	0.173
Groundnut	high-2	individ.	no	0	average	average	0.100	0.100
Groundnut	high-3	individ.	no	0	average	average	0.056	0.056
R.Sorghum/Cowpea	high-1	common	yes	4000	1	average	0.258	0.000
							3.52	2.88

	t=8	t=9	t=10	t=11	t=12	t=13
Production (kg)						
- Maize	123	41	0			
- R.Sorghum	0	155	155			
- W.Sorghum	0	0	0			
- Millet	0	472	472			
- Groundnut	67	136	0			
- Cowpea	28	28	0			
Consumption (kg)						
- Maize	117	39	0	0	0	0
- R.Sorghum	8	8	9	50	50	25
- W.Sorghum	0	0	0	0	0	0
- Millet	0	82	122	325	332	396
- Groundnut	7	14	13	36	37	44
- Cowpea	11	0	0	0	0	0
Sales (kg)						
- R.Sorghum				132		
- Cowpea				38		
Purchases (kg)						
- R.Sorghum	8			0	0	
- Millet	0			0	378	
Stocks of produce (kg)						
- Maize	0	0	0	0	0	0
- R.Sorghum	0	144	287	102	51	25
- Millet	0	376	708	371	34	16
- Groundnut	46	138	123	83	44	0
- Cowpea	15	40	39	0	0	0
Finances (CFA)						
- EAR-EAE	7500	7500	0	-25000	20000	20000
- Sales				10356		
- Purchases	768				0	40480
Consumption						
- in 1000 kilocalories	520	520	520	1472	1501	1673
- in proteins (1000gm)	7	8	8	23	24	26
Food shortage						
- in 1000 kilocalories	0	0	0	0	0	0
- in proteins (1000gm)	0	0	0	0	0	0

Annual net revenues: 0 CFA

Source: Based on calculations in the model defined in Appendix 1 (equations (1), (2'), (3), ... (38))

its strategies. Nonetheless, it produces less groundnuts on the common fields, more red sorghum (by applying *zaï*), part of which is sold, and maize is no longer grown by applying a dose of 8000 kg/ha of organic manure, but a dose of 4000 kg/ha. The results of the 'Exploitation Centrale' are only slightly better. In fact, the net revenues are always 0, but the stocks at the end of the target consumption year have somewhat increased. Therefore it may be concluded that the effects seem only marginal.

In this model the principal constraint limiting the practice of *zaï* is the availability of organic manure. Organic manure is in the first place used for growing maize, because they have to meet their needs of food in the harvesting period, but the rest is to a large extent devoted to *zaï*. We have repeated our calculations while assuming that the availability of organic manure is increased by 1000 and 2000 kg (see Table 5.5). It is clear that the practice of *zaï* benefits considerably from a greater availability of organic manure. The areas where the *zaï* is practised show an increase by 54% and 158%, respectively. In addition, the ability to meet their food needs improves considerably, although the net revenues stay at zero.

TABLE 5.5: Some Results of the Linear Programming Model on Different Doses of Organic Manure

Availability of organic manure	Surface where organic manure has been applied (ha)		Shadow price of organic manure (FCFA)	Covering of nutritive requirements at the end of the target consumption year[1]		Net revenues (FCFA)
	no *zaï*	*zaï*		in 1000 kilocalories	in proteins kilocalories	
2000 kg	0.56	0.26	14	9%	11%	0
3000 kg	0.91	0.40	12	35%	39%	0
4000 kg	0.58	0.67	11	58%	64%	0

Note:
[1] By the stocks that remain at the end of the target consumption year

The digging of holes, according to our model, is not a limiting factor (the shadow prices are zero for the constraints of labour spent on digging holes). This is not surprising. Digging holes is very hard work and probably this burden of hard labour is the true constraint, not the time available during this period. Moreover, a subtle distinction must be made in the assumption that farmers do not have much to do in the periods before the season. Other activities, such as animal husbandry, activities other than farming (commerce, repair of houses, etc.), the fight against erosion in the off-season (construction of dikes), participation in social and religious activities and the effect of fasting in the fasting period on the work accomplished may play a big role.[10]

---

[10] On the basis of village studies in Donsin, Lowenberg-de Boer et al. (1993) calculate that the opportunity cost of labour for extra-agricultural activities in the dry season are 50 FCFA/hour; other authors have also found non-agricultural income to be important to the households in Burkina Faso, especially to those in the north (Reardon et al., 1988, 1992).

The 'Exploitation Centrale' practises *zaï* on the higher grounds. The *zippélé* do not enter in the solution. This conforms to the observations on the practice of *zaï* by the farmers in Donsin (Kaboré *et al.*, 1994). Yet, in the province of Yatenga *zaï* is especially practised on deteriorated grounds on condition that the run-off is not too high (Vlaar, 1992). A possible explanation is that the availability of land is more limiting in Yatenga than it is on average on the Central Plateau. Still, even if one diminishes the availability of high and low land in the 'Exploitation Centrale', bringing the *zippélé* into cultivation does not bring immediate results. The availability of high and low land at the 'Exploitation Centrale' must be decreased by more than 50% before it enters in the solution. On the other hand, in the case of Yatenga, it is not simply the high density of the rural population which reduces the availability of cultivable land, the process of land deterioration also plays an important role and reduces the fertility and cultural aptitude of the available land. The different aptitudes of the cultivable high grounds and the ever smaller differences between cultivable high land of lesser quality and the *zippélé* make the choice of sites to practice *zaï* more complicated. Moreover, the importance of regenerating soil through the practice of *zaï* must not be ignored, when the farmer has no cultivable land at all.[11] In the linear programming model discussed above, we have only included the increase in land area during one year; even if the regenerated part is small from one year to the next, the cumulation of regenerated plots over several years may not be negligible.

SECTION 4

CLASSIFICATION

Agro-ecological conditions and socio-demographic factors (densities, ethnic groups) vary considerably from place to place in the Central and North-West regions. These differences considerably influence farmers' strategies and have a strong effect on the parameters (rainfall, length of the growing season, availability and quality of cultivable land, yields, working hours). The distinction of zones based on agro-ecological, socio-economic and demographic criteria is important for a better understanding of farmers' strategies. Moreover, even within a zone notable differences may exist between one household and another. The analyses in the previous sections, and other sensitivity analyses show the strong influence of the availability of land, labour and manure on the strategies and the levels of food security in the 'Exploitation Centrale'.

A proper classification strongly depends on the objectives of the study one wants to conduct. For example, if one wants to study farmers' possibilities of intensifying the struggle against erosion, one will concentrate more on criteria such as the availability of land, the degree of deterioration of the land and the availability of labour (if the work demands a considerable investment in labour). In this section we concentrate on some key criteria which can guide the sectoral regional analysis discussed in the next section. In the field of the application of linear programming in agricultural production systems studies various authors have formulated criteria which would result in a classification which is as precise as possible (see e.g. Day (1963), Sheehy & McAlexander (1965), Lee (1966), Miller (1966), Buckwell & Hazell (1973), Paris & Rausser (1973)). These mathematical criteria deal with technical coefficients, quantities of

---

[11] The maintenance of user's rights can also play a role in the decisions to cultivate a field that has not been cultivated for some years (Sawadogo et al., 1995).

available resources and coefficients in the objective function of the linear programming model. In practice, these criteria are difficult to apply. They require detailed data about the individual households. Such data do not exist in our case. We have therefore followed a more practical approach, which will be illustrated for the North-West region. Based on agro-ecological and demographic criteria the North-West region is divided into zones and sub-zones; then, a number of production systems, which are well discernable, are distinguished on the basis of socio-economic characteristics. A representative household corresponds to a certain production system in a certain sub-zone.

The Different Zones in the North-West Region

The North-West region covers an area of 30817 km², equal to 11% of the national territory and is situated in the Soudano-Sahelian Zone which has an average rainfall of 400 to 800 mm from the north to the south with very significant variations in space and time. With regard to the soil there are different formations according to the location, which results in different levels of deterioration (more pronounced in the north and in the centre) and of fertility. The region has permanent water flows where irrigation and horticulture are practised. The region has an average population density of 41.1 inhabitants/km² with an average growth rate of 1.9%. This density is higher in the provinces of Passoré, 54.9 inh./km², of Bam, 40.9 inh./km² and of Yatenga, 40.6 inh./km². The province of Sourou is thinly populated with 28 inh./km² and is the zone where migrants from Passoré and Yatenga come to. The main crops are millet (which prevails in the north) and sorghum (white sorghum in the central part and red in the south). Millet and sorghum by themselves cover 80 to 90% of the cultivated areas. Irrigated fields do not cover more than 1% of the total of cultivated areas and are concentrated around the valley of Sourou, the Bam lake and the wells in the larger urban districts.

Raising of small ruminants and poultry takes up an important place in the region, in particular in the north. The distribution of livestock per household shows figures that are higher for the region than at national level for cattle (0.8 against 0.6), goats (1.1 against 0.8) and poultry (2.1 against 2) according to INERA (1994). In each province there are urban markets and they form the link between the rural zone and the towns for commercial transactions. Because of the proximity of Mali there are some important markets near the border in Yatenga (Thiou, Nogodoum) and in Sourou (Di, Gouran). These are mostly cattle markets and cereal markets and these are the places where the trade of animals for the interior of the country and for Cote d'Ivoire and Ghana takes place.

For the division in sub-zones we have based ourselves on the work of the 'Institut d'Etudes et Recherches Agricoles' (INERA, 1994). The criteria applied are: rainfall, morphology and pedology, the degree of degradation of natural resources, the crops and animal husbandry, the density of the population and the major ethnic groups; the region can be split up in three, more or less homogeneous zones. Within each zone, sub-zones can be demarcated on the basis of more detailed differences in soil and rainfall conditions. The following demarcation can be proposed for the North-West region (Figure 5.5):

A.   The northern zone, which is made up of the districts of Banh, Kaïn ad Sollé shows rather pronounced differences along a north-south gradient. This zone is characterized by a low rainfall (440 to 500 mm), a peneplain ground profile with formations of sandy dunes. When one considers the natural resources one notices that the deterioration is light on the one hand because of the low density of population (15 inh./km²) which is dominated by the Peuhl and the Rimaïbé, and on the other hand because of the low degree of erosion. The production systems are dominated by pastoral semi-transhumant animal husbandry

106

and a cropping system based on millet grown on sandy soil. The shortage of food crops is chronic in this zone where one finds *pastoralists* and *agro-pastoralists*. The pastoralists practise transhumance either in the direction of Mali or in the direction of the provinces of Sourou, Sanguié or the Sissili.

B. The central and south zone (Ouahigouya-Yako-Kongoussi). This zone is characterized by a high population density, mostly Mossi (50 inh./km²), and a high degree of deterioration of natural resources. The production systems are characterized by the dominant cultivation of cereals (millet and sorghum) in combination with the raising of small ruminants. The deficit in food crops is chronic. When considering the rainfall, the soil and the water resources near the surface, the zone can be sub-divided in five sub-zones:

— The sub-zone of Ouahigouya has an average rainfall of 500 to 600 mm and laterite crusts; the deterioration of the cultivable grounds is very serious.

— The sub-zone of Tikaré, Guibaré and Saboé where the rainfall is the same as in the Ouahigouya sub-zone has hills formed in an early geological period that have allowed the formation of ravines and shallows. There are brown eutrophic tropical soils.

— The sub-zone of Kongoussi, is formed by the Bam lake basin. The soil is either hydromorphous and little humified or little developed by alluvial hydromorphous deposits. The lake makes it possible to irrigate crops (horticulture).

— The sub-zone of Yako shows characteristics similar to those of Ouahigouya, but with an average rainfall of 600 to 700 mm per year and a deficit in food crops which, however, is not permanent.

— The sub-zone of Samba and Pilimpikou has 'birrimian' hills and a rainfall between 600 and 700 mm. Apart from the food crops cultivated in the zone (millet and sorghum) some groundnuts and cotton are cultivated. On the shore of the Bam lake the *horticultural farmers* grow French beans and other vegetables (onions, tomatoes, cabbages) under the management of the two large cooperatives: the SOCOBAM and the COMAKO.

C. The south-west zone. This zone is the most successful in agriculture. The average rainfall is between 600 and 700 mm with a low deterioration of natural resources. This zone is the one where migrants from Yatenga and Passoré have settled. Three sub-zones can be distinguished:

— The sub-zone of Tougan-Di-Toma: the peneplain ground profile shows a levelling-off of laterite crusts. The soil is ferruginous and tropical and occasionally there are lithosols on the crusts. Apart from millet, sorghum and groundnuts, the raising of small ruminants with a transit trade of cattle characterises this part of the zone. The density of the population is 30 inh./km² and the main ethnic groups are the Samo and the Dafing.

— The sub-zone of the Sourou valley. The soils are either entrophic brown, or tropical vertisols or hydromorphous with little humus formation. Irrigation is applied and the pluvial cereal crops (sorghum) are intercropped with irrigation crops, which are sold (rice and horticultural products). It is a zone where cattle-raisers send their cattle. The

107

population density is 27.5 inh./km² on average and the zone is favoured by migrants. The state has arranged perimeters which are maintained by the farmers. The cultivable potenthe land in the zone is 7250 ha.

— The sub-zone of Yé-Gossina. This consists of very flat plains with a hydrographic network that has hardly been mapped. The soil is hydromorphous and little humified. It consists mostly of soil that is little developed by alluvial hydromorphous deposits. There are good water resources and cereal crops are cultivated together with cotton. The average population density, dominated by Dafing is low (20 inh./km²).

Like in the previous zone there are, in the entire zone, manual farmers and farmers who apply animal traction. At the level of the second sub-zone horticulture and pluvial and irrigated rice are practised by producers set up by the state.

Classification of the Agricultural Households

In each zone of the North-West region (Figure 5.5), five types of production systems can be observed which are quite different. They are:

(I)   pastoralists who live on the products of cattle breeding and the occasional sale of a few animals, often practising transhumance;

(ii)   agro-pastoralists who cultivate fields of millet and keep a few cows and many small ruminants;

Figure 5.5: Different Zones of the North-West Region of Burkina Faso

(iii)   manual subsistence farmers;

(iv)   farmers with draught animals (traction by oxen, donkeys and horses);

(v)   horticulturalists who have their farms around the established perimeters of Sourou and the Bam lake.

The differences between these five production systems are evident. The survey of the North-West region above has established which production systems can be found in each sub-zone. In our research a representative household corresponds to a certain production system in a certain sub-zone. The structure of the linear programming models for the representative households differs for the different production systems, differences between sub-zones lead to different values of parameters in the models. The important question remains whether our choice of representative households is detailed enough. To a large extent this depends on the availability of information and data, on time and costs of collecting data and on the capacity of managing the complexity of the analysis. Certainly, other criteria to distinguish representative households could be defended as well. For instance, Dugué (1989) distinguishes in Yatenga between households with a high emigration rate and those with a low emigration rate. Households with a low emigration rate are supposed to invest more in the intensification of their agricultural production system (provided that they are capable of doing so). We have decided to restrict ourselves to the classification outlined above. Other differences between households are analysed by applying sensitivity analyses.

In order to illustrate the appropriateness of the classification we present here some data of the villages of Baszaïdo, Kalamtogo and Lankoé (see Figure 5.1), collected by the research team INERA/RSP in Tougan. The population of Baszaïdo and Kalamtogo consists of the Mossi, and in Lankoé live four ethnic groups: Mossi, Samo, Peulh and Rimaïbé. In the villages we do not find the production systems (I) and (v). In Baszaïdo and Kalamtogo we find only crop farmers, in Lankoé both crop farmers and agro-pastoralists. The last ones are Peulh and Rimaïbé. Between the households a great difference exists between the autochthones (Peulh, Rimaïbé and Samo) and the immigrants (Mossi).

Table 5.6 shows some differences between the households in the three villages as regards equipment and ethnic group. This table shows the clear distinction between production systems (ii), (iii) and (iv) indeed. Ethnic differences coincide with different production systems. The table shows as well that some important characteristics (e.g. size of the household, arable surface per household, cattle) differ significantly between production systems.

As can be seen, the households with equipment have more resources in land, labour and cattle than those without (in most of the cases this difference is significant). For the village of Lankoé differences between ethnic groups and between levels of land and cattle resources imply different agricultural production strategies. For instance, the Samo, although they are autochthonous and landowners, have less land than the Mossi who are immigrants and are at the demand end where land is concerned. This situation can be explained by the large number of active members of the Mossi households and the poor fertility of the fields they have on loan, which favours extensification. They grow much more millet (which is easy to grow) inter-cropped with cowpea. The Samo, on the other hand, have the most fertile fields and grow tobacco with a lot of manure on the fields close to their compounds in the dry season so that in the rainy season these fields are very fertile and produce good yields of maize and sorghum. In this way the Samo can cultivate their smaller fields intensively. The Peuhl and the Rimaïbé, who are traditionally cattle farmers have a larger number of TLU than the other ethnic groups, but less cultivable land which they fertilize with the droppings of their animals.

109

TABLE 5.6: Some Data on the Households in the Villages of Baszaïdo, Kalamtogo and Lankoé in the North-West Region of Burkina Faso

	Average population	Average number of actives	Emigration rate	Livestock (in TLU[1])	Average surface (ha)	Surface-/active
Baszaïdo						
- not equipped households	8.4	3.7	13%	1.06	3.5	0.94
	(0.51)	(0.58)		(1.32)	(0.10)	
- equipped households	18.9	7.9	8%	6.69	7.0	0.89
	(0.70)	(0.69)		(1.34)	(0.06)	
Kalamtogo						
- not equipped households	9.4	4.4	4%	1.11	7.5	1.70
	(0.40)	(0.40)		(0.93)	(0.16)	
- equipped households	23.3	10.5	37%	3.48	13.2	1.25
	(0.70)	(0.50)		(0.79)	(0.28)	
Lankoé (Mossi households)						
- not equipped households	13.8	5.7	15%	0.73	3.2	0.60
	(0.62)	(0.81)		(1.34)	(0.10)	
- equipped households	24.5	9.2	12%	6.89	6.2	0.67
	(0.65)	(0.64)		(1.50)	(0.05)	
Lankoé (Samo households)						
- not equipped households	11.0	5.2	9%	0.03	2.7	0.51
	(0.49)	(0.56)		(2.29)	(0.13)	
- equipped households	17.7	9.6	7%	1.55	3.3	0.34
	(0.45)	(0.50)		(1.14)	(0.14)	
Lankoé (Peulh + Rimaïbé households)						
- not equipped households	13.4	7.2	10%	3.80	1.7	0.27
	(0.60)	(0.90)		(1.37)	(0.11)	
- equipped households	13.8	5.8	10%	18.59	2.2	0.38
	(0.36)	(0.57)		(0.96)	(0.21)	

Note: Variation coefficient between brackets
1 TLU = Tropical Livestock Unit: 1 TLU = 1 cow, 10 goats, 10 sheep, 2 donkeys or 2 horses.

SECTION 5

REGIONAL AGRICULTURAL MODEL

The model presented in the foregoing is a micro-economic model which describes one representative household. In order to study interventions in the field of functioning of markets, price formation and credit programmes we make use of an agricultural sector model for the region presented in the section labelled 'Classification'. In such a model all representative households for the region are included. In this section we do not discuss the detailed structure of the model, but rather some ideas on which the construction of the model is based.

Regional Sector Models

The regional model of the agricultural sector deals with the strategies of production, consumption and marketing by the farmers, the strategies of non-producing consumers (who do not produce themselves or only a little bit) and the strategies of the merchants.

Basically, our regional sector model is a combination of the following elements:(1) the models for the representative households distinguished in the section labelled 'Classification';

110

(2) a model to describe the consumption by non-producing consumers; (3) a description of the strategies of the merchants, dependant on supply and demand of the producers and consumers, the agricultural prices, transaction costs (e.g. storage and transport costs) and the condition of communication networks; and (4) an objective function representing at the same time the objectives of the various actors.

The description of the merchants' strategies is the main element of the sector model. In most of the years, the region is a shortage area, which means that merchants have to import cereals. In the model, the merchants receive a certain net margin for each transaction and maximize their profit. When constructing the model for the merchants, we assumed that there is, per sub-zone, a single market and a single aggregated merchant who purchases agricultural products right after the harvest, sells the purchased products during the lean period, and is limited by his capital availability. In the literature concerning sector studies, it is in general assumed that the markets are perfect (i.e. homogenous products, no market barriers, and perfect knowledge of prices) and competitive (i.e. a lot of consumers and suppliers who have no influence on price formation). In the present literature these suppositions are gradually relaxed (see e.g. Janvry, Fafchamps and Sadoulet (1991)). Probably, these suppositions are satisfied on some urban markets. However, the situation is often different in the rural markets of our region. We observe differences during the year: after the harvest, a lot of producers supply a part of their harvest at the market to acquire a certain revenue; nevertheless, the number of merchants will be limited because the supply will be limited; during the lean period, the demand on the rural markets increases, which attracts merchants; however, also now the number of merchants will be limited because the purchasing power of the consumers is limited; between these two periods, the demand and supply will be restricted, which causes the number of merchants on the market to be limited. Yonli (1988) notifies that the character of the rural markets has become rather oligopolistic because of the disappearance of a lot of producers on the markets and since the supply on the markets is provided by a reduced number of local merchants (Yonli, 1988). A consequence of this phenomenon is that the merchants in the North-West region of Burkina Faso can influence the prices they demand or pay.

These observations explain why we follow another approach than many other authors. Merchants have a certain freedom in determining a price. This freedom is, however, limited, i.e. consumers will move towards other markets when prices at a particular market are too high. The merchant will demand and pay a price that avoids clients to go to other markets and that gives him a profit that is reasonable. The possibility to go to other markets engenders the prices of all markets in the region, as well as prices on markets outside the region (e.g. the national market in Ouagadougou) are linked. The differences between the prices on the different markets will not be higher than the costs of a client to go to another market. The differences will not be less than the transaction costs to import the products from another market.

The regional agricultural sector model combines models for the different actors. The important element of the interaction between these models is the price formation. Per sub-zone, producer and consumer prices are introduced. The consumer price is obtained by adding to the producer price the storage and transport costs and a profit margin of the merchant. We introduced the concept of a reasonable profit, which is equal to the opportunity costs of invested capital. This reasonable profit allows the merchant to obtain an acceptable profit level. When introducing endogenous prices, the model is not linear. That is why equilibrium prices (consumer and producer prices) are determined by an iterative optimisation process.

Conclusions

If we are to take the results of our models prudently, it is clear that there is no simple solution to attain higher and more sustainable food security levels. In fact, sensitivity analysis seems to indicate that it will be difficult, yes, even impossible, for most households in the Central and North-West regions of Burkina Faso to meet the food needs of the members of the household and to preserve the natural resources (land, water, vegetation) without help from outside. Intensification of agriculture (i.e. increasing yields) is necessary to meet the needs of the rural (and urban) population in the future. Such intensification can only be achieved by stopping erosion of the soil and by improving its fertility significantly. The investments necessary (for instance, construction of permeable dikes, transport of stones for anti-erosion works, chemical fertilizers, carts, animal traction and equipment for ploughing and ridging, etc.) do not seem to be within the reach of the farmers.

The challenge which agricultural research can take up is working out not only techniques to improve the agricultural production in the short term, but also scenarios for a sustainable intensification of agriculture in the long term. Intensification of the current farming systems can only be based on improvement (or intensification) of the 'traditional' techniques (integration agriculture — animal husbandry, agro-forestry, local methods of preserving water and soil, etc.) and the careful integration of other techniques such as chemical fertilization. Solutions may vary from one location to the next and also depend on the type of household, which shows the importance of (countrywide) agricultural research oriented towards these activities in different regions of Burkina Faso. Moreover, agricultural research should not concentrate on technical questions only. The intensification of production systems is also, and maybe first of all, a matter of organising the local population (including organisation of credit systems, supply of inputs). This is the reason why in all studies the cooperation of the rural population is of vital importance. A lesson we learned from the past is that we must not make any plans without involving the rural population. By discussing the various options and by sharing the knowledge of everyone involved it will be possible to arrive at ideas for a more permanent agricultural development and food security.

The linear programming models presented in this document are the means used to investigate and analyse the farmers' strategies. They represent an important means to profit from secondary data collected in several village studies and offer great possibilities of analysis, both at the household level and at the level of the entire sector. This type of model which has been used for several years in agricultural research in Africa (Eicher and Baker, 1982) has only recently been introduced in the national agricultural research programme in Burkina Faso (Ouédraogo, 1995). The linear programming models defined in this chapter seem to describe in a fairly satisfactory manner the average socio-economic conditions in the households in the Central Plateau, the farming methods and techniques available to the farmers and the factors and criteria steering and limiting their production strategies. Nevertheless, there are many important factors that have not been included in this model. The study of farmers' strategies and food security is very complex and can only be approached by integrated studies of an interdisciplinary nature (Schweigman et al., 1990; Maatman et al., 1992). This study is only one component of such an approach. The results of the calculations should be interpreted with great caution. In our opinion they are not the most important results of the present study. When formulating a linear programming model the researchers are compelled to reflect profoundly on the concepts to be used (for instance, what exactly is a strategy), the decisions that have to be taken into account, the relations between decisions and factors, etc. During this process, step by step, a better understanding of the food security problems is obtained. For that reason linear programming seems to be a useful instrument for analysing food security problems and for the search for appropriate solutions.

REFERENCES

Agbessi Dos-Santos, H., Damon, M. 1987. "Manuel de nutrition africaine: éléments de base appliqués". Institut Panafricain pour le Développement (IPD) and Editions Karthala.

Agripromo 1993. "Témoignage: le *zaï* nous permet de manger". In: "Gérer la fertilité des sols", Agripromo no. 83 Octobre 1993. 1993 "Fiche Technique: comment faire le zaï". In: "Gérer la fertilité des sols", Agripromo no. 83 Octobre 1993.

Ancey, G. 1975. "Niveaux de decision et functions objectif en milieu Africain". INSEE, AMIRA Note de Travail No. 3. Paris.

Bakker-Frijling, M.J., Konaté, G. 1988. "La demande alimentaire, la consommation alimentaire et l'état nutritionnel de la population du Plateau Mossi, Burkina Faso". Projet CEDRES/A-GRISK, Université de Ouagadougou et Université de Groningen.

Bleiberg, F.M., Brun, T.A., Goihman, S., Gouba, E. 1980. "Duration of activities and energy expenditure of female farmers in dry and rainy seasons in Upper Volta". *British Journal of Nutrition* 43: 71-82.

Broekhuyse, J.Th. 1982. "Production et productivité agricole dans la savanne sèche". CEBEMO, 's-Hertogenbosch, Pays-Bas. Institut Royal des Tropiques, Amsterdam. Pays-Bas.

Broekhuyse, J.Th. 1983. "Economie rural des Mossi: Tomes I-II". Institut Royal des Tropiques, Amsterdam. Pays-Bas.

Broekhuyse, J.Th. 1988. "De la survivance à la suffisance; étude des problèmes et des perspectives du développement aboutissant à une idéologie paysanne au Plateau Nord des Mossi". Projet CEDRES/AGRISK, Université de Ouagadougou et Université de Groningen.

Brun, Th.A., Bleiberg, F., Goihman, S. 1981. "Energy expenditure of male farmers in dry and rainy seasons in Upper Volta". *British Journal of Nutrition* 45: 67-75.

Buckwell, A.E., Hazell, P.B.R. 1972. "Implications of aggregation bias for the construction of static and dynamic linear programming supply models". *American Journal of Agricultural Economics* 54: 119-134.

Day, R.H. 1963. "On aggregating linear programming models of production". *Journal of Farm Economics* 45: 797-813.

Delgado, C.L. 1978. "Livestock versus foodgrain production in Southeast Upper-Volta: a resource allocation analysis". Monograph in the Entente Livestock Project Series, Center for Research on Economic Development (CRED), University of Michigan, USA.

Dugué, M.J. 1987. "Variabilité régionale des systèmes agraires au Yatenga. Consequences pour le développement". INERA/DSA/CIRAD, Projet de Recherche - Développement, Ouagadougou, Burkina Faso.

Dugué, P. 1989. "Possibilités et limites de l'intensification des systèmes de cultures vivriers en zone Soudano-Sahelienne: le cas du Yatenga (Burkina Faso)". Collection 'Documents Systèmes Agraires', No. 9. Institut d'Etudes et de Recherches Agricoles (I.N.E.R.A.), Burkina Faso et Département Systèmes Agraires du CIRAD, France.

Eicher, G.K. and Baker, D.C. 1982. "Research on Agricultural Development in Sub-Saharan Africa: a Critical Survey". Michigan State University, East Lansing. International Development Paper 1.

Feldstein, H.S., Poats, S.V. (Eds.) 1989. "Working Together: Gender Analysis in Agriculture". Volume 1: Case Studies. Kumarian Press. Connecticut, USA.

Gué, N.J. 1995. "Stratégies paysannes de consommation et vulnérabilité saisonnière dans la zone Nord-Ouest du Burkina Faso". INERA/RSP Zone Nord-Ouest.

Imbs, F. 1987. "Kumtaabo, une collectivite rurale Mossi et son rapport à l'espace". Editions de l'ORSTOM, Collection Atlas des Structures Agraires au Sud du Sahara No. 21, Paris.

INERA 1994. "Diagnostic des contraintes et potentialités: définition des axes de recherche pour le CRRA Nord-Ouest". INERA, Juin 1994.

Jaeger, W.K. 1984. "Agricultural Mechanisation: the Economics of Animal Traction in Burkina Faso". Ph.d. Thesis, Stanford University.

Jaeger, W.K. 1987. "Agricultural Mechanisation: The Economics of Animal Draft Power in West Africa". Westview Press, Boulder, Colorado.

Janvry, A. de, Fafchamps, M., Sadoulet, E., 1991. "Peasant Household Behaviour with Missing Markets: Some Paradoxes Explained". *The Economic Journal*, 101: 1400-1417.

Kaboré, D., Kambou, F., Dickey, J., Lowenberg-DeBoer, J. 1994. "Economie des cordons pierreux, du paillage et du zaï dans le nord du plateau Central du Burkina Faso, une perspective préliminaire". In: Lowenberg-DeBoer et al. (Eds.), 1994.

Kambou, N.K., Taonda, S.J.-B., Zougmoré, R., Kaboré, B., Dickey, J. 1994. "Effet des pratiques de conservation des sols sur l'évolution de la sédimentation, des états de surface et des rendements de mil d'un site érodé à Yilou, Burkina Faso". In: Lowenberg-DeBoer et al. (Eds.), 1994.

Kohler, J.M. 1971. "Activités agricoles et changements sociaux dans l'ouest-mossi, Haute Volta". Editions de l'ORSTOM, Mémoires ORSTOM No. 46, Paris.

Konaté, G. 1988. "Introduction a l'étude des structures démographiques et de l'organisation sociale Mossi". Projet CEDRES/AGRISK, Université de Ouagadougou et Université de Groningen

Lang, M.G., Roth, M., Preckel, P. 1984. "Risk Perceptions and Risk Management by Farmers in Burkina Faso". Paper presented at the Farming Systems Research Symposium, Kansas State University, Manhattan, Kansas, October, 1984.

Lee, J.E. 1966 "Exact Aggregation: A Discussion of Miller's Theorem". Agricultural Economics Research 18: 58-61.

Lowenberg-DeBoer, J., Tahirou, A., Kaboré, D. 1993. "The Opportunity Cost of Capital for Agriculture in the Sahel: Case Evidence from Niger and Burkina Faso". INERA/RSP. Projet ARTS.

Lowenberg-deBoer, J., Boffa, J.-M., Dickey, J., Robins, E. (Eds.) 1994. "Recherche intégrée en production agricole et en gestion des ressources naturelles: projet d'appui à la recherche et à la formation agricoles (ARTS)". Rapport Technique. Purdue University et Winrock International.

Maatman, A., Schweigman, C., Thiombiano, T., van Andel, J. 1992. "Food security and Sustainable Agriculture on the Central Plateau in Burkina Faso: Observations on a Research Agenda". Tijdschrift voor Sociaal Wetenschappelijk Onderzoek van de Landbouw (TSL), jaargang 7, no. 2.

Maatman, A. et C. Schweigman 1995a. "Etude des systèmes de production agricole du Plateau Central au Burkina Faso: application de la programmation linéaire: Tome 1" (à paraître). Document de Recherche du Projet "Analyze des Stratégies Paysannes", Réseau SADAOC. INERA/RSP Tougan, Burkina Faso. Université de Ouagadougou, Burkina Faso et Université de Groningen, Pays-Bas.

Maatman, A. et C. Schweigman 1995b. "Etude des systèmes de production agricole du Plateau Central au Burkina Faso: application de la programmation linéaire: Tome 2" (à paraître). Document de Recherche du Projet "Analyze des Stratégies Paysannes", Réseau SADAOC. INERA/RSP Tougan, Burkina Faso. Université de Ouagadougou, Burkina Faso et Université de Groningen, Pays-Bas.

Marchal, J.Y. 1983. "Yatenga, Nord Haute-Volta: la dynamique de l'espace rural soudano-sahélien". Traveaux et documents de l'ORSTOM No. 167. ORSTOM, Paris.

Matlon, P.J., Cantrell, R., King, D., Benoit-Cattin, M. (Eds.) 1984. "Coming Full Circle. Farmer's Participation in the Development of Technology". International Development Research Centre (IDRC), Ottawa, Canada.

Matlon, P.J. 1987. "The West African semi-arid tropics". In Mellor, Delgado, Blackie (Eds.).

Matlon, P.J., Fafchamps, M. 1988. "Crop Budgets for three Agroclimatic Zones for the West African Semi-Arid Tropics". ICRISAT, Resource Management Program, Economic Group, Progress Report 85. ICRISAT, Patancheru, India.

Matlon, P.J. 1990. "Improving Productivity in Sorghum and Pearl Millet in Semi-arid Africa". Food Research Institute Studies, Vol. XXII, No.1.

McIntire, J. 1981. "Two aspects of farming in Upper Volta: animal traction and intercropping". ICRISAT, West Africa Economics Program, Village Studies Report No. 7. ICRISAT, Ouagadougou, Upper Volta.

McIntire, J. 1983. "Crop Production Budgets in Two Villages of Central Upper Volta".

ICRISAT, West Africa Economics Program, Progress Report No. 4. ICRISAT, Ouagadougou, Upper Volta.

Mellor, J.W., Delgado, C.L., Blackie, M.J. 1987. Accelerating food production in Sub-saharan Africa". Baltimore, MD: Johns Hopkins University Press.

Miller, T.A. 1966. "Sufficient Conditions for Exact Aggregation in Linear Programming Models". Agricultural Economics Research 18:52-57.

Ministère de l'Agriculture et de l'Elevage (MAE) 1988. "Enquête d'envergure de statistique agricole dans l'Ex-Ord du Centre-Nord, Campagne 1986/1987". Rapport Final. Ouagadougou, Burkina Faso.

Nagy, J.G., Ohm, H.W., Sanders, J.H. 1986. "Cereal Technology Development - West African Tropics: a Farming Systems Perspective". SAFGRAD/FSU, End of Project Report. International Programs in Agriculture, Purdue University, West Lafayette, Indiana.

Nagy, J.G., Ohm, H.W., Sanders, J.H. 1989. "A Case Study of the Purdue University Farming Systems Project". In: Feldstein et Poats (Eds.).

Norman, D.W. 1974. "Rationalising Mixed Cropping under Indigenous Conditions: The Example of Northern Nigeria". *Journal of Development Studies,* Vol. 11, No. 1: 3-21.

Ohm, W.H., Nagy, J.G. (Eds.) 1985. "Appropriate Technologies for Farmers in West Africa". Proceedings of a Workshop Conducted under the Auspices of the Semi-arid Food Grain Research and Development Program (Safgrad). Safgrad/fsu, International Programs in Agriculture, Purdue University, West Lafayette, Indiana.

Ouédraogo, S. 1995. "La modélisation comme outil de simulation intégrée". In: Interprétation agronomique des données du sol: un outil pour la gestion des sols et le développement agricole". AB-DLO Thema's 2. Séminaire BUNASOLS/AB-DLO/INERA, 14-16 Mars 1995.

Prudencio, Y.C. 1983. "A Village Study of Soil Fertility Management and Food Crop Production in Upper-Volta: Technical and Economic Analysis". Ph.D. thesis, Department of Economics, University of Arizona, USA.

Prudencio, Y.C. 1987. "La gestion paysanne des sols et des cultures au Burkina Faso: implications pour la recherche et le développement agricole". Edition préliminaire, Mars, 1987. Organisation de l'Unité Africaine (OAU)/Commission Scientifique Technique et de la Recherche (CSTRC)/Semi-Arid Food Grain Research and Development (SAFGRAD). Ouagadougou, Burkina Faso.

Paris, Q., Rausser, G.C. 1973. "Sufficient Conditions for Aggregation of Linear Programming Models". *American Journal of Agricultural Economics* 55:659-666.

Reardon, T., Matlon, P., Delgado, C. 1988. "Coping with Household-level Food Insecurity in Drought-affected Areas of Burkina Faso". *World Development*, Vol. 16, No. 9: 1065-1074.

Reardon, T. and P. Matlon 1989 "Seasonal Food Insecurity and Vulnerability in Drought-affected Regions of Burkina Faso". In: Sahn (Ed.).

Reardon, T., Delgado, C., Matlon, P. 1992. "Determinants and Effects of Income Diversification Amongst Households in Burkina Faso". *Journal of Development Studies*, January, 1992.

Robins, E., Sorgho, M.C. 1994. "Enquête d'opinion: évaluation paysanne de la confection de zaï à Donsin en 1993". INERA. Programme RSP. Station de Recherches Agronomiques de Kamboinsé.

Roth, M., Abott, P., Sanders, J., McKinzie, J. 1986. "An Application of Whole Farm Modelling to New Technology Evaluation, Central Mossi Plateau, Burkina Faso". SAFGRAD/FSU, International Programs in Agriculture, Purdue University, West Lafayette, Indiana.

Roth, M. 1986. "Economic Evaluation of Agricultural Policy in Burkina Faso, a Sectoral Modelling Approach". Ph.D. Thesis, Department of Agricultural Economics, Purdue University, West Lafayette, Indiana.

Ruthenberg, H. 1980. "Farming Systems in the Tropics". 3rd ed. Clarendon Press, Oxford University.

Sahn, D.E. (Ed.) 1989. "Seasonal Variability in Third World Agriculture: The Consequences for Food Security". Published for the International Food Policy Research Institute (IFPRI). The Johns Hopkins University Press.

Sawadogo, H., Ouédraogo, M., Kaboré, D., Maatman, A. 1995. "Le *zaï*, une technique de conservation et de récupération des sols; aperçu des résultats des études villageoises dans les zones Centre et Nord-Ouest du Burkina Faso". INERA/RSP/Zone Nord-Ouest.

Schweigman, C., Snijders, T.A.B., Bakker, E.J. 1990. "Operations Research as a Tool for Analysis of Food Security Problems". *European Journal of Operational Research* 49 (1990): 211-221. North-Holland.

Shaner, W.W., Philipp, P.F., Schmehl, W.R. 1982. "Farming Systems Research and Development: Guidelines for Developing Countries". Boulder: Westview.

Sheehy, S.J., McAlexander, R.H. 1965. "Selection of Representative Benchmark Farms for Supply Estimation". *Journal of Farm Economics* 47: 681-695.

Sherman, J.R., Shapiro, K.H., Gilbert, E. 1987. "The Dynamics of Grain Marketing in Burkina Faso: Volume 1: An Economic Analysis of Grain Marketing". Final Report: Burkina Faso Marketing Development Research Project. Center for Research on Economic Development, University of Michigan and International Agricultural Programs, University of Wisconsin.

Singh, R.D., Kehrberg, E.W., Morris, W.H.M. 1984. "Small Farm Production Systems in Upper Volta: Descriptive and Production Function Analysis". SAFGRAD/FSU, Department of Agricultural Economics, Purdue University, West Lafayette, Indiana.

Singh, R.D. 1988. "Economics of the Family and Farming Systems in Sub-saharan Africa: Development Perspectives". Westview Press, Boulder, Colorado.

Steiner, K.G. 1984. "Intercropping in Tropical Smallholder Agriculture with Special Reference to West-africa". Deutsche Gesellschaft für Technische Zusammenarbeit (GTZ), West Germany.

Tallet, B. 1985. "Genèse et evolution des exploitations agricoles familiales dans les milieux de savannes ouest-africaines". Thèse de Doctorat 3ème cycle, Juin 1985. Université de Paris.

Thiombiano, T. 1987. "Le rôle des prix dans les décisions de produire et de vendre". CEDRES-Etudes, Revue Economique et Sociale Burkinabé. Spécial, Dixième Anniversaire.

Thiombiano, T., Soulama, S., Wetta, C. 1988. "Systèmes alimentaires au Burkina Faso". Série no. 1. Résultats de Recherche, Centre d'Etudes, de Documentation et de Recherches Economiques et Sociales (CEDRES). Juin, 1988. Université de Ouagadougou.

Vlaar, J.C.J. (Ed.) 1992. "Les techniques de conservation des eaux et des sols dans les pays du Sahel". Comité Interafricain d'Etudes Hydrauliques (CIEH), Burkina Faso et Université de Wageningen (UAW), Pays-Bas.

Yonli, E. P. 1988. "Marchés et prix dans l'approvisionnement du Plateau Mossi en sorgho et mil". Projet CEDRES/AGRISK, Université de Ouagadougou et Université de Groningen.

APPENDIX 1: Linear Programming Model for The 'Exploitation Centrale'

In the model the concept plot refers to a piece of land:
— which is cultivated with maize (*MA*), red sorghum (*RS*), white sorghum (*WS*), millet (*MI*) or groundnut (*GN*) as monocrop, or sorghum or millet intercropped with cowpea (*CP*).
— which is sown in period 1, 2, 3, 4 or 5 (see below)
— which is weeded intensively or extensively
— which is of a certain soil type (see below)
— where a certain dose of organic manure is applied (0, 800, 2000, 4000 or 8000 kg/ha).

Two nutrient types are taken into account: energy (*KC*) and proteins (*PR*). The planning period is divided in the following periods, which are numbered t=1, 2, ..., 13:

1   the first half of the month of May (1 - 15 May)
2   the second half of the month of May (16 - 31 May)
3   the first half of the month of June (1 - 15 June)
4   the second half of the month of June (16 - 30 June)
5   the first half of the month of July (1 - 15 July)
6   the second half of the month of July (16 -31 July)
7   the month of August (1 - 31 August)
8   the month of September (1 - 30 September)
9   the month of October (1 - 31 October)
10   the month of November (1 - 30 November)
11   the months of December, January and February (1 December - 28 February)
12   the months of March, April and May (1 March - 31 May)
13   the months of June, July and August (1 June - 31 August)

Eight categories of land ("soil types") are distinguished in section entitled "Linear Programming Model":

s_1:   common fields situated at a short distance from the dwellings (less than 100 m). These are in general small fields which are supposed not to be at the lower parts of the toposequence;
s_2:   common fields at sandy soils situated at the higher and medium parts of the toposequence, at some distance from the compound (100m - 1000m);
s_3:   common fields at sandy soils situated at the higher and medium parts of the toposequence, at a larger distance from the compound (more than 1000m);
s_4:   common fields at clayey soils situated at the lower parts of the toposequence, at some distance from the compound (100m - 1000m);
s_5:   common fields at clayey soils situated at the lower parts of the toposequence, at a larger distance from the compound (more than 1000m);
s_6:   individual fields situated close to the dwellings (less than 100 m);
s_7:   individual fields at sandy soils situated at higher and medium parts of the toposequence, at some distance from the compound (100m - 1000m);
s_8:   individual fields at sandy soils situated at higher and medium parts of the toposequence, at a larger distance from the compound (more than 1000m).

Under the sub-heading "Linear Programming Model", three other categories are added:

s_9:   degenerate fields (*zippélé*), situated close to the compound (less than 100m);
s_{10}:   degenerate fields (*zippélé*), at some distance from the compound (100m - 1000m);

119

s_{11}: degenerate fields (*zippélé*), at a larger distance from the compound (more than 1000m).

We define the following sets:

P	$=$	$\{ MA, RS, WS, MI, GN, CP \}$
$PCER$	$=$	$\{ MA, RS, WS, MI \}$
S	$=$	$\{ s_1, s_2, s_3, s_4, s_5, s_6, s_7, s_8 \}$ (in section 2.2)
S	$=$	$\{ s_1, s_2, s_3, s_4, s_5, s_6, s_7, s_8, s_9, s_{10}, s_{11} \}$ (in section sub-titled "Linear Programming Model")
J	$=$	{ all cultivated plots to be considered }
$J(s)$	$=$	{ all cultivated plots of soil type s}
JC	$=$	{ all plots corresponding with common fields}
JI	$=$	{ all plots corresponding with individual fields}
N	$=$	$\{ KC, PR \}$

The variables are, with $j \in J$, $p \in P$, $s \in S$, $n \in N$

$SUR(j)$	surface of cultivated plot j,
$FAL(j)$	surface of fallow land corresponding to cultivated plot j,
$PROD(p,t)$	harvest of product p in period t (in kg), $t = 8, 9, 10$,
$SEEDS(p,t)$	quantity of product p (in kg) to be reserved in period t as seeds for the next season, t = 8, 9 and 10,
$PROD'(p,t)$	the harvest of product p (in kg) in period t, available for consumption or sales in the target consumption year, t = 8, 9, 10
$CON(p,t)$	consumption of product p in period t (in kg), $t = 8 ,..., 13$.
$SAL(p,t)$	sales of product p in period t (in kg), $t = 8, ..., 13$,
$PUR(p,t)$	purchases of product p in period t (in kg), $t = 8, ..., 13$,
REV	net revenues during the target consumption year in FCFA,
$STOCK(p,t)$	stock of product p at the end of period t (in kg), $t = 8, ..., 13$,
$STOCKR(p)$	stock of product p (in kg) put aside at the end of period 13, to contribute to the food needs in the harvest period of the next growing season, $p \in PCER$,
$SAFST(p)$	volume of the safety stock of product p (in kg) reserved at the end of period 13 to meet food requirements after the harvest of the next growing season, if the harvest proves disappointing, $p \in PCER$,
$FIN(t)$	financial means (in FCFA) at the end of period t, $t = 8, ..., 13$,
$CONS(n,t)$	consumption of nutrient n, in period $t = 8, ..., 13$,
$CONS'(n,t)$	consumption of nutrient n by the consumption of base cereals, $t = 8, ..., 13$,

DEF(n,t)	deficit in nutrient *n*, in period $t = 8,..., 13$,
DEFR(n)	deficit in nutrient *n*, in the harvest period of the next growing season, if the consumption was only based on the stocks of agricultural produce at the end of period 13.

The parameters are, $p \in P, s \in S, j \in J, n \in N$:

AV(s)	available area of soil type *s* (in ha),
λ(j)	the ratio between the area of the fallow supplement of plot *j* and the area of the cultivated plot *j*,
AVMAN	quantity of manure available to the head of the household in kg,
MAN(j)	quantity of manure applied on plot *j* in kg/ha,
AVLABC(t)	labour available in period *t* for agricultural activities on the common fields (in hours), $t = 1, ..., 10$,
AVLABI(t)	labour available in period *t* for agricultural activities on the individual fields (in hours), $t = 1, ..., 10$,
LAB(j,t)	required labour in period *t* to cultivate 1 ha of plot *j*, $t = 1, ..., 10$,
SOWDAYS(t)	favourable days in period t for preparation of the fields and for sowing (in hours), $t = 1, ..., 5$,
DUR(t)	duration (in days) of period *t*, $t = 1, ..., 5$,
LABSOW(j,t)	required labour in period *t* for preparing and sowing 1 ha of plot *j*, $t = 1, ..., 5$,
YLD(j,p,t)	yield of product *p* in period *t* if 1 ha of plot *j* is cultivated (in kg/ha), $t = 8, 9, 10$,
γ(j,p,t)	quantity of product *p* to be reserved as seed per ha of plot *j* in period *t* , $t = 8, 9, 10$,
STOCK(p,7)	stock of product *p* at the end of period 7 (beginning of the harvest period),
f(p,t)	fraction of the stock of product *p* lost in period *t* because of storage losses, $t = 8, ..., 13$,
FIN(7)	financial means (in FCFA) at the end of period 7,
EAR(t)	extra-agricultural revenues (in FCFA) during period *t*, $t = 8, ..., 13$,
EAE(t)	extra-agricultural expenses (in FCFA) during period *t*, $t = 8, ..., 13$,
PRS(p)	selling price that the household expects to receive for the sale of 1 kg of product p (in FCFA),
PRP(p)	purchasing price that the household expects to have to pay when buying 1 kg of product p (in FCFA),
ρ(t)	interest rate on capital deposited in period *t*, $t = 8, ..., 13$,
DEM(n,t)	demand of nutrient *n* in period $t = 8, ..., 13$,
VAL(p,n)	contents of nutrient *n* in 1 consumed kg of product *p*,
MAXSR(t)	maximum quantity of red sorghum allowed to be consumed per period *t*, $t = 8,...,13$,
$\Theta_1(n)$	desired fraction of the total demand of nutrient *n* that can be satisfied by the consumption of product *p*,
$\Theta_2(n)$	minimal fraction of the demand of nutrient *n* that always has to be satisfied in every period,
$\Theta_3(n)$	fraction of the consumption of nutrient *n*, based on daily meals, to be satisfied by the consumption of base cereals $p, p \in P_{cer}, p \neq SR$

DEMR(n)	demand of nutrient *n* in the harvest period in the following growing season,
α	the fraction of the production of base cereals (*MA, SB, MI*) that has to be produced by the household itself,
β	fraction of the net revenues that has to be invested in the creation of a safety stock,
$\omega(n)$	weight coefficients,
$\omega_R(n)$	weight coefficients.

The model consists of the following elements:

$$Maximise:\ REV - \sum_{n \in N} \sum_{t=8..13} \omega(n) * DEF(n,t) - \sum_{n \in N} \omega_R(n) * DEFR(n) \tag{1}$$

where the variables have to satisfy the following conditions:

Land constraints

$$\sum_{j \in J(s)} (SUR(j) + FAL(j)) \le AV(s),\ s \in S \tag{2}$$

$$FAL(j) = \lambda(j) * SUR(j) \tag{3}$$

Organic manure constraint

$$\sum_{j \in J} MAN(j) * SUR(j) \le AVMAN \tag{4}$$

Labour constraints

$$\sum_{j \in JC} LAB(j,t) * SUR(j) \le AVLABC(t) \tag{5}$$

$$\sum_{j \in JI} LAB(j,t) * SUR(j) \le AVLABI(t) \tag{6}$$

$$\sum_{j \in JC} LABSOW(j,t) * SUR(j) \le SOWDAYS(t) * AVLABC(t)/DUR(t) \tag{7}$$

$$\sum_{j \in JI} LABSOW(j,t) * SUR(j) \le SOWDAYS(t) * AVLABI(t)/DUR(t) \tag{8}$$

Definition of the production and the seeds to be reserved

$$PROD(p,t) = \sum_{j \in J} YLD(j,p,t) * SUR(j) \tag{9}$$

$$SEEDS(p,t) = \sum_{j \in J} \gamma(j,p,t) * SUR(j) \tag{10}$$

$$PROD'(p,t) = PROD(p,t) - SEEDS(p,t) \qquad \text{for } t = 8, 9, 10 \tag{11a}$$

$$PROD'(p,t) = 0 \qquad \text{for } t = 11, 12, 13 \tag{11b}$$

Balances of the stocks of agricultural produce

$$STOCK(p,t) = (1-f(p,t)) * STOCK(p,t-1) + (1 - f(p,t)/2) * \tag{12}$$
$$(PROD'(p,t) + PUR(p,t) - SAL(p,t) - CON(p,t))$$

$$STOCK(p,13) = STOCKR(p) + SAFST(p) \quad \text{for } p \in PCER \tag{13}$$

$$PUR(p,t) = 0, \text{ for } t = 9, 10, 11 \tag{14}$$

$$SAL(p,t) = 0, \text{ for } t = 8, 9, 10, 12, 13 \tag{15}$$

Financial balances

$$FIN(t) = (1+\rho(t)) * FIN(t-1) + (1+\rho(t)/2) * \tag{16}$$
$$(EAR(t) - EAE(t) +$$
$$\sum_{p \in P} PRS(p)*SAL(p,t) - \sum_{p \in P} PRP(p)*PUR(p,t))$$

$$REV = FIN(13) - FIN(7) \tag{17}$$

Definition of food deficits and conditions for consumption

$$CONS(n,t) = \sum_{p \in P} CON(p,t) * VAL(p,n) \tag{18}$$

$$DEF(n,t) \geq \Theta_1(n) * DEM(n,t) - CONS(n,t) \tag{19}$$

$$DEFR(n) \geq \Theta_1(n) * DEMR(n) - \sum_{p \in Pcer} VAL(p,n)*STOCKR(p) \tag{20}$$

$$CON(SR,t) \leq MAXSR(t) \tag{21}$$

$$STOCK(SR,13) \leq MAXSR(8) + MAXSR(9) + MAXSR(10) \tag{22}$$

$$CONS(n,t) \geq \Theta_2(n) * DEM(n,t) \tag{23}$$

$$CONS'(n,t) = \sum_{\substack{p \in PCER \\ p \neq SR}} CON(p,t) * VAL(p,n) \qquad (24)$$

$$CONS'(n,t) \geq \Theta_3(n) * (CONS(n,t) - CON(SR,t)*VAL(SR,n)) \qquad (25)$$

Condition of agricultural autosubsistence production

$$\sum_{\substack{p \in PCER \\ p \neq SR}} \sum_{t=8..10} PROD'(p,t) \geq \alpha * \sum_{\substack{p \in PCER \\ p \neq SR}} \sum_{t=8..13} CON(p,t) \qquad (26)$$

Balance between safety stock and net revenues

$$\sum_{p \in PCER} PRP(P) * SAFST(P) \geq \beta * REV \qquad (27)$$

Non-negativity of the variables

$$SUR(j) \geq 0 \qquad (28)$$
$$CON(p,t) \geq 0 \qquad (29)$$
$$SAL(p,t) \geq 0 \qquad (30)$$
$$PUR(p,t) \geq 0 \qquad (31)$$
$$STOCK(p,t) \geq 0 \qquad (32)$$
$$STOCKR(p) \geq 0 \qquad (33)$$
$$SAFST(p) \geq 0 \qquad (34)$$
$$FIN(t) \geq 0 \qquad (35)$$
$$DEF(n,t) \geq 0 \qquad (36)$$
$$DEFR(n) \geq 0 \qquad (37)$$

In the section named Linear Programming Model under *zaï*, constraints (2) has been replaced by the following constraint:

$$\sum_{j \in J(s1)} (SUR(j) + FAL(j)) \leq AV(s_1) + 1/3 \sum_{j \in J(s9)} SUR(j) \qquad (2')$$

$$\sum_{j \in J(s2)} (SUR(j) + FAL(j)) \leq AF(s_2) + 1/3 \sum_{j \in J(s10)} SUR(j)$$

$$\sum_{j \in J(s3)} (SUR(j) + FAL(j)) \leq AV(s_3) + 1/3 \sum_{j \in J(s11)} SUR(j)$$

$$\sum_{j \in J(s)} (SUR(j) + FAL(j)) \leq AV(s), \quad s = s_4, ..., s_8$$

124

$$\sum_{j \in J(s9)} (SUR(j) + FAL(j)) \le AV(s_9) - 1/3 \sum_{j \in J(s9)} SUR(j)$$

$$\sum_{j \in J(s10)} (SUR(j) + FAL(j)) \le AF(s_{10}) - 1/3 \sum_{j \in J(s10)} SUR(j)$$

$$\sum_{j \in J(s11)} (SUR(j) + FAL(j)) \le AV(s_{11}) - 1/3 \sum_{j \in J(s11)} SUR(j)$$

Next the labour constraint (38) has been added:

$$\sum_{j \in JC} LABDI(j,t) * SUR(j) \le AVLABDI(t), \quad t = March \ and \ April, \ 1, \ 2, \ 3 \tag{38}$$

where: $LABDI(j,t)$ and $AVLABDI(t)$ are the parameters that represent respectively the labour requirements to dig the holes per plot j and per period t (including the periods March and April) and the availability of labour for these activities in each period.

BASE VALUES OF SOME OF THE PARAMETERS OF THE MODEL[12]

TABLE 2.1 Estimations of the average yields for some prevailing conditions

Crop	Soil type	Level of manuring (kg/ha manure)	Yield
Maize	Higher fields	8000	1100
Red Sorghum	Higher fields	800	475
White Sorghum	Higher fields	800	425
Millet	Higher fields	0	330
Groundnut	Higher fields	0	400

TABLE 2.2 Average agricultural calendars for maize and groundnut, on higher and lower fields for red sorghum, white sorghum, millet and for intercropping of sorghum or millet with cowpea

MAIZE
preparation
fertilization
sowing
1st weeding
2nd weeding
harvest

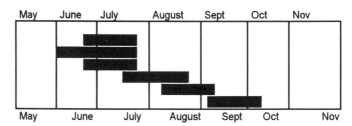

Red sorghum/
White sorghum/
Millet +
Cowpea
Higher fields
preparation
fertilization
sowing
1st weeding
2, 3rd weeding
harvest RS/WS/MI
harvest Cowpea

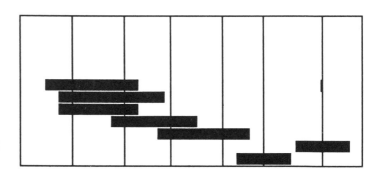

---

[12] Linear Programming Model.

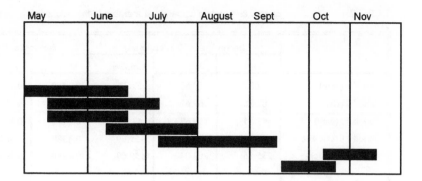

Red sorghum/
White sorghum/
Millet +
Cowpea
Lower lands
preparation
fertilization
sowing
1st weeding
2,3rd weeding
harvest RS/WS/MI
harvest Cowpea

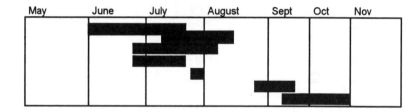

Groundnut
preparation
fertilization
sowing
1st weeding
2nd weeding
harvest

TABLE 2.3   Estimations of labour inputs in hours/ha of the `Exploitation Centrale' for some prevailing conditions

Crop	Soil type	Manure kg/ha	Prep	Sowing	Fert.	1st weeding	2nd weeding	Harvest
MA	High	8000	350	125	200	300	35	215
RS	High	800	50	105	30	295	265	130
WS	High	800	45	105	30	270	250	125
MI	High	0	30	100	-	255	190	115
GN	High	0	230	170	-	300	20	375

TABLE 2.4   Energy and protein demand per member of the `Exploitation Centrale' per period

	Energy (in 1000 kilocalories)			Proteins (in 1000 grammes)		
	Men	Women	Children	Men	Women	Children
Period 8 (Sept)	87.35	69.67	53.23	0.1460	0.1217	0.0730
Period 9 (Oct)	87.35	69.67	53.23	0.1460	0.1217	0.0730
Period 10 (Nov)	87.35	69.67	53.23	0.1460	0.1217	0.0730
Period 11 (Dec-Febr)	222.74	198.55	159.69	0.4380	0.3651	0.2190
Period 12 (March-June)	235.85	202.04	159.69	0.4380	0.3651	0.2190
Period 13 (July-Aug)	301.36	229.90	159.69	0.4380	0.3651	0.2190
Sept-Nov	262.05	209.00	159.69	0.4380	0.3651	0.2190

TABLE 2.5   Value in nutrient N of a meal based on 1 kg of product P

	Maize	Red sorghum	White sorghum	Millet	Ground-nut	Cow-pea
1000 kilocalories	3.57	3.39	2.57	3.41	5.46	3.42
proteins (grams)	42	49	37	52	104	104

CHAPTER 6

FARM HOUSEHOLD STRATEGIES FOR FOOD SECURITY IN NORTHERN GHANA: A COMPARATIVE ANALYSIS OF HIGH AND LOW POPULATION FARMING SYSTEMS

Al-Hassan, R., Famiyeh, J.A. and *Jager, A. de*

Introduction

In many economically deprived regions of Sub-saharan Africa the farm-household is, predominantly, a production-consumption unit (Low, 1986). Consequently, its theoretical analysis requires a framework that is significantly different from mainstream microeconomic theory. For example, the primary goal of farm households in northern Ghana is food self-sufficiency (Fey, 1992; Baur, 1992; Runge-Metzger, 1993; ICRA/NAES, 1993). This goal is justified under the existing high climatic risk. In such households food entitlement is gained through own food production and cash incomes from the sale of crops, livestock and non-farm products. Paid casual employment and remittances from emigrated relatives sometimes provide additional sources of food entitlement. It follows from here that rural subsistence households have multiple objectives toward increasing household food security (McKee, 1986; Herbon, 1990). However, variations are expected to exist in the strategy, i.e. the means and criteria used to achieve the multiple objectives (White, 1969). Therefore, the issue of strategy for achieving household food security is important in research and policy meant to address peasant farmers' economic problems.

A review of the literature on food security in northern Ghana reveals a rather wide dimension of the subject covering issues of the physical environment, socio-economic factors, farmer crop production decisions, storage, marketing, and the issue of modelling to aid policy interventions (see Baur, 1992; Runge-Metzger, 1992). A significant contribution to the analysis of northern Ghana farmers' food security decisions has been provided by de Haan and Runge-Metzger (1990). They refer to issues such as utilization of income increases and its effects on nutrition; sustainability of production systems; intra-household distribution of income; external effects and locational disadvantages limiting access to resources; and key socio-economic influences on household and regional food security.

The information gap in this literature is the identification of differences in strategies adopted by farm-households to achieve food security. The study of farm-household food security strategies will reveal farm-household resource levels; decisions that farmers make consistent with their resources; and the criteria used to arrive at those decisions. These details about the farm-household provide a logical basis for classifying households into recommendation domains.

Although the literature on the subject of food security is quite extensive with diverse opinions often expressed on issues, a single most important factor affecting food security (whether at household, community or regional level) is the high population growth and its influence on the availability of labour and land.

A feature of areas with high population growth rates is a high dependency ratio and in subsistence economies, this means a high consumer/worker ratio. Unless agricultural land use becomes more intensive in terms of frequency of cultivation, and the use of productivity enhancing factors such as fertilizers and irrigation, as well as adoption of improved soil management practices, the contribution of own food production to the attainment of a household's food security may be low. Therefore in order to attain food security, households experiencing declining per capita land availability, but who have no access to use productivity enhancing factors may tend to rely on non-food productive activities more than households with relatively abundant land, to gain entitlement to food.

This chapter compares farmers' strategies for achieving food security under high and low population density farming systems in northern Ghana. The importance of this chapter is seen within the context of the farming systems approach to agricultural research adopted recently in Ghana. The classification of households into recommendation domains in Farming Systems Research (FSR) is an essential first step because this research methodology evolved from the realisation that results of agricultural research did not diffuse equally among farmers (Rhoades, 1989).

The objectives of the chapter are therefore to provide a descriptive analysis of farm household strategies for achieving food security and to identify limitations of, and opportunities presented by existing strategies for policy intervention, emphasising differences in strategies presented by relative land availability. Such a description may subsequently be used to determine a framework in terms of resource levels, productive activities, and constraints for quantitative analytical modelling of strategies of classes of farm-households.

METHODOLOGY

Farming Systems Research

The Farming Systems Research (FSR) approach is used as a basic methodology to describe and classify the prevailing farming systems in Northern Ghana, to describe resource endowment, allocation of resources and their efficiency and rentability. Existing information from previous FSR at the Savannah Agricultural Research Institute (SARI) at Nyankpala was screened, alongside a general review of literature and secondary data. The general classification of farming systems from SARI was followed and given the limited resources available initially two of the identified farming systems in Northern Ghana were selected for further study. The village Dulugu (Upper-East) is representing a high density population farming system (about 200 persons per square kilometre), while Kpongu represents a low density population area (50 persons per square kilometre). Both villages are less than 10 kilometres away from the regional capitals of Bolgatanga and Wa. Kpongu has a relatively higher annual rainfall of 1050 mm compared to 900 mm in Dulugu. Dulugu is characterised by sole compound system of farming while the bush system is the only one practised in Kpongu. In March 1993, Dulugu consisted of 210 compounds (dwellings) while Kpongu had 119 such dwellings. In all, 60 farm-households in the two villages were involved in the study.

Half of the population consisted of households already involved in previous FSR at SARI, the remaining 30 (15 per village) were selected at random from compiled sampling frames. In close collaboration with the FSR teams at SARI the following activities have been undertaken:

— rapid rural appraisal in the study villages;

— inventory and pre-season questionnaire (separate survey for collective household and individual activities);

— monthly seasonal questionnaire on inputs/outputs;

— yield measurements;

— post-season questionnaire;

— group discussion on evaluation of strategies.

The acquired data formed the input for a quantitative descriptive analysis.

Identification of Strategies

Following the classification and description of existing farming systems, the identification of farm-household objectives with regard to attaining food security, major constraints, available strategies and instruments to achieve the objectives are determined. Basically the primary data collection activities mentioned above are used to identify these objectives and strategies. Since it is assumed that within one farming system the objectives and strategies may vary, various 'farm styles' will be identified using correspondence analysis and group discussion with the farmers.

Systems Approach

Based on the identified farming systems and 'farm styles' major farm-household groups are formulated representing the agricultural sector in Northern Ghana. At farm level the elements which determine farm-household decisions are disentangled and quantified using Linear Programming (LP). The farm level model consists of a set of goals or objectives, available or potential available activities, existing constraints and the technical coefficients linking them. The set of spreadsheet farm models developed by SARI (Runge-Metzger, 1993) and the LP-models developed for the central-plateau in Burkina Faso (Maatman and Schweigman, 1994) form the basic framework of the farm level models in Northern Ghana. The developed models are tested in an iterative process confronting model results with actual data and farm-household opinions. The identified groups of farm-households and their farming systems are studied in terms of their productivity, stability and sustainability.

Identification of Technical and Policy Options

The various instruments available to improve the food security situations are to be identified. Within the context of this project two main instruments are studied in more detail:

— technical innovations;
— agricultural policy options.

The influence of public goods and services and trade and distribution channels on farm-houshould strategies are studied in close cooperation with other sub-projects in SADAOC.

The close linkage with the SARI-FSR-teams provides the project with basic information on potential innovations applicable at farm-household level. Monitoring OFTs at farm-household level provides information on essential factors in adoption of new technologies and serve as research guidance for future technical research.

Existing agricultural policies, objectives and instruments relevant to farm-household decision making processes are identified and described in an agricultural policy logical framework. This framework addresses the effectiveness of available instruments to national and regional policy makers to realise specific objectives taking into account the difference in objectives that exists at policy and farm-household level.

Development Scenarios

The impact of identified options to improve food security situation at farm-household level is tested through running the options in the developed farm-household models. Various sets of combinations of instruments and interventions are formulated and their impacts at farm level are assessed.

The preliminary results presented in this chapter only refer to the first step of the described methodology: description of farming systems with specific emphasis on existing strategies relating to food security.

Farming Systems and the Study Area

The present description of farmers' strategies is based on two representative villages in Northern Ghana (Figure 6.1), which is the study area of the Analysis of Farmers' Strategies (AFS) project[1]. The two villages were selected on the basis of the concept of 'farming system window' (Chambers, 1983). The concept postulates that for the study of a wide but reasonably homogenous area, one representative village unit (the window) may be selected for detailed study of the entire area. The Savannah Agricultural Research Institute, which has the mandate for agricultural research in Northern Ghana, has identified six farming systems windows (Figure 6.1). The 'farming systems window' concept was used in this study.

Population density is one of the factors of the farming systems window classification in Northern Ghana. The two windows are represented by the villages of Dulugu (in the Upper East region) and Kpongu (in the Upper West region) in high and low population densities respectively. The reasons for this high population density is not documented but could include in-migration from Burkina Faso, and a clustering of people in areas of relatively high soil fertility. Differences in population pressure are often represented by the compound and bush systems of farming, reflecting relative land availability. Compound farming is characterised by high population density, while the bush farming system is characterised by low population density. There are, of course, other farming systems which have variously been classified mainly on the basis of the dominant cash crop produced; for example, cotton-based, tobacco-based, rice-based, and yam-based farming systems.

The compound farming system has the distinct feature that all the farmland actually lies

---

[1] The AFS project is one of the projects under the broader research programme, Sustainable Food Security in West and Central Africa, commonly known under the French acronym, SADAOC. The AFS is devoted to the analysis - of-farmers strategies, for food security with respect to organisation of production, consumption and marketing.

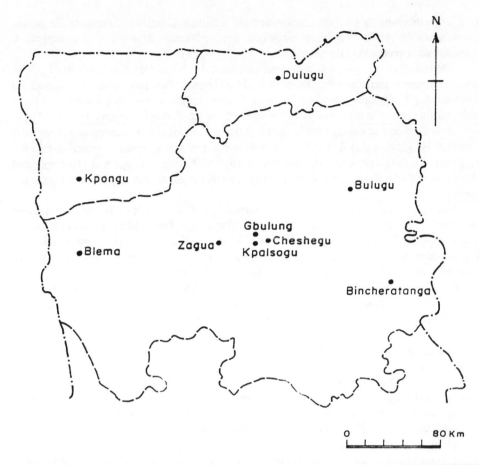

System (Village)	Rainfall (mm)	Crops	Population density
Western Dagbon (Cheshegu, Gbulung, Zagua, Kpalsogu)	1060	Maize, sorghum, groundnuts, rice	50 -100 (H)
Eastern Dagbong (Bulugu)	1050	Maize, millet, groundnuts, yam	20 - 50 (M)
Nanumba (Bincheratanga)	1250	Yam, maize, groundnuts, rice	10 - 50 (M)
West Gonja (Blema)	1100	Sorghum, maize, cowpea, yam	10 - 20 (L)
Upper West (Kpongu)	1050	Sorghum, maize, cowpea, yam	10 - 50 (M)
Upper East (Dulugu)	890	Millet, sorghum, groundnuts	50 - 200 (H)

Source: Baur (1992)
   H = High; M = Medium; L = Low

Figure 6.1: Farming Systems Windows of Northern Ghana (Classification by SARI)

next to the dwelling (or compound) of household members. Compound farmlands are farmed permanently on year-to-year basis. Since one or more households may live in a compound, farmland can become very fragmented.

The bush farming system is characterised by farmlands located often between a half and several kilometres away from the village of clustered houses. Farmlands under this system are fallowed after 4 - 6 years continuous cropping. Fallow periods are usually between 3 - 5 years although this period may be shortened considerably as population increases (Baur, 1992).

Both Dulugu and Kpongu are located less than 10 km away from the respective regional capitals, Bolgatanga and Wa. Dulugu is characterised by sole compound system of farming while the bush farming system is the only one practised in Kpongu. Dulugu had 210 compound dwellings in 1993. Kpongu had 119 dwelling units within the clustered settlement during the same period[2].

Northern Ghana, lies within the Guinea savannah belt of West Africa. The area experiences one rainfall season lasting from May to October (Papadakis, 1966). Many areas of the region are prone to food insecurity because of the vicious circle of low productivity, low income and low investments. Social services and delivery are generally poor; besides, many farmers cannot gain easy access to them on account of distance and finance. There is a higher incidence of poverty[3] in Northern Ghana than Southern Ghana (UNICEF, 1984; Ewusi, 1976; 1984, Boateng et al., 1989). This region also contributes[4] nearly 36 per cent to national poverty (Boateng et al., 1989 p. 15). Although agricultural productivity is very low on account of very poor soils, low rainfall and general environmental degradation, between 63 and 75 percent of the economically active population of northern Ghana are engaged in agriculture (Ghana Population Census, 1984).

Annual mean rainfall ranges from 800 mm in Dulugu to 1000 mm in Kpongu. Inter-year variability in timing and amounts of rain are a source of worry to farmers (ICRA/NAES, 1993). Drought spells occur during the rainy season, often interrupting normal plant growth (Runge-Metzger and Diehl, 1993; Freeman, 1994). Although temperatures are high, they are not abnormal for plant growth. They are, however, high enough to reduce potential moisture through evapotranspiration. The annual mean evapotranspiration ranges between 1600 mm in the south and 1750 mm in the north-east (Benneh, 1971).

The organic matter content of the Upper `A' horizon is generally around only 2 per cent (Quansah et al., 1991). The adverse climatic and soil conditions favour only annual crops with growing period of up to 150 days. The two farming systems are consequently based on cereals, legumes and, to some extent, roots and tubers. Cassava production in the bush farming system has been increasing since the early 1980s due to its ability to grow reasonably well on impoverished soils and its high calorific value (MOFA, 1990). Cereals, legumes, leafy vegetables and oils account for less than 15 per cent of the total energy supply (Baur, 1992). Legumes provide substantial dietary protein. In many places fish constitutes the major source of animal protein except where hunting is important.

---

[2] Profiles of these villages including history, social structure, available services are available as RG-AFS Working Document 1.

[3] The incidence of poverty is estimated as a simple head count of number of individuals below a poverty line representing an annual expenditure below 32,981 Cedis, and a hard core poverty line of 16,491 Cedis. The poverty lines are expressed in September 1987 constant prices.

[4] The contribution to poverty is the weighted share of a the region in the national poverty measure, Pα developed by Foster, Greer and Thorbeck (Boateng et. al., 1988).

There is a wide variation in the estimate of calorific balance[5] among farm-households. Alderman and Higgins (1992) estimate that the median calorie availability in both rural and urban northern savannah zone is lower than other regions of Ghana. In spite of the low calorific intake there are significant inter-and intra-year variability in the amounts and types of food consumed. Seasonal variability in available food is reflected by the well-known 'hunger gap' between new crop harvests and when food stocks are depleted. This period is characterized by a negative dry season energy balance because it coincides with the period of increased energy requirement for a new cropping season.

A shortfall of food energy of about 300 calories per capita per day may occur in the lean season (NORRIP, 1983). Ewusi (1984) reports that both children and adults receive only about 60 percent of calorie and other nutritional requirements. The seasonal food shortage is one major factor responsible for retarded children's 'catch-up' growth. A child in the interior savannah is more likely to suffer from chronic and transitory malnutrition (Alderman and Higgins, 1992).

PRELIMINARY RESULTS

Cropping Systems

The practice of growing two or more crops simultaneously on the same plot of land (i.e. mixed- or intercropping) is predominant in both the compound and bush farming systems (Table 6.1). This age-old food cropping technique has several advantages including reduction in crop output variability, weed suppression (Norman, 1974) and higher total energy output.

TABLE 6.1: Major Cropping Systems in Compound and Bush Farming Systems in Dulugu and Kpongu Respectively on Collective[6] Farm Holdings in Cropping Season 1993 (% of Total Cropped Area)

Cropping system	Compound farming (Dulugu)	Bush farming (Kpongu)
Mono cropping	2.9%	3.7%
All cereal mix	55.7%	26.9%
Cereal-legume mix	10.4%	43.8%
All-legume mix	31.0%	8.0%
Root/tuber crop	—	17.6%

Source: SADAOC/AFS/Ghana

---

[5] Calorie balance is the difference between total calories produced and calories consumed.

[6] In many parts of Northern Ghana, the distinction is made between collective and individual productive activities. All working members of the household participate in collective activities under the leadership of the household head, for their common consumption. In addition, individuals may undertake productive activities, o n which they make all decisions and control income.

Apart from aspects like soil type, limited farmland and dietary preference, the differences in the cropping systems can also be explained by the roles of individual crops in the food system, and subsequently, production strategies in the two farming environments. In the compound system, the 'hunger-breaking' crop is the early millet found everywhere soon after land preparation in May. Early millet is frequently intercropped with sorghum which constitutes the main cereal crop consumed throughout the year. Hence the all-cereal mixed crop is the most important cropping pattern in the compound farming system. Similarly, the cereal-legume mixture assumes much importance in the bush farming system as cowpea plays the role of 'hunger-breaker' after the 'hunger-gap' occurring between April and June. Sorghum and/or maize would frequently constitute the cereal component in the cereal-legume cropping pattern.

In both farming systems, substantial amounts of legumes are produced either in the all-legume crop mixture (Dulugu) or in the cereal-legume cropping mixture (Kpongu). This aspect of crop production provides the main source of protein in household diets.

It is observed that sole cropping constitutes between 3 and 4 percent of the cropped area. On the individually owned plots in the low-population system a higher percentage of sole crops may be found due to the existence of cotton and yam on these fields.

Animal Husbandry

About 70 percent of all livestock in Ghana are located in Northern Ghana (MOFA/PPMED, 1991). For example, in 1986, 77.4 percent of cattle in Ghana were in the northern Guinean/Sudan savannah zone comprising Northern, Upper West and Upper East Regions (ISNAR R52, 1991).

Livestock Management

Livestock management in Northern Ghana has remained undeveloped over the years. The free range system which is practised is limited by feed and water inadequacy especially in the dry months. Animals suffer considerable weight loss during the period (Runge-Metzger, 1993).

In many localities, cattle owners entrust their animals to Fulani herdsmen who live close to the villages and maintain kraals. Herdsmen walk animals often for long distances in search of pasture and water. They inform owners about births, disease outbreaks and deaths. In return they are paid in grain and they retain revenues from milk sale. In other localities, male children (usually sons) take on the responsibility of herding the cattle. Generally, bullocks are cared for better than other livestock. They are fed on grain, *pito* mash and groundnuts during land preparation. Sheep and goats merely roam about in villages. They are however, tethered during the cropping season to prevent them from destroying crops. Pigs are also kept in pens and fed on household waste and pito mash. Fowls move about freely except at night and are fed on grain and termite larvae.

There is greater livestock concentration in bush system households than in compound system households (Table 6.2). High livestock concentration in the compound system will have a negative effect on crop production as free-range livestock compete with crops for land. This is not the case in the bush system.

TABLE 6.2   Average Number of Livestock per Farm-household
in Dulugu and Kpongu in 1993

Type of livestock	Compound farming (Dulugu)	Bush farming (Kpongu)
Cattle	1.6	3.8
Goats	3.6	5.6
Sheep	2.3	1.3
Guinea fowls	2.5	6.8
Fowls	6.0	6.1
Pigs	0.4	0.4
Rabbits	0.1	-
Donkeys	0.1	-
Total (TLU)	1.9	3.6

Source: SADAOC/AFS/Ghana
TLU = Tropical Livestock Units[7]

The Role of Livestock

Most farm-households in Northern Ghana do not use institutional banking facilities which are available only in the regional capitals and a few sub-urban towns such as Bawku, Navrongo, Tumu, Lawra and Yendi. Consequently, livestock provide a means to keep cash reserves. Although this way of keeping money does not expose rural households to institutional banking services and the benefits to be derived, it is an important means of ensuring that, in high inflationary conditions, the value of household wealth remains stable, and does not depreciate.

The various livestock types play reasonably specific roles in the farm-household. Roles that have direct effect on food security include insurance against crop failures, source of cash for food purchases during the hunger gaps, and for acquisition of farm inputs. Livestock also provide a means for non-food expenditures such as school fees, medical expenses; and socio-cultural functions such as marriages, funerals, and gifts. Some cattle are trained to provide animal traction services such as ploughing and transporting.

Livestock sales are highest during the periods November-December and March-April. The first peak sale provides the cash for grain purchases for storage against the hunger season in May-June. The second peak sale is meant for financing farm input purchase and to a limited extent, food. The compound system households on average sold about ¢16,000 (approximately $20) of livestock during 1993-94. The Figure for Kpongu was ¢52,800 (or $66). Clearly, livestock sale is much more important in the bush system, which has the higher concentration of livestock, than in the compound system.

Given the role of livestock in supplementing food intake either directly through purchases or indirectly through input purchases, the lower concentration of livestock in the compound system is probably a further limitation to the attainment of food security. It also presents a difficult choice in land use for either crop or livestock, and begs for research into the optimal integration of crop and livestock especially in the compound farming system.

---

[7] Livestock is aggregated using a weighting system based on value and standard cow = 1

Resource Use and Allocations

Available Farmland

The principal indicator of land availability is the population density. However, other indicators of immediate relevance are `average parcel[8] size per household or per capita', `average parcel size per man-equivalent labour[9]', and the R value computed as percent of cultivated and fallow land actually cultivated.

Table 6.3 provides information on these land-availability indicators for the two farming systems. Consistent with the differences in population density in the two farming systems, there is a drastic difference in the parcel size per household and per capita among the two systems with land availability being more restricted in the compound system. The smaller parcel sizes in the compound system is not because of difficult usufruct conditions as found in southern Ghana; it is because there is simply not enough compound farmland for every household member as households expand in size. This is true because in both farming systems, 87-97 percent of tenure is owner-occupier.

Available land (including fallow land) per household in the bush system is 4.5 times that in the compound system; and land per man equivalent is about twice larger in the bush system than in the compound system. Nevertheless, crop yields under both systems are similar[10]. While households in both systems experience periodic food deficit each year (Fortes and Fortes, 1936; Hunter, 1967; ICRA, 1993) the shortages are more severe in the compound system because of a restricted access to land.

The implication then is that households in the compound system must either encourage significant out migration of their members or endeavour to make major decisions involving the following alternative sources of gaining food entitlement:

(i)   intensive cultivation of compound farmlands

(ii) gaining access to distant farmlands

(iii) paying major attention to non-farm economic activities

(iv) undertaking intensive small livestock production.

Baur (1992) provides a number of indicators of land use intensity which show much greater frequency of cropping farm land in the compound system (Table 6.4). Unless productivity enhancing inputs are used in the compound system, soils are likely to be exhausted and crop yields poor. Our surveys supplement Baur's land use intensity indicators with information on use of inputs such as fertilizers and mechanised land preparation.

---

[8]   A parcel is the total farmland owned and available for farming by the household in any cropping year. This will include land deliberately left to fallow by the household.

[9]   A man-equivalent of labour is the unit of labour determined by aggregating labour of all workers with appropriate conversion factors based on age and sex (Appendix 3)
Total man-equivalent labour = Σ [cFi * Ai]
      where I = 1...4 age classes
cFi = conversion factor for age class I
Ai = number of household members in age class i

[10] The AFS yield measurement data indicate that plant densities are slightly higher under the bush farming system than under the compound system. However, legumes tend to yield slightly more under the compound system.

TABLE 6.3: Land availability indicators

Indicator	Compound farming	Bush farming
Population density 1)	200 pers/km^2	50 pers/km^2
Av. parcel size/hh 2)	0.75 ha	3.51 ha
Av. parcel size/caput 3)	0.15 ha	0.35 ha
Av. parcel size/me 2)	0.23 ha	0.50 ha
R-value 4)	99%	48%

Source: (1) 1984 Population Census
          (2) SADAOC/AFS/Ghana
          (3) Average household size   = 5 for compound system
                                       =10 for compound system
          (4) SARI farm surveys (Baur, 1992)
             hh = household
             me = man-equivalent labour

TABLE 6.4: Land Use Intensity

Indicator	Compound farming (Dulugu)	Bush farming (Kpongu)	All northern Ghana
R-value	99%	48%	30-99%
Trees/ha farmland	6	64	6-64
Continuously cultivated > 7 years	98%	6%	2-26%
Relative frequency of cow-dung use[1])	100%	4%	n.a.
Relative frequency of fertilizer use [1])	9%	38%	n.a.

Source: SARI Farm Surveys (Baur, 1992)

     [1])   SADAOC/AFS/Ghana
     na = Not Available

The results indicate that because households in the compound system cannot increase their food production capacity by increasing the existing cultivated area the land is used intensively. Households in the bush system can increase production capacity through expansion in land area if labour does not become limiting. If family members can afford to travel long distances (often up to 15 km), more farmlands can be obtained. Analysis of data on the distribution of parcel type indicates that only 7 percent of households in the compound system own bush farms. This implies that distant farming is not a significant phenomenon under the compound farming system as land is limited everywhere.

Farm Labour Utilisation

For both farming systems, the dominant source of farm labour is the household itself. Over 80 percent of labour utilised on farms is family labour, about 10 and 8 percent are hired and communal labour; the remaining 2 percent is labour from minor sources such as reciprocal (or exchange) labour and visitors. Communal labour is often important for weeding.

Family labour is heterogenous in terms of its age, gender and traditional responsibility. So as to ensure productivity of available family labour, decisions as to who does what cannot be avoided. The AFS pre-season surveys indicate that generally, men in the bush system households are scheduled to perform most of the on-farm work. In the compound system, both men and women perform the same on-farm activities.

From gender perspective, male adult family members devote more time to crop production than adult females (Runge-Metzger and Diehl, 1993). Adult females predominantly take part in specific operations of the crop production, mainly planting and harvesting. The reason is that:

— The quantity of food produced each season depends, essentially, on how much planting is done and how much crop is actually harvested. The time specificity of planting and harvesting in the cropping season means that all available labour must be mobilised to perform them;

— Women must necessarily spend time on non-farm income-generating activities to supplement household income. They can do this during the non-critical operations.

The largest portion of the time of adult women in spent on the management of the household including food processing, cooking and child care.

Total labour utilisation in the two systems (Table 6.5), combined with an average land area per household of about 1.5 hectares in the compound system and 4 hectares in the bush system, gives a per hectare labour utilisation during one cropping season as 785 hours in the compound system and 457 hours in the bush system. This is yet another piece of evidence of higher agricultural production intensity in the compound system than in the bush system.

TABLE 6.5: Time Spent on Activities (Average Adult Hours/day)

	Compound farming (Dulugu)			Bush farming (Kpongu)		
	F	NF	O	F	NF	O
Farming season	6.7	3.7	1.1	6.6	2.6	2.5
Off-season	1.6	7.3	1.1	2.6	4.7	2.6

Source: SADAOC/AFS/Ghana

F   =   farm activity
NF   =   non-farm activity
O^{11} =   other activities

Individual adult household members in the compound system spend 1.5 times more time on non-farm income-generating activities during the off-season than individual adults in the bush farming system. This is consistent with the crucial need to generate non-farm income to compensate for shortfalls in food production due to limited availability of land in the compound system.

In spite of the marked difference in the average number of persons per household between

---

[11] The estimates of time spent on other activities seems low considering a day length of 12 hours. Runge-Metzger and Diehl (1993, pp. 125-131) for example estimate an average of 3-4 hours for these activities.

the two farming systems (7 in Dulugu and 13 in Kpongu), the consumer-worker ratio[12] shows no significant difference between the two systems. Runge-Metzger and Diehl (1993) compute a higher ratio of 1.6 for Navrongo — a high population density semi-urban town in the Upper East Region. However, their estimate of the ratio for Sawla — a large village within the low population density area of the Upper West Region — is 1.3 (Runge-Metzger and Diehl, 1993, p. 58). The data also show little variation in the consumer-worker ratio between households within either system, with coefficient of variation ranges of 1.1 - 1.3 in the compound system and 1.2 - 1.3 in the bush system.

With a consumer-worker ratio of about 1.3, each adult worker is required to feed an equivalent of only one infant[13] in addition to himself. Whether households can meet this requirement depends on the land and capital resources available and as well as the rainfall and edaphic conditions, ill-health, and the inadequate use of off-season periods for non-farm economic activities.

Ownership of Farm Equipment Stock

Another factor which would affect the ability of households to produce enough to feed themselves is the nature of the stock of equipment owned and/or used. It appears that most households do not own any major farm equipment such as tractors and bullocks (Table 6.6).

On collective ownership basis, households in Kpongu do not own bullocks and implements. On the contrary, about 16 percent of households in Dulugu own bullocks and implements. Of the interviewed family members in Dulugu 11% owned bullocks and implements individually, while in Kpongu this was only 2%.

Given the very close proximity of plots in Dulugu and the available bullocks and tractors in the village the majority of the households have a potential access to mechanised land preparation. Of course lack of cash is the major bottleneck to actual access of these services. The bush farms in Kpongu are farmed in much more extensive way, with hardly any mechanised input.

---

[12] Indicator of the extent to which households are able to cater for the food needs of their members. Calculated is the ratio between men equivalent labour (me) and consumer units (cu)

$$\text{Total consumer units (cu)} = \sum_{j=1}^{4} (cf_j * A_j)$$

where $j = 1....4$ age class
cf_j = conversion factor for age class j
A_j = No. of household members

[13] The consumption weight for an infant is 0.4

TABLE 6.6: Percentage of Farm-households Owning Types of Farming Equipment (Excluding Equipment Owned by Individual Farm-household Members)

Type of equipment	Compound farming (Dulugu)	Bush farming (Kpongu)
Bullock and implements	16%	—
Tractor and implements	3%	—
Cutlasses	100%	100%
Hoes	100%	100%
Axes, mattocks	—	16%

Source: SADAOC/AFS/Ghana

Non-Farm and Non-Food Economic Activities

Non-farm economic activities are very important in the compound system due to population pressure and land limitation. On the contrary, under the bush system, cash crop production plays the role that non-farm activity plays within the compound system.

In Dulugu, about 50 percent of all household members have farming as their primary occupation and about 47 percent have crafts[14] (basketry or blacksmithing) as their secondary occupation. In Kpongu, 81 percent of persons in the household are engaged primarily in farming, and as low as 8 percent (mainly female) have any secondary occupation. What this means is that supplementary economic activities must still be in the farm sector in the form of cash crop production. Indeed, most households in Kpongu produce cotton and yam, and to a lesser extent, groundnut and bambara nuts as cash crops.

From gender perspective, basketry, like farming in Dulugu, is not gender specific. However, trading and firewood-charcoal production in Kpongu are predominantly female occupations.

Trade and Stocks

Throughout Northern Ghana, rural markets exist for both agricultural and basic industrial products. Markets are often scheduled for three- or six-day intervals but most villages would have one or two small shops from which households can buy most of their regular requirements. The basic agricultural products found on most rural markets are cereals, legumes, dried root crops and fresh vegetables and fruits. In Dulugu, about 96 percent of household crop purchases are cereals of which maize is about 60 percent. A similar situation exists in Kpongu where 80 percent of crop purchases by households is cereal with sorghum and maize constituting 40 and 30 percent respectively. Households purchases of cereals and legumes show a specific pattern over the year (Figure 6.2). Especially in Kpongu a peak is observed to bridge the 'hunger gap'.

In Dulugu households do not sell food crops at all. This confirms the point that all crops produced here are meant for own consumption. In Kpongu, yams constitute about 76 percent of crops sold by households. Cotton sales constitute about 6 percent of the total crop sales.

---

[14] It is expected that in other high population density areas, non-farm activities which dominate as additional source of entitlement to food will depend on the local conditions, including indigenous knowledge, available resources and market availability.

These results differ from the NAES survey finding maize as the most important cash crop. This probably reflects a change in the cropping system over the last four years, particularly with respect to cotton production.

In relation to crop storage, the AFS monthly household crop stocks data in 1993 indicate that the four most important crops for which stocks are kept for most part of the year are sorghum, maize, millet and, to a limited extent, groundnuts in order of importance. Stock levels are generally high in September (almost immediate post-harvest) and in February, just before the hunger season (Figure 6.3). The pattern is the same for both farming systems households. This suggests that contrary to the view that farmers will sell immediately after harvest when prices are lowest, grain farmers in Northern Ghana do not sell at that time unless difficulties in household finances make it unavoidable. Indeed, many households would rather buy stocks early when prices are low than sell early only to buy later at high prices (Baur, 1992; Abu, 1992). The case of yam farmers is different due to the perishable nature of the crop. Its transport requirement which cannot be provided easily during the rains when prices are attractive makes it quite imperative to sell immediately after harvest.

On average, a Kpongu household stored a total amount of 2.90 tonnes of grain in 1993. The average Dulugu household stored about 1.25 tonnes of grain during the same year. Of course, these figures relate directly to household size. In terms of consumer units, Kpongu households are twice as large as Dulugu households. It is estimated that whereas about 80 percent of crops stored by Kpongu households came from own production, only about 40 percent of crops stored by Dulugu households came from the same source (Baur, 1992).

Concerning trade of animals it is observed that a peak in the number of animals sold occurs in the period November/December and May/April (Figure 6.4). In these periods cash needs are high for buying gifts and purchase of food before the 'hunger gap'.

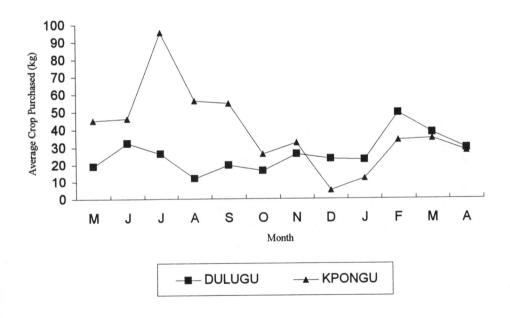

Figure 6.2: Cereal and Legume Purchases in Dulugu and Kpongu in the Period 1993-94

Source: SADAOC/AFS/Ghana

143

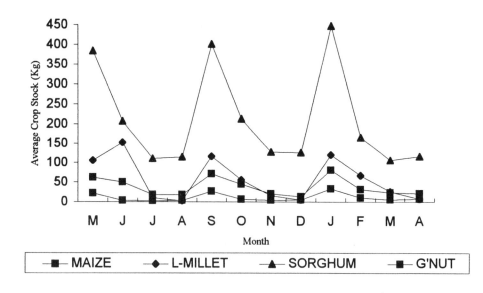

Figure 6.3: Average Stock of Food Crops per Households in Kpongu in the
Period, 1993-94

Source: SADAOC/AFS/Ghana

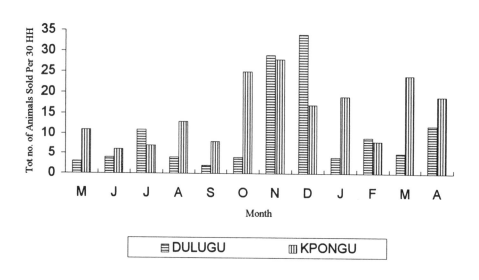

Figure 6.4: Total Number of Animals Sold Per 30 Households in Dulugu
and Kpongu in the Period 1993-1994

Source: SADAOC/AFS/Ghana

Crop production, non-farm economic activities and livestock trading dominate the sources of food entitlement in the two farming systems, and therefore provide the basis of any assessment of the extent of achievement of food security.

The 1993 performance indicators suggest that the bush system farm-households are better performers than the compound system households (Table 6.7). Farm-households in the bush system farm larger areas, apply more farm labour and produce greater energy output per hectare and per man-equivalent labour than the high population density compound households.

The cropped area per man-equivalent labour is higher in the low population density system than in the high population density compound system. The difference is quite significant particularly because the variation in cropped area per household is far greater in the compound system than in the bush system. This fundamental difference in land use per unit of labour is manifested directly in energy and protein production by households. Energy and protein output per hectare and per labour unit are higher in the bush system. But within the systems, the productivity of labour is higher than the productivity of land in the bush. Within the compound system, it is the productivity of land which is higher. It appears therefore that households in either system are making the best of the more limiting resource.

TABLE 6.7: Performance Indicators of Farm-households in Dulugu and
Kpongu in 1993 in Average Values per Farm-household
(Cedi Values Are at 1993 Prices; Between Brackets the CV)

Indicator	Compound farming (Dulugu)		Bush farming (Kpongu)	
Cropped area (ha)	1.4	(1.5)	4.0	(0.6)
Farm labour used (me)	92	(0.5)	209	(0.3)
Energy production/yr (MJ)	9963	(2.2)	49240	(0.6)
Protein production/yr (kg)	80	(2.7)	270	(0.5)
Food stock/yr (kg)	1248	(0.7)	2902	(0.7)
Crop sales (¢)	-		18000	(1.0)
Crop purchases (kg)	314	(1.1)	470	(1.1)
Livestock sales (¢)	16300	(1.6)	52800	(1.5)
Livestock purchases (¢)	11800	(2.9)	3800	(1.5)
Non-farm income (¢)	308500	(2.1)	67500	(0.6)
Non-farm expenditures (¢)	233900	(2.5)	32100	(0.7)
Total cash expenditures(¢)	167200	(0.7)	145300	(0.4)
Potential labour (me)	3.3		7.1	
Consumer units (cu)	4.2		8.4	

Source: SADAOC/AFS/Ghana

The energy output per hectare in Kpongu is even considerably higher than in Dulugu: 12310 MJ/ha versus 7116 MJ/ha. This, among others, may contribute to the higher average annual rainfall in Kpongu compared to Dulugu. The relatively lower output levels of the compound system is due to land scarcity, but because non-farm income is substantially higher in the compound system than in the bush system, shortfalls in household food production are met with non-farm income.

Crop storage offers an avenue for ensuring food security, at least, in the short-run. The indicator, crop purchase-stock ratio suggests that for each kilogram of crop stored, compound households would purchase 0.25 kg — about one and a half times what bush system households would purchase. This is consistent with the different levels of crop production that occur in the two settings.

From the nutritional perspective, the energy indicators (Table 6.8) show that bush system households are able to meet their energy requirement from their own production. Compared to the minimum energy requirement of 3,869 MJ/cu/year (White, 1976), households in Kpongu are able to produce 5,862 MJ/cu/year while those in Dulugu produce 2,372 MJ/cu/year. This means that the rest of energy required by Dulugu households must come from food purchases if they do not reduce consumption. Indeed, the NAES Food Consumption survey shows that whereas Kpongu households were 93 percent energy self-sufficient in 1990, Dulugu households were 63 percent self-sufficient[15]. However, the same survey indicates that there is no significant difference between the daily calorie consumption per consumer unit in the two villages, the calorie deficit being made up through purchases. About 30 percent of energy produced by Kpongu households is actually sold in the form of edible cash crop (yam) to generate income for non-food expenditure.

TABLE 6.8   Food Security Indicators in Dulugu and Kpongu
in 1993 in Average Values Per Farm-Household

Indicator	Compound farming (Dulugu)	Bush farming (Kpongu)
Cropped area/me (ha)	0.42	0.56
Energy prod./me/yr (MJ)	3019	6935
Energy prod./cu/yr (MJ)	2372	5862
Energy requirement/cu/yr	3869	3869
Supply-Demand Ratio	61%	151%
Protein/me/day (gr)	67	104
Protein/cu/yr (gr)	52	88
Protein requirement/cu/day	37	37
Supply-Demand Ration	141%	237%
Stock/Crop purchase ratio	4.0	6.25

Source: SADAOC/AFS/Ghana

Although the protein production information in Table 6.8 indicates that Kpongu households produce more protein than Dulugu households, Fey's analysis (Fey, 1992) shows that Dulugu households consume more protein than Kpongu households. This contradiction needs further investigation. The minimum daily protein requirement for the average consumer is set by FAO at 37 g. Given this figure households in the two systems will be meeting their protein needs adequately.

In order to assess the food security situation at farm level a number of whole farm performance indicators are required, for instance, total net family income (crop production,

---

[15] The report indicates that apart from compound farming households represented by Dulugu, many Northern Ghana farm households are energy self-sufficient.

livestock activities, non-farm income), cash flow situation during the year, food stock situation during the year. Such type of food security indicators at farm-household level are currently being developed and calculated.

Conclusion

Food production among households in both the low and high population density farming systems in Northern Ghana is still done within the traditional subsistence economic system. Households produce much of what they eat. Poor rainfall and soil condition as well as severe resource constraints cause considerable instability in food production and, consequently, expose farm-households to food insecurity.

Major differences in the resource base are observed between farming systems in low and high density population areas. Land is the basic limiting factor in high density population farming systems and labour in the low density population farming systems. In both systems capital is scarce factor for most of the farm-households.

Although farm-households in the two farming systems try to achieve self-sufficiency in food production, most of them are not food self-sufficient throughout the year. In particular, the food situation during the period just before the next harvest is a very difficult one. To be able to live through the period, households find other avenues to generate income for purchasing the additional food required. This chapter makes it clear that for farmers in high-population land-scarce areas like Dulugu in the Upper East Region, the opportunity to achieve this lies in non-farm economic activity and, to a limited extent, livestock production. For farmers in low-population density systems like Kpongu in the Upper West Region, the opportunity lies in producing cash crops for sale and/or livestock production.

Therefore, the two identifiable strategies for achieving household food security are:

(I) maximising own-food production and non-farm income;

(ii) maximising own-food production and cash crop income.

In both strategies, livestock sale is an in-built decision variable aimed at providing greater access to food, either directly through purchases or indirectly through farm input application. Variations in these two strategies arise from differences in the type of cash crop adopted and/or the type of non-farm activity undertaken.

Cash generating activities are another important strategy both to make up for food shortages and to enable the households to take part in the monetary economy. In high density population areas households appear to be dependent mainly on non-farm activities, while in low population density systems cash crops are cultivated for that purpose. Given the limited opportunities for non-farm activities within Northern Ghana, food security situation in especially the high population density areas is critical.

Achievement of food security is given by a household's ability to meet its target consumption level on a regular basis. Consequently, issues of unstable production, income, and prices are extremely important in the households' ability to produce and/or acquire food. The recurrence of hunger during a definite period in the year implies that many households experience regular transitory food insecurity due to fluctuations in household income and food consumption. There is the urgent need for households to, at least, keep grain safety stocks to be assured of food throughout the hunger period. If their safety stock cannot be met during the year, then they either intensify non-farm economic activity or cash crop production to generate greater cash income for food purchases or simply eat less for sometime during the year.

REFERENCES

Abu Katie/GGAEP, 1992. GGAEP Target Group Survey of Dagbon Area in Northern Region. GGAEP, Tamale.

Alderman, Harold and Paul Higgins, 1992. Food and Nutritional Adequacy in Ghana. Working Paper 27. Cornell Food and Nutrition Policy Program. Cornell University.

Baur, H., 1992. Farm Household Systems in Northern Ghana (Draft Report) PARTICIP GmbH Wehingen.

Benneh, George, 1971. Water Requirements and Limitations Imposed on Agricultural Development in Northern Ghana. In Proceedings of International Conference on Factors of Agricultural Growth in West Africa, Held in Legon, April 1971, ISSER, University of Ghana, Legon.

Boateng, 1989. 'Ghana Poverty Profile', ISSER, University of Ghana, Legon.

Chambers, R., 1983. 'Rural Development: Putting the Last First, Harlow U.K.

Eyeson, J.K. & Ankrah, E.K. 1975. 'Composition of Foods Commonly used in Ghana' Food Research Institute. CSIR, Accra.

Ewusi, K., 1976. Disparities in Levels of Regional Development in Ghana. Technical Publication No. 43, University of Ghana, Legon.

Ewusi, K., 1984. The Dimensions and Characteristics of Rural Poverty in Ghana. Technical Publication No. 43, University of Ghana, Legon.

Fey, Martin., 1992. Farm Household Food Consumption in Northern Ghana: Case Study in Eight Villages. Preliminary Report. Gottingen.

Fortes, M. and Fortes, S.I. 1936. 'Food in the Domestic Economy of the Tallensi' 9, Africa.

Freeman, Ade H. 1994. Population Pressure, Land Use, and the Productivity of Agricultural Systems in the West African Savannah. In Issues in Rural African Development 2. Winrock International, Morrillton, Arkansas.

Ghana Statistical Service, 1986. Country Report of 1984 Population Census. Accra.

Herbon, D. 1990. 'Ensuring Lasting Economic and Social Security' in Cammann, L. (ed.) 1990, 'Peasant Household Systems: Proceedings of an International Workshop, Feldafing.

Herbon, D. 1990. 'The Relative Importance of Farming' in Cammann, L. (ed.) 1990, 'Peasant Household Systems: Proceedings of an International Workshop. Feldafing.

Hunter, J.M. 1967. 'Seasonal Hunger in a Part of the West African Savannah: A Survey of Bodyweight in Nangodi, North-East Ghana' Transactions: Institute of British Geographers, 41.

International Course Centre for Development Oriented Research in Agriculture and Nyankpala Agricultural Experiment Station, 1993. Coping with Uncertainty: Challenges for Agricultural Development in the Guinea Savannah Zone of the Upper West Region, Ghana. Working Document Series 28.

ISNAR R52, 1991. 'Review of the Ghana Agricultural Research System' CSIR/ISNAR Report. Vol. 11 Annexes. CSIR, Accra, Ghana.

Low, A. 1986. 'On-farm Research and Household Economics' in Moock, Joyce L .(ed.) 1986 'Understanding African Rural Households and Farming Systems', Westview Press London.

Mckee, Katharine. 1986. 'Household Analysis as an Aid to Farming Systems Research: Methodological Issues' in Moock, Joyce, L. (ed.) (1986).

Maatman A. And C. Schweigman 1993. A Study of Farming Systems in the Central Plateau in Burkina Faso. Application of Linear Programming. Working Paper WP- AFS-93-01. INERA; University of Ouagadougou and University of Groningen.

Ministry of Agriculture Ghana; Policy, Planning, Monitoring and Evaluation (PPMED), 1991. 'Agriculture in Ghana: Facts and Figures'. Accra, Ghana.

Ministry of Agriculture, 1990. Medium Term Agricultural Development Programme, Accra, Ghana.

Norman, D.W. 1974. Rationalising Mixed Cropping under Indigenous Conditions: the Example of Northern Nigeria. Journal of Development Studies, Vol. 11.

Norrip, 1983. The Northern Region, Ghana. Vol. 2 Regional Development Strategy. NORRIP Technical Unit, Tamale, Ghana.

Papadakis, J. 1966. 'Crop Ecology Survey in West Africa' Vol. 1. FAO, Rome.

Prudencio, Coffie Y. 1983. A Village Study of Soil Fertility Management and Food Crop Production in Upper Volta. Technical and Economic Analysis. PhD. Thesis. Department of Economics, University of Arizona.

Quansah, C., Bonsu, M., Gana, B. K. and Basta, H. P. 1991. Preparatory Work: Land and Water Resource Management Study. Department of Crop Services, Ministry of Food and Agriculture, Accra.

RG-AFS Note 1., 1993. Village Profiles.

Rhoades, Robert E., 1989. Evolution of Agricultural Research and Development since 1950: Toward an Integrated Framework. Gatekeeper Series No. SA12 International Institute for Environment and Development.

Runge-Metzger, A. and De Haen, 1990. 'Farm Households in the Process of Development as Partners in Project Work' in Cammann, L. (ed.) (1990). 'Peasant Household Systems': Proceedings of an International Workshop, Feldafing.

Runge-Metzger, A. and L. Diehl, 1993. Farm Household Systems in Northern Ghana. A Case Study in Farming Systems Oriented Research for the Development of Improved Crop Production Systems. Nyankpala Agricultural Experiment Station Report No. 9, Nyankpala, Ghana.

UNICEF, 1984. Ghana Situation Analysis of Women and Children. Accra, Ghana.

Watson, D.J. 1971. 'The Nutritive Value of Some Ghanaian Foodstuffs' Ghana J. Agric. Sc. 4 95-111.

White, G.F., 1969. 'Strategies of American Water Management', Ann Arbor. Univ. of Michigan Press.

APPENDIX

CONVERSION FACTORS

(i) Man-equivalent Labour

Age	Male	Female
0-3	0.0	0.0
4-8	0.25	0.25
9-15	0.7	0.6
16-49	1.0	0.8
50 +	0.7	0.5

Source: Fey, 1992

(ii) Consumer unit

Age	Male	Female
0-8	0.4	0.4
9-15	0.7	0.7
16-49	1.0	0.9
50 +	0.8	0.8

Source: Fey, 1992

(iii) Livestock units

Animal	au.
Cattle	1.0
Oxen (trained)	1.25
Sheep/goats	0.1
Pigs	0.2
G-fowl/chicken/duck	0.04
Turkey	0.05
Rabbits	0.04

(iv) Equipment units

Equipment	eu.
Tractor	3.0
Plough	0.1
Ridger	0.1
Bullock	1.25
ATP	0.05
ATR	0.05
Hoe	0.01
Cutlass	0.005

Source: SADAOC/AFS/Ghana

Energy and Protein Content of Crops

Crop	Calorie (MJ/kg)	Protein(g/kg)
Rice	9.66	64
Maize	15.06	95
Millet	15.48	90
Early millet	15.69	86
Late millet	15.48	90
Sorghum	15.06	96
Sweet potato	6.19	8
Cassava	7.20	7
Yam	5.94	18
Tiger nut	11.80	30
Edible dry bean	14.64	225
Cowpea	14.64	225
Pigeon pea	12.97	198
Groundnut	25.02	205
Soybean	17.65	325
Hot pepper	3.05	28
Okro	1.26	16

Source: Watson (1971), Eyeson and Ankrah (1975)

CHAPTER 7

AGRICULTURAL SUPPLY RESPONSE AND STRUCTURAL ADJUSTMENT IN GHANA AND BURKINA FASO — ESTIMATES FROM MACRO-LEVEL TIME-SERIES DATA

K. Y. Fosu, N. Heerink, K. E. Ilboudo, M. Kuiper, and *A. Kuyvenhoven*

Introduction

Liberalisation of prices is a major element of structural adjustment programmes which have been implemented in many African countries since the beginning of the 1980s. One of the main objectives of price liberalisation is to increase the relative prices of agricultural products with the purpose of stimulating agricultural production. Measures taken to this effect include the abolition of government control of most agricultural prices, devaluation of the exchange rate, and reduction of protective import tariffs for industrial products.

The extent to which (relative) price increases of agricultural products lead to increases in agricultural production in sub-Saharan Africa is highly debated. According to some authors, price responsiveness in low-income countries is often constrained by the poor state of the transport and marketing facilities, communication and information services, and social infrastructure. Many of these constraining 'nonprice' factors typically fall into the category of public goods, and would not be forthcoming simply through higher prices (e.g. Beynon, 1989 and Chhibber, 1989). Others argue that both private and public resources devoted to agriculture are highly responsive to price changes. When relative prices change in favour of agriculture, private investments that remove nonprice constraints may become attractive and changes in government expenditures and public investment in rural infrastructure may be more forthcoming (e.g. Peterson, 1979; Jaeger, 1992).

Empirical research on supply response to price changes in countries in sub-Saharan Africa is needed to help to clarify this discussion. Little empirical research has been done thus far on the supply response of agriculture to price changes in sub-Saharan Africa. Reviews of the (scanty) available evidence by Bond (1983) and Pütz (1992) conclude that, with few exceptions, own-price elasticities of cash crops are positive and significant. The size of the short-run elasticities is not large, however, ranging between 0.4 and 0.6. Long-run price elasticities tend to be larger and of fairly sizable magnitude (ranging from 0.5 to 1.2), particularly for perennials like coffee and cocoa.

Available studies of supply response in sub-Saharan Africa have usually concentrated on cash crop production. Production of food crops, particularly root crops, is generally assumed to show little or no response to price changes. More recent studies, however, pay increasing attention to food production. A review by Pütz (1992: Table A.1) of available econometric estimates of supply responses of individual food crops in sub-Saharan Africa indicates that own-price generally has a small but significant positive effect on acreage (with short-run elasticities generally between 0.1 and 0.3) as well as output (with elasticities ranging from 0.1 to 0.8[1]).

A comparison of the responses to price changes of total cereal production and total root crop production has been made in a recent study on Ghana (World Bank, 1993). The estimated short-run elasticity equals 0.53 for cereals and 0.37 for root crops, whereas the long-term elasticities equal 2.94 and 0.37 respectively. On the basis of these results it was concluded that farmers in Ghana, just like farmers in other parts of the world, respond promptly and significantly to price changes. The analysis suffers, however, from a number of shortcomings. In particular, the effects of structural factors and of policy changes are disregarded.

It is important to distinguish the supply response of individual crops from the supply response of the agricultural sector. The aggregate supply response of agriculture is smaller, because individual crops compete for the same scarce resources. Resources are fairly fixed for agriculture as a whole, but can easily be shifted between different crops. A macro-level empirical analysis of aggregate supply response to price changes in nine African countries (Bond, 1983) found that the estimated coefficient of the price variable differs significantly from zero in Ghana and Kenya only (with short-run elasticities equal to 0.1 and 0.2, respectively). In an analysis of total agricultural production in The Gambia, Pütz (1992) found a small and nonsignificant supply response of aggregate production to price changes for a micro-level data set, but a statistically significant short-run elasticity of 0.7 for a macro-level data set.

Empirical analyses of supply response in sub-Saharan Africa based on time-series data may have to control for major policy switches that occur in the course of time. Prices are usually not liberalised in isolation, but are part of a whole package of adjustment measures. Many of these measures, e.g. privatisation of input supply or cuts in government investments, are likely to have an impact on the strength of the response of agricultural production to price changes. It may therefore be necessary to distinguish between the pre-adjustment period and the period after adjustment programmes have been implemented.

The purpose of this chapter is twofold. One is to present a detailed analysis of the supply response of root crops, cereals, and non-food crops to price changes and structural factors in Ghana and Burkina Faso. The results are used to make inferences about the extent to which policies aimed at price liberalisation may be successful in stimulating agricultural production in Central-West Africa. The second purpose is to examine the extent to which the supply response (of both food and cash crops) to prices has changed as a result of the implementation of adjustment programmes in Ghana and Burkina Faso. This is accomplished by examining the differences in supply response between the pre-adjustment period and the whole time range for which data are available.

The chapter consists of three parts. First, the Nerlove model of supply response is presented and an extension of the Nerlove model that may be used to analyse supply response in low-income countries is discussed. Second, the specifications of the two models that are used in this chapter are given, and choices with respect to the variables used are motivated. Third, the results of applying these two models to time-series data for Ghana and Burkina Faso are presented and discussed. Finally, conclusions are drawn.

The Nerlove Model and Supply Response in Low-income Countries

The model originally developed by Nerlove to estimate supply response has been so widely adopted that it can be considered the standard approach to the determination of agricultural

---

[1] Larger elasticities are found in some studies on Senegal, but the methodologies applied in these studies are questionable (Pütz 1992: 34).

supply response. The model can be used to examine the effect of price expectations and adjustment lags in production on agricultural supply. It consists of three equations:

$$P_t^e = P_{t-1}^e + \beta(P_{t-1} - P_{t-1}^e) \tag{7.1}$$

$$Q_t^d = a_0 + a_1 P_t^e + a_2 Z_t + u_t \tag{7.2}$$

$$Q_t = Q_{t-1} + \gamma(Q_t^d - Q_{t-1}) \tag{7.3}$$

with

P_t^e	=	expected price in period t
P_t	=	actual price in period t
Q_t^d	=	desired output in period t
Q_t	=	actual output in period t
Z_t	=	non-price variable
u_t	=	random disturbance term
β	=	coefficient of expectation ($0 \leq \beta \leq 1$)
γ	=	area adjustment coefficient ($0 \leq \gamma \leq 1$)
a_0, a_1, a_2	=	unknown coefficients.

By substitution and rearrangement, the reduced-form equation can be derived (see e.g. Askari and Cummings, 1976:33) which can be estimated by means of maximum likelihood procedures or least-squares techniques:

$$Q_t = \pi_0 + \pi_1 P_{t-1} + \pi_2 Q_{t-1} + \pi_3 Q_{t-2} + \pi_4 Z_t + \pi_5 Z_{t-1} + v_t \tag{7.4}$$

with:

$\pi_0 = a_0 \beta \gamma$
$\pi_1 = a_1 \beta \gamma$
$\pi_2 = (1-\beta) + (1-\gamma)$
$\pi_3 = -(1-\beta)(1-\gamma)$
$\pi_4 = a_2 \gamma$
$\pi_5 = -a_2(1-\beta)\gamma$
$v_t = \gamma u_t - (1-\beta)\gamma u_{t-1}$

Equation (7.4) is over-identified, since there are six reduced-form coefficients π, and only five structural parameters, $a_0, a_1, a_2, \beta,$ and γ (Sadoulet and de Janvry, 1995:87). Some sort of restriction thus has to be imposed in order to get a unique solution. One of the most frequently imposed constraints is the restriction $\beta=1$, thus assuming that the expected normal price is equal to the last period's price, i.e. $P_t^e = P_{t-1}$. This leads to a much simpler, exactly defined, reduced-form equation:

$$Q_t = c_0 + c_1 P_{t-1} + c_2 Q_{t-1} + c_3 Z_t + v_t \tag{7.5}$$

with:

$c_0 = a_0 \gamma$
$c_1 = a_1 \gamma$ $\quad (c_1 \geq 0)$
$c_2 = 1-\gamma$ $\quad (0 \leq c_2 \leq 1)$
$c_3 = a_2 \gamma$
$v_t = \gamma u_t$

155

Coefficients estimated by this model provide the following information:

— the area adjustment coefficient: $1-c_2$ ($=\gamma$)
— the short-term supply response: c_1 ($=a_1\gamma$)
— the long-term supply response: $\dfrac{c_1}{1-c_2}$ ($=a_1$)

Alternatively, a weighted average of the prices in the last two or more periods may be used to represent expected normal price (see e.g. Bapna, 1981 and Tsibaka, 1986).

An important extension of the model, which is meant to better reflect the specific conditions of agriculture in low-income countries, has been developed by Behrman (1968) in a study of annual crop production in Thailand. Behrman extends the general model with expectations about yield and with two proxies for risk. The essence of the model can be represented as follows (see Behrman, 1968:169 and Sadoulet and De Janvry, 1995:92):

$$P^e_t = c_0 + P^e_{t-1} + \beta(P_{t-1} - P^e_{t-1}) + u_{1t} \tag{7.6}$$

$$A_t = b_0 + A_{t-1} + \delta(A^d_t - A_{t-1}) + u_{2t} \tag{7.7}$$

$$A^d_t = a_0 + a_1 P^e_t + a_2 Y^e_t + a_3 \sigma_{pt} + a_4 \sigma_{yt} + u_{3t} \tag{7.8}$$

$$Y_t = d_0 + d_1(R_t - R^*) + d_2 T + d_3 T^2 + u_{4t}, \tag{7.9}$$

$$Y^e_t = Y_t - d_1(R_t - R^*) - u_{4t} = d_0 + d_2 T + d_3 T^2 \tag{7.10}$$

with:

δ = adjustment coefficient for the planted area
A_t = actual planted area
A^d_t = desired planted area
Y^e_t = expected yield (expected production per unit planted area)
σ_{pt} = standard deviation of crop price over three preceding periods relative to standard deviation of index of prices of alternatives over preceding three periods
σ_{yt} = standard deviation of actual crop yields over three preceding production periods
R_t = annual rainfall
R^* = mean annual rainfall
T = time-trend
$u_{1t}, u_{2t}, u_{3t}, u_{4t}$ = random disturbance terms
$a_0, a_1, a_2, a_3, a_4, b_0, c_0, d_0, d_1, d_2, d_3$ = unknown coefficients.

The first two equations are similar to equations (7.1) and (7.3) of the standard Nerlove model.[2] In the second equation, output is limited to planted area. The third equation contains a number of extensions of the standard model. The first new element is the expected yield (Y^e_t). The idea behind the inclusion of this variable is that not only the value per unit of production (P^e_t), but also the number of units one expects to be able to harvest influences the total acreage devoted to a crop (Behrman, 1968:157-158). The second addition is the inclusion of proxies for yield and price risk. Especially for subsistence farmers, not only the expected values of

---

[2] Notice that in Behrman's version of the model, a constant and an error term are added to these two equations.

prices and output are important but other characteristics of the probability distributions of these variables as well. Price risk is approximated by the standard deviation of the price of the crop of concern relative to the standard deviation for the price index of alternatives in the last three periods (σ_{pt}). Yield risk is approximated by the standard deviation of the yields in the last three periods (σ_{yt}). These approximations may be justified on the ground that the subjective probability distributions of the farmers probably are based in large part on previous experience (Behrman, 1968:158). The fourth equation gives yield as a function of the difference between the actual rainfall in the period and mean annual rainfall, a linear time-trend and a quadratic time-trend. The latter two variables are meant to reflect secular trends in yields (that result from soil depletion, expansion of irrigation facilities, spread of new seeds, and so on). Expected yield, which is the variable used to explain desired planted area in the third equation, can be derived from this equation by assuming that expected rainfall equals mean rainfall (and the error term equals zero). This gives the last equation.

Behrman (1968) estimated the model for 50 administrative units in Thailand. It was found that the area adjustment and price expectation coefficients (β and δ) play only minor roles, whereas expected yield and (to a lesser extent) price and yield risk are strong determinants of the area planted.

Estimated Models for Ghana and Burkina Faso

Two models are estimated for the supply response of major crops in Ghana and Burkina Faso. In this section, we will present these models and discuss the choices that were made with respect to variable selection. The first model is a (traditional) Nerlove model, with the assumption that $\beta=1$:

$$A_t = c_0 + c_1 P_{t-1} + \Sigma c_{2i} P_{ci,t-1} + c_3 A_{t-1} + c_4 D_t + c_5 T + v_t \qquad (7.11)$$

with:

A_t = harvested area of the crop per head of the agricultural population in year t (in hectares per capita, x 1000)

P_{t-1} = producer price of the crop at t-1 (in local currency per metric ton), divided by the average price of all crops in that year

$P_{ci,t-1}$ = producer price of competitive crop i at t-1 (in local currency per metric ton), divided by the average price of all crops in that year

D_t = dummy for the occurrence of a drought in year t (equals 0 in the case of no drought, and 1 in the case of a drought)

T = time-trend

$c_0,..,c_4$ = unknown coefficients

v_t = disturbance term with standard properties.

In the literature, acreage (either planted or harvested acreage) is commonly used as a proxy for output. This is preferred to the use of harvested crop weight or volume, since the latter are partly dependent upon variables that are not under control of the farmer, like weather conditions (Rao, 1989:5). In our model, output is proxied by harvested area, as reliable data on planted area are not readily available. The harvested area is divided by the size of the agricultural population. In Ghana and Burkina Faso, the total harvested area is strongly related to the size of the agricultural population. Expansion of agricultural production takes place to a large extent through the extension of the cultivated area. An analysis of the effects of price changes and nonprice constraints on agricultural production should make proper adjustments for the impact of these demographic changes.

Prices used in the model are producer prices. They are defined as the average annual price received by the farmers in the most important production regions. Agricultural price data for countries in sub-Saharan Africa are often not of the desired quality. Market price data are unavailable for most African countries. Because many African governments fix (or used to fix) official prices for food and export crops, only these prices are generally reported. In countries like Tanzania, Cameroon, Zaire, and Ethiopia these official prices differ to a large extent from actual market prices for the major food crops. Reported farmgate prices in these countries are two to seven times higher than official prices in some years. In such cases, it would clearly be inappropriate to use official prices for estimating supply response. For Ghana and Burkina Faso, however, it is commonly presumed that differences between official and market prices are relatively small. Available evidence for Burkina Faso is consistent with this presumption (see UNDP and World Bank, 1992: Table 8-1).

Models of agricultural supply response equations should be homogeneous of degree zero in prices. In other words, supply should not change if the prices of all crops increase at the same rate. This is based on the assumption that it is not the absolute price of the crops that matters for production decisions, but the price of a crop relative to other crops. To satisfy this homogeneity constraint, the crop price in year t is divided by the average price of the crops (weighted by the harvested area of each crop) in that specific year. Thus, if all the crop prices increase by the same percentage, total supply does not change.

The coefficient of own-price can be positive or negative. In the latter case, supply response to the price of the crop in question is said to be 'perverse'. The response to price changes of competitive crops should generally be positive. When intercropping is the dominating system of cultivation, however, the coefficient of a competitive crop may be positive. When the price of a crop is known in advance of the growing season, as is for example the case with cocoa in Ghana, the price in the same period should be taken instead of the price in the period before.

The nonprice variables in the model consist of a weather indicator and a trend variable. The weather indicator is included to account for the effect of droughts on harvested area. Droughts will affect output mainly through changes in yields. In cases of extreme droughts, however, harvested area may also be negatively affected. No time-series data on rainfall in Ghana and Burkina Faso are available from the data sources that were consulted for this chapter (see section headed 'Estimation Results'). A dummy variable indicating the occurrence of a drought is used instead. A negative coefficient is expected for this variable. The trend variable reflects secular trends in area harvested. Such secular trends may occur as a result of technological change, improvements in rural infrastructure, environmental degradation, or other structural factors. The coefficient of the trend variable may be positive or negative.

It would be desirable to incorporate separate variables indicating the state of the rural infrastructure into the model. The available information is very scanty, however, and is often not of the desired quality. For this reason, a trend variable is used as a proxy for the impact of infrastructural improvements and other structural changes.

The second model that we use in our analysis is based on the variant of the Nerlove model developed by Behrman (see section headed the, 'Nerlove Model and Supply Response in Low-Income Countries'):

$$A_t = d_0 + d_1 P_{t-1} + \Sigma d_{2i} P_{ci,t-1} + d_3 \sigma_{pt} + d_4 \sigma_{yt} + d_5 Y_t^e + d_6 D_t + d_7 A_{t-1} + d_8 T + u_{1t} \qquad (7.12)$$

$$Y_t = d_8 + d_9 D_t + d_{10} T + d_{11} T^2 + u_{2t} \qquad (7.13)$$

$$Y_t^e = d_8 + d_{10} T + d_{11} T^2 \qquad (7.14)$$

with:

σ_{pt} = standard deviation of the crop price over the three preceding periods relative to the

158

standard deviation of the average price of other crops over the three preceding periods

σ_{yt} = standard deviation of actual crop yields (in hectogrammes per hectare) over the three preceding production periods

Y_t = crop yield per unit of land (in hectogrammes per hectare)

Y_t^* = expected crop yield per unit of land

$d_0,...,d_{11}$ = unknown coefficients

u_{1t}, u_{2t} = disturbance terms with standard properties.

One can expect the farmers to be risk-averse, since especially for subsistence farmers the severe consequence of a return below the expected value probably does not offset the rewards of a return above the expected value (Behrman, 1968:96). Two-sided tests will be applied for the price and yield variability variables, however, to test whether farmers are actually risk-averse or tend to be risk-takers. Expected yield is likely to have a positive impact, since probably not only the return (reflected in the price) but also the output per hectare influences planting decisions. On the other hand, when farmers have a target production level of a certain (subsistence) crop, then an increase in expected yield may imply that farmers will cultivate fewer acres of the crop in question.

The expected yield variable is estimated from actual yields in the following way. First the actual yield is regressed on the drought dummy variable, the time-trend, and the square of the time-trend. Expected yield is derived from this actual yield equation by assuming that there will be no drought (and that the disturbance term equals zero). This gives expected yield as a function of the trend variables only. For many crops the impact of these trend variables may not differ significantly from zero. In most countries in sub-Saharan Africa, a trend variable appears to be of minor importance for capturing technological change. There have been few technological breakthroughs in Africa, compared to the dissemination of the Green Revolution in Asia for example. The expansion of draft animal equipment, hybrid crop varieties (cocoa, maize, cassava), fertilizer, and crop protection in selected programmes are exceptions to this rule (Pütz, 1992:31). In cases where the two trend variables do not differ significantly from zero, expected yield will be excluded from the regression equation.

It should be noted that the price and yield variability variables in the model are likely to depend on the availability of public facilities such as research institutes, irrigation facilities, rural roads, and other components of rural infrastructure. Risk tends to decrease when more public services and facilities become available. As a result, these two variables are likely to capture part of the effect of rural infrastructure on agricultural production.

Estimation Results

The two models presented above are estimated on the basis of nation-wide data for Ghana and Burkina Faso. Data are obtained from the data files of the FAO and the World Bank, and from international publications (UNDP and World Bank, 1992 and World Bank, 1995). For each country, the five major annual crops are selected on the basis of their harvested areas. The analysis concentrates on annual crops only, because these have been least researched. Supply response of perennials, particularly cocoa in Ghana, has received relatively more research attention (see e.g. Bateman, 1965; Fosu, 1992 and 1994; Frimpong-Ansah, 1989; and Strycker, 1990). In addition, the acreage under cultivation of perennials hardly changes in the short run. Modelling supply response of perennials requires a more complex modelling approach, which includes the modelling of the stock of trees through e.g. a vintage approach.

Regressions are made for two different periods in both countries. Firstly, regressions are

159

made for the period 1966 -1989. This is the period for which time-series data are available for all variables. Secondly, regressions are made for the pre-adjustment period only, i.e. the period 1966-1982 in both Ghana and Burkina Faso. In Ghana an Economic Recovery Programme (ERP) was introduced in 1983, followed by the implementation of a Structural Adjustment Programme (SAP) from 1986. In Burkina Faso, 1983 was the year of the start of the so-called 'autonomous adjustment' policy reform initiated by the Sankara government. Its purpose was to improve the rate and pattern of economic growth, not so much to redress macroeconomic imbalances (Savadogo and Lariviere, 1994; Sedogo and Michelsen, 1995). It was followed by the introduction of a formal SAP in 1991. By comparing the results for the entire period with those for the period until 1983, the impact of adjustment programmes on the strength of supply response can be assessed. Unfortunately, it is not (yet) possible to compare pre-adjustment and adjustment periods, because the number of observations for the adjustment period is too small at the moment to allow for proper statistical analysis. However, it was possible to assess the difference between the 1966 - 1982 and 1983 - 1989 periods with a Chow break-point test[3].

REGRESSION RESULTS FOR THE STANDARD NERLOVE MODEL

Ghana

For Ghana the major annual crops selected for the analysis include three cereals (maize, millet and sorghum) and two root crops (cassava and yam). Cocoa, the most important export crop, is not included in the analysis, because the production of cocoa involves (long-term) planting decisions. For the same reason plantain, which is an important food crop in Ghana, is excluded. The prices of cocoa and plantain, however, do serve as explanatory variables in the supply response equations of the selected annual crops.

Ghana can roughly be divided into two regions: a savanna zone in the North and a forest and coastal zone in the South. The Northern Zone is mainly a grain zone. The crops produced here include sorghum, millet, maize, and yam. The Southern Zone can be considered as a root and tuber crop zone. The major crops here are cassava, yam, plantain, maize and cocoa. Evidently, only the prices of crops grown in the same region are relevant for explaining the supply response of a crop.

Maize

The estimation results of the standard Nerlove model for maize are given in Table 7.1. Since maize is an important crop in the north as well as in the south, the prices of the other four selected crops as well as cocoa and plantain are included as competitive prices in the model.

The first two columns give the results for the pre-adjustment period (1966 - 1982), while columns three and four give the results for the entire sample period (1966 - 1989). For each period, the first column presents the estimation results of the full equation, while the second column gives the results when all variables with coefficients that are either statistically insignificant (at the 10 percent significance level) or have the wrong sign are excluded from the

---

[3] The Chow break-point test partitions the data set into two subsets and tests whether the coefficients can be regarded as constant over both subsets. Each subset must contain more observations than the number of coefficients in the test equation (Hall *et al.*, 1990:15-19).

model (except for the own-price, which is always retained in the model). The resulting model is termed the 'reduced model'.

The estimated coefficients for the own-price are positive for both periods. But only for the whole 1966-1989 period does the estimated coefficient differ significantly from zero. The area harvested (lagged one period) proves not to be significant in either variant of the model. As a result, the long-run supply elasticities equal the short-run elasticities for both periods. The short-run supply elasticities equal 0.64 for the whole period, and 0.24 for the pre-adjustment period (but the estimated coefficient is insignificant in the latter case). These elasticities are substantially larger than the elasticity of 0.03 that was estimated for maize in Ghana from cross-section micro data for 1987 by Heerink *et al.* (1995). The difference may be explained to some extent by the fact that no lagged price variables could be included in the latter study.

The drought dummy is insignificant. Only in the case of an extremely severe drought can one expect the area harvested to be affected. The time-trend is also insignificant in both versions of the model. Structural factors therefore do not seem to have much impact on acreage decisions with respect to maize in Ghana.

Table 7.1: Regression Results for Harvested Area of Maize:
Standard Nerlove Model

Explanatory variables		Period 1966-1982		Period 1966-1989	
		Entire Model	Reduced Model	Entire Model	Reduced Model
P_{t-1}	Maize	32.69	9.53	18.44	29.19
		(1.39)	(0.79)	(1.54)	(3.91)
A_{t-1}	Maize	-0.39	-	0.26	-
		(-0.64)		(1.26)	
$P_{ci, t-1}$	Plantain	1.42	-	15.01	-
		(0.03)		(0.97)	
	Cassava	-173.70	-64.00	-70.04	-86.55
		(-1.93)	(-1.99)	(-2.17)	(-3.63)
	Yam	-81.19	-	-45.98	-43.08
		(-1.63)		(-1.65)	(-2.35)
	Sorghum	-36.17	-39.16	-30.50	-40.87
		(-1.15)	(-2.64)	(-1.81)	(-3.17)
	Millet	-27.90	-	-28.58	-22.65
		(-1.02)		(-2.59)	(-2.34)
$P_{ci, t}$	Cocoa	-10.90	-	-5.44	-4.61
		(-0.95)		(-1.56)	(-1.40)
T	Time-trend	-0.12	-	0.01	-
		(-0.04)		(0.01)	
D	Dummy	31.32	-	12.95	-
		(0.91)		(1.57)	
adjusted R^2		0.40	0.51	0.69	0.71
LM [probability]		0.64 [0.55]		0.03 [0.97]	
Chow F-value [probability]				0.76 [0.64]	
Number of observations		16	16	23	23

Note: The numbers in parentheses are t-values; an intercept term was included in all equations

The high probabilities of the LM test statistic[4] indicate that there does not seem to be a problem with serial correlation in either of the models. The break-point Chow test does not indicate a significant difference between the pre-SAP and SAP period.

Millet

Regression results for millet are presented in Table 7.2. Since millet is an important crop in the savanna zone, only the prices of sorghum, maize, and yam are included as competitive prices in the model.

TABLE 7.2:   Regression Results for Harvested
Area of Millet: Standard Nerlove Model

Explanatory variables		Period 1966-1982		Period 1966-1989	
		Entire Model	Reduced Model	Entire Model	Reduced Model
P_{t-1}	Millet	5.41	-6.68	-4.26	-7.80
		(0.33)	(-1.25)	(-0.77)	(-1.99)
A_{t-1}	Millet	0.13	-	0.20	0.30
		(0.35)		(0.79)	(1.59)
$P_{ci, t-1}$	Sorghum	-5.21	-	-4.97	-
		(-0.39)		(-0.50)	
	Maize	-7.98	-	-1.23	-
		(-0.49)		(-0.17)	
	Yam	16.60	-	12.43	-
		(0.54)		(0.76)	
T	Time-trend	-0.67	-	-0.25	-
		(-0.62)		(-0.57)	
D	Dummy	-0.70	-	-0.90	-
		(-0.06)		(-0.15)	
adjusted R^2		-0.37	0.04	-0.02	0.17
LM [probability]		0.81 [0.46]			0.81 [0.46]
Chow F-value [probability]					0.36 [0.79]
Number of observations		16	16	23	23

Note: The numbers in parentheses are t-values; an intercept term was
included in all equations

---

4 Since the model contains harvested area as a dependent as well as an explanatory variable (lagged one period), the Durbin-Watson test statistic for first order serial correlation (or autocorrelation) will be biased towards 2, and thus unable to detect serial correlation. Therefore the results of an LM test for first order serial correlation is given, instead of the DW statistic. The LM test statistic is the Breusch-Godfrey Lagrange multiplier test statistic (nR^2), which is asymptotically $\chi^2(\rho)$ unbiased under general conditions. A low probability indicates that the null hypothesis of the absence of serial correlation should be rejected.

Only a small proportion in the variation of the harvested area of millet is explained by the model, as indicated by the low values of the adjusted R-squared. The own-price of millet has a negative coefficient for both periods, indicating a perverse supply response. None of the prices of competitive crops has a significant influence on decisions with regard to the area devoted to growing millet. The lagged harvested area variable has a significant effect in the whole sample, not in the pre-adjustment period sample. The short-run own-price elasticity equals -0.45, while the long-run elasticity equals -0.64 for the whole sample. As in the case of maize, the time-trend and the drought dummy do not differ significantly from zero.

Sorghum

The results of the traditional Nerlove model for sorghum, the third cereal that is analysed, are given in Table 7.3. The own-price of sorghum has a negative coefficient, which differs significantly from zero for the pre-adjustment period. The lagged harvested area variable has a significant positive coefficient for both periods. The corresponding short-run and long-run price elasticities equal -0.67 and -1.74 respectively for the pre-adjustment period.

For both samples, the price of millet is the only price of a competitive crop that seems to affect the harvested area of sorghum. A positive time-trend is found for the pre-adjustment period, indicating that structural factors had a significant positive effect on the (per capita)

TABLE 7.3: Regression Results for Harvested Area of Sorghum:
Standard Nerlove Model

Explanatory variables		Period 1966-1982		Period 1966-1989	
		Entire Model	Reduced Model	Entire Model	Reduced Model
P_{t-1}	Sorghum	-11.27	-17.59	-11.73	-7.47
		(-0.68)	(-3.09)	(-1.41)	(-1.40)
A_{t-1}	Sorghum	0.63	0.61	0.44	0.34
		(2.27)	(2.68)	(1.51)	(1.60)
$P_{ci, t-1}$	Millet	-20.67	-18.45	-6.10	-6.92
		(-2.24)	(-2.81)	(-1.15)	(-1.87)
	Maize	-2.12	-	-6.07	-
		(-0.23)		(-0.85)	
	Yam	-11.27	-	17.88	-
		(-0.68)		(1.32)	
T	Time-trend	1.46	1.07	0.11	-
		(2.23)	(2.40)	(0.25)	
D	Dummy	-4.06	-	3.68	-
		(-0.61)		(0.74)	
adjusted R^2		0.43	0.53	0.07	0.10
LM [probability]		0.34 [0.72]		0.19 [0.83]	
Chow F-value [probability]		1.69 [0.21]			
Number of observations		16	16	23	23

Note: The numbers in parentheses are t-values; an intercept term was included in all equations

163

harvested area of sorghum before the economic policy reforms started in 1983.

The value for the Chow break-point test is considerably higher than it is for maize and millet. The differences between the coefficients for the pre-adjustment and the adjustment period, however, are not statistically significant (at the five percent significance level).

Cassava

Turning to the root crops, the results for cassava are presented in Table 7.4. The own-price of cassava has a negative impact on harvested area in both samples. Only for the whole period does the estimated coefficient differ significantly from zero. The lagged harvested area variable also differs significantly from zero for the whole period. The corresponding short-run and long-run price elasticities of cassava are -0.33 and -0.51, respectively. A similar negative supply response is found for cassava in Ghana in the study by Heerink *et al.* (1995). It should also be noted that the Chow test indicates a structural break in 1983, which is statistically significant (at the five percent significance level).

A positive time-trend is found for the whole sample period, but not for the pre-adjustment period. None of the variables in the model has a significant effect on the harvested area of cassava in the pre-adjustment period, resulting in a very low R-squared (and a negative adjusted R-squared) for the estimated equation.

TABLE 7.4: Regression Resultsf for Harvested Area of Cassava: Standard Nerlove Model

Explanatory variables		Period 1966-1982		Period 1966-1989	
		Entire Model	Reduced Model	Entire Model	Reduced Model
P_{t-1} Cassava		1.02	-5.48	-9.21	-21.81
		(0.06)	(-0.61)	(-0.77)	(-1.96)
A_{t-1} Cassava		0.59	-	0.24	0.35
		(1.66)		(1.07)	(1.70)
$P_{ci, t-1}$	Plantain	-11.86	-	-9.70	-14.33
		(-1.08)		(-1.42)	(-2.20)
	Maize	-3.08	-	0.44	-
		(-0.51)		(0.11)	
	Yam	-5.93	-	1.15	-
		(-0.49)		(0.11)	
$P_{ci, t}$	Cocoa	4.16	-	2.18	-
		(1.00)		(2.07)	
T	Time-trend	1.21	-	0.68	0.87
		(1.25)		(2.11)	(3.02)
D	Dummy	-2.37	-	-2.69	-
		(-0.22)		(-0.70)	
adjusted R^2		-0.19	-0.04	0.59	0.51
LM [probability]		0.34 [0.72]		0.39 [0.68]	
Chow F-value [probability]				3.04 [0.05]	
Number of observations		16	16	23	23

Note: The numbers in parentheses are t-values; an intercept term was included in all equations

The other root crop analysed is yam. The results of the traditional Nerlove model are given in Table 7.5. As with cassava (and millet), the response of harvested area to changes in the own-price is negative and significantly different from zero for the whole sample period. The lagged harvested area variable differs significantly from zero. The resulting short-run and long-run price elasticities for the whole sample period equal -0.66 and -1.16, respectively. The results of the Chow test, however, do not indicate a significant structural break-point in 1983.

TABLE 7.5: Regression Results for Harvested Area of Yam:
Standard Nerlove Model

Explanatory variables		Period 1966-1982		Period 1966-1989	
		Entire Model	Reduced Model	Entire Model	Reduced Model
P_{t-1}	Yam	-10.98	-9.31	-25.44	-17.51
		(-0.60)	(-0.91)	(-2.43)	(-2.99)
A_{t-1}	Yam	-0.03	0.55	0.30	0.43
		(-0.06)	(2.84)	(1.42)	(2.12)
$P_{ci, t-1}$	Plantain	-10.04	-	1.00	-
		(-0.66)		(0.19)	
	Cassava	-8.96	-	-26.30	-18.89
		(-0.31)		(-2.31)	(-1.70)
	Maize	-5.50	-7.11	1.56	-
		(-0.54)	(-2.03)	(0.38)	
	Sorghum	-14.49	-	-2.70	-
		(-0.98)		(-0.42)	
	Millet	1.14	-	-10.20	-11.23
		(0.10)		(-2.68)	(-2.80)
$P_{ci, t}$	Cocoa	-3.13	-	-1.38	-1.84
		(-0.60)		(-1.08)	(-1.44)
T	Time-trend	-0.56	-	0.21	0.43
		(-0.42)		(0.83)	(2.58)
D	Dummy	1.36	-	6.34	-
		(0.10)		(2.09)	
adjusted R^2		0.38	0.61	0.64	0.56
LM [probability]		1.02 [0.39]		0.94 [0.42]	
Chow F-value [probability]				1.04 [0.47]	
Number of observations		16	16	23	23

Note: The numbers in parentheses are t-values; an intercept term was included in all equations

The time-trend has a significant positive effect on the harvested area of yam. But the drought dummy does not differ significantly from zero (despite its significance in the full equation for the period 1966-1989), as is the case with all the regressions for Ghana presented above.

Conclusion

A few general conclusions emerge from the results presented in Tables 7.1 to 7.5. Firstly, the supply response to prices is positive for maize (for the whole period) only. For the other four crops, it is either negative or insignificant. This result may be explained from the fact that cassava, yam, millet and sorghum are food crops that are grown with traditional technologies in Ghana. In maize production, on the other hand, improved technologies are used on a relatively large scale. It should further be noted that price control has characterised agriculture in Ghana during most of the period under consideration. Few consumer goods were available to farmers during this period which may have caused the perverse supply response (see e.g. Azam and Besley, 1989 and Bevan *et al.*, 1990).

Secondly, the estimated coefficients for the own-price variable tend to be higher in absolute value and more significant for the whole sample period as compared to the pre-adjustment period. The only exception to this rule is sorghum. Yet, it is only for cassava that the Chow break-point test indicates that the results for the pre-adjustment period differ significantly from the results for the adjustment period.

Finally, structural factors seem to have had a limited impact on supply response in Ghana during the sample period. In only three cases — sorghum (for the pre-adjustment period), cassava (whole sample), and yam (whole sample) — do the estimated coefficients for the time-trend variable differ significantly from zero.

Burkina Faso

For Burkina Faso three cereals (sorghum, millet, and maize), and two cash crops (groundnuts and cotton) are selected on the basis of harvested area. Cereals are the most important crops throughout Burkina Faso. Burkina Faso can roughly be divided into three zones: a Sahel Zone in the north, a Dry Savanna Zone in the middle, and a Humid Savanna Zone in the south-west (and South). Sorghum and millet are the major cereals of the Sahel and Dry Savanna Zone. In the Savanna Zone they are grown in combination with groundnuts. In the Humid Savanna Zone in the south-west, sorghum and maize are the major cereals, while cotton is the major cash crop.

The price series for millet and sorghum in Burkina Faso are exactly the same, except for three years (1983, 1985, 1987). The perfect correlation which thus exists for the period 1966-1982 prevents a separate estimation of the coefficients of the prices of millet and sorghum. Therefore, in all models for Burkina Faso, the average of the prices of millet and sorghum is used rather than separate prices for these two crops.

Sorghum

Regression results for sorghum are presented in Table 7.6. The own-price used in the model is the average price of sorghum and millet, as explained above. Since sorghum is an important crop in all regions in Burkina Faso, the prices of groundnuts as well as maize and cotton are included as prices of competitive crops in the model.

None of the prices of competitive crops has a significant effect on the harvested area of sorghum, nor has harvested area lagged one period. The own-price of sorghum has a negative effect on harvested area for both sample periods, but this effect is not significantly different from zero (at the 10 percent significance level). Only the time-trend has a significant negative

166

TABLE 7.6: Regression Results for Harvested Area of Sorghum: Standard Nerlove Model

Explanatory variables		Period 1966-1982		Period 1966-1989	
		Entire Model	Reduced Model	Entire Model	Reduced Model
P_{t-1}	Sorghum/Millet	222.46	-73.36	5.70	-153.48
		(0.43)	(-0.22)	(0.02)	(-1.41)
A_{t-1}	Sorghum	-0.48	-	-0.51	-
		(-1.62)		(-2.10)	
$P_{ci, t-1}$	Maize	45.22	-	34.50	-
		(0.36)		(0.39)	
	Groundnuts	20.13	-	-10.22	-
		(0.46)		(-0.29)	
	Cotton	42.40	-	47.40	-
		(0.90)		(1.24)	
T	Time-trend	-2.94	-4.09	-2.10	-3.02
		(-0.74)	(-2.19)	(-0.96)	(-4.51)
D	Dummy	29.18	-	14.90	-
		(1.48)		(1.13)	
adjusted R^2		0.48	0.42	0.53	0.48
LM [probability]		1.50 [0.27]			3.96 [0.06]
Chow F-value [probability]					0.30 [0.87]
Number of observations		16	16	23	23

Note: The numbers in parentheses are t-values; an intercept term was included in all equations

effect on the harvested area of sorghum, indicating that structural factors tend to lower the per capita harvested area of sorghum.

Millet

The second cereal that is analysed is millet. Since it is an important crop in the Sahel and the Dry Savanna Zone but not in the Humid Savanna Zone, the price of groundnuts is the only price of a competitive crop included in the model.

Neither the price of groundnuts nor the harvested area lagged one period has a significant effect on the harvested area of millet (Table 7.7). The own-price has a negative effect, which is significantly different from zero for the whole sample period. The corresponding short-run (and long-term) price elasticity is -1.28. The drought dummy has a significant effect in the regression for the entire period, indicating that the harvested area of millet has decreased during periods of extreme drought in Burkina Faso.

167

TABLE 7.7: Regression Results for Harvested Area of Millet:
Standard Nerlove Model

Explanatory variables		Period 1966-1982		Period 1966-1989	
		Entire Model	Reduced Model	Entire Model	Reduced Model
P_{t-1}	Sorghum/Millet	162.16	-191.29	-139.93	-206.32
		(0.50)	(-1.19)	(-1.02)	(-2.48)
A_{t-1}	Millet	-0.01	-	0.15	-
		(-0.04)		(0.65)	
$P_{ci, t-1}$	Groundnuts	37.24	-	7.33	-
		(1.30)		(0.34)	
T	Time-trend	-2.69	-	-0.10	-
		(-1.53)		(-0.16)	
D	Dummy	-15.24	-	-18.75	-17.50
		(-1.04)		(-1.85)	(-1.96)
adjusted R^2		0.03	0.03	0.17	0.27
LM [probability]		0.21 [0.81]			0.43 [0.66]
Chow F-value [probability]					1.02 [0.41]
Number of observations		16	16	23	23

Note: The numbers in parentheses are t-values; an intercept term was included in all equations

Maize

The results for maize are given in Table 7.8. The prices of sorghum/millet and cotton are included in the model as competitive prices because maize is an important crop in the Humid Savannah-Zone only.

The own-price of maize has a positive effect on the harvested area, but the effect is not significantly different from zero for both sample periods. Harvested area (lagged one period), however, is significant for both periods. Neither the time-trend nor the drought dummy has a significant effect on the harvested area of maize in Burkina Faso.

Groundnuts

Two cash crops are analysed for Burkina Faso, with results for groundnuts given in Table 7.9. Despite the fact that the own-price can be expected to be quite important for cash crops, this price is not significant in either variant. As in the case of maize, however, the estimated coefficients are positive for both sample periods. It should further be noted that a significant negative time-trend is found for the pre-adjustment period.

TABLE 7.8: Regression Results for Harvested Area of Maize:
Standard Nerlove Model

Explanatory variables		Period 1966-1982		Period 1966-1989	
		Entire Model	Reduced Model	Entire Model	Reduced Model
P_{t-1}	Maize	44.35	36.70	44.76	18.78
		(1.15)	(1.64)	(1.83)	(1.25)
A_{t-1}	Maize	0.58	0.46	0.50	0.45
		(2.26)	(2.17)	(2.79)	(2.59)
$P_{ci, t-1}$	Sorghum/millet	17.36	-	-52.05	-54.92
		(0.12)		(-1.14)	(-1.92)
	Cotton	-7.76	-	-1.60	-
		(-0.47)		(-0.18)	
T	Time trend	0.09		0.49	-
		(0.05)		(0.83)	
D	Dummy	11.02	-	4.31	-
		(1.37)		(0.94)	
adjusted R^2		0.36	0.41	0.43	0.42
LM [probability]		0.73 [0.51]			0.58 [0.57]
Chow F-value [probability]					0.38 [0.77]
Number of observations		16	16	23	23

Note: The numbers in parentheses are t-values; an intercept term was included in all equations

TABLE 7.9: Regression Results for Harvested Area of Groundnuts:
Standard Nerlove Model

Explanatory variables		Period 1966-1982		Period 1966-1989	
		Entire Model	Reduced Model	Entire Model	Reduced Model
P_{t-1}	Groundnuts	6.96	5.06	3.85	4.42
		(1.04)	(0.98)	(0.68)	(1.25)
A_{t-1}	Groundnuts	0.51	-	0.27	0.47
		(1.53)		(0.96)	(2.46)
$P_{ci, t-1}$	Sorghum/Millet	8.19	-	-28.62	-
		(0.11)		(-0.76)	
T	Time-trend	-0.20	-0.55	-0.17	-
		(-0.44)	(-2.61)	(-1.10)	
D	Dummy	4.73	-	-0.64	-
		(1.21)		(-0.24)	
adjusted R^2		0.25	0.27	0.17	0.21
LM [probability]		0.34 [0.72]			0.05 [0.95]
Chow F-value [probability]					0.45 [0.72]
Number of observations		16	16	23	23

Note: The numbers in parentheses are t-values; an intercept term was included in all equations

Cotton

The last crop analysed for Burkina Faso is cotton. The results are presented in Table 7.10. As expected, the estimated coefficients for the own-price are positive and significantly different from zero. In addition, the harvested area lagged one period has a significant positive effect, which indicates that farmers adjust the cultivated area with a certain time lag. The corresponding short-run and long-run price elasticities are 0.38 and 0.64 for the pre-adjustment period, and 0.61 and 1.23 for the whole period, respectively. Furthermore, a positive time-trend is found for the whole sample period.

The results of the LM-test indicate a problem with serial correlation in the regression for the whole sample period[5]. Furthermore, the Chow break-point test gives a high value which is highly significant. Hence the regression results for cotton differ fundamentally as between the pre-adjustment period and the adjustment period.

TABLE 7.10: Regression Results for Harvested Area of Cotton:
Standard Nerlove Model

Explanatory variables		Period 1966-1982		Period 1966-1989	
		Entire Model	Reduced Model	Entire Model	Reduced Model
P_{t-1}	Cotton	1.72	2.88	4.37	5.34
		(0.80)	(3.07)	(1.75)	(3.16)
A_{t-1}	Cotton	0.25	0.41	0.47	0.50
		(1.10)	(2.24)	(2.68)	(3.11)
$P_{ci, t-1}$	Sorghum/Millet	18.01	-	-10.75	-
		(0.93)		(-0.58)	
T	Time-trend	-0.22	-	0.31	0.35
		(-1.14)		(2.39)	(3.31)
D	Dummy	-0.77	-	0.04	-
		(-0.64)		(0.04)	
adjusted R^2		0.57	0.61	0.78	0.80
LM [probability]		0.53 [0.60]			5.01 [0.02]
Chow F-value [probability]					4.45 [0.01]
Number of observations		16	16	23	23

Note: The numbers in parentheses are t-values; an intercept term was included in all equations

Conclusion

From the results presented above, a number of conclusions can be drawn. Firstly, the supply response to own-price changes is insignificant for most crops. The only exceptions are cotton and millet. Supply response to price is positive for cotton (for both sample periods), which

---

5 It may be noted, however, that the results of the extended Nerlove model are very similar to the results obtained here. As will be seen below (the Appendix), serial correlation is not a problem in the extended Nerlove model for cotton.

is the major cash crop in Burkina Faso. For the other cash crop, groundnuts, the estimated supply response is positive but insignificant. The supply response for food crops does not differ significantly from zero, except for millet (for the whole sample period).

Secondly, the estimated coefficients for the own-price are higher, in an absolute sense, for the whole period as compared to the pre-adjustment period for sorghum, millet and cotton. Only for cotton, however, does the Chow test indicate a significant break-point in 1983.

Thirdly, structural factors seem to have had a significant negative impact on the cultivation of some crops. Negative coefficients for the time-trend were found for sorghum (both sample periods) and for groundnuts (before 1983). A positive coefficient was estimated, on the other hand, for the time-trend in the equation for cotton (whole sample period).

Finally, drought does not seem to have had much effect on the harvested areas of different crops in Burkina Faso. Only for millet (whole sample period), a significant negative impact of drought on the harvested area was found.

RESULTS OF THE MODEL WITH RISK AND EXPECTED YIELD

As indicated in the section entitled 'Estimation Models for Ghana and Burkina Faso', the model with risk and expected yield adds three variables to the standard model. The price and yield risk variables are defined as standard deviations of the prices and yields in the preceding three years. The expected yield depends on the time-trend and the square of this trend. This variable is only included in case the time-trend variables are significant in the regression of the actual yield. If the expected yield is included, the time-trend is dropped from the equation in order to prevent multicollinearity problems. Thus, the effects of infrastructure for example, as captured by the time-trend in the traditional Nerlove model, are now reflected in the expectations concerning the yield. The results of estimating the extended Nerlove model are presented in the Appendix for both Ghana (Tables 7A.1 to 7A.5) and Burkina Faso (Tables 7B.1 to 7B.5).

In Ghana, price variability is found to affect the harvested areas of maize and yam (during the pre-adjustment period) only. As expected, price variability has a negative impact in both cases. Yield variability, however, has a positive effect on the harvested area of cassava and yam (whole sample and pre-adjustment period) and of maize and sorghum (pre-adjustment period). Apparently, farmers in Ghana often decided to grow larger land areas of food crops in response to a high variation in yield, particularly during the pre-adjustment period. A possible explanation for this result may be the need to meet a minimum food production level. When the risk of low yields is relatively high, a larger land area is cultivated.

Expected yield is included in the regressions for three crops. The yield of the other two crops, sorghum and cassava, proved to be fairly stable over time. Expected yield is found to affect harvested area in two cases only: millet (during the pre-adjustment period) and yam (whole sample period). In both cases, the impact is found to be negative.

The regression results for Burkina Faso indicate that price variability does not affect the acreage decisions of farmers. None of the estimated coefficients for the standard deviation of price differs significantly from zero. As regards yield variability, a positive coefficient is found for cotton (whole sample period) while a negative coefficient is found for maize (pre-adjustment period). Likewise, expected yield is found to affect the harvested area in two cases only: groundnuts (pre-adjustment period) and cotton (whole sample period). The estimated coefficients are negative for the former and positive for the latter.

The results for the other variables in the extended Nerlove model are roughly consistent with the results obtained from the standard Nerlove model. The general conclusions that were drawn above still hold for Ghana as well as for Burkina Faso. One notable difference is the significant positive supply response for maize in Burkina Faso for the whole sample period. When estimated for the pre-adjustment period, however, the supply response of maize is not significantly different from zero.

Conclusion

In this chapter, a detailed empirical analysis has been made of supply response of major annual crops in Ghana and Burkina Faso. The results indicate that supply response to the own-price is positive only for those crops that are grown with nontraditional technologies, i.e. maize in Ghana and cotton in Burkina Faso. Supply response of traditional food crops is either insignificant or (particularly in Ghana) negative. A possible explanation for the perverse supply response in Ghana may be the price control in agriculture and the limited availability of consumer goods that characterised most of the period under examination (see e.g. the studies of Azam and Besley, 1989, and Bevan *et al.*, 1992 on this subject). The results for Ghana differ greatly from the results obtained by the World Bank study mentioned in the introduction (World Bank, 1993). In that study, long-run elasticities of 2.94 and 0.37 were estimated for cereals and root crops in Ghana, respectively. The long-run elasticity for cotton in Burkina Faso lies within the range of 0.5 to 1.2 that is found for cash crops in the literature (see Bond, 1983, and Pütz, 1992).

The estimated supply responses tend to be stronger in magnitude and more significant for the whole sample period as compared to the pre-adjustment period. Not only are the estimated positive coefficients for maize in Ghana and cotton in Burkina Faso higher when estimated on the basis of the whole sample instead of the pre-adjustment period only, also the negative coefficients for food crops like millet in Ghana and Burkina Faso and cassava and yam in Ghana are higher (in an absolute sense) and more significant when they are estimated on the basis of the entire sample. The results of the Chow-test, however, show a structural break in 1983 for two crops only: cassava in Ghana and cotton in Burkina Faso. In Ghana, the own-price supply response became negative after 1983, while the own-price supply response of cotton in Burkina Faso became more strongly positive.

These results indicate that the introduction of adjustment policies has had only a limited impact on supply reponse in Ghana and Burkina Faso. Price liberalisation policies aimed at stimulating agricultural production may therefore be only partially successful. In fact, the production of traditional food crops may respond negatively to relative price improvements, particularly in Ghana. Only cash crops or food crops grown with improved technologies seem to respond positively to relative price improvements.

Price risk, as measured by the degree of price variability, is found to have a limited impact on the harvested areas of agricultural crops in Ghana and Burkina Faso. Significant (negative) coefficients were found for maize and yam in Ghana only. Yield variability, on the other hand, is found to have a positive effect on the harvested area of various crops in Ghana. This holds true in particular for the pre-adjustment period. It should be noted that the degree of variability of prices and yields is related to the availability of irrigation facilities, rural transport infrastructure, agricultural technologies, and other structural factors. By reducing price and yield variability, public investment in rural infrastructure may therefore have important effects on food production, particularly in Ghana.

Finally, population growth is probably the most important structural factor affecting the harvested area of most crops in Ghana and Burkina Faso. For many crops, harvested area increases proportionally with population growth. In order to correct for demographic influences, it was decided to concentrate the analysis on harvested area per capita. In the resulting regression equations, the time-trend variable differed significantly from zero in some equations only. For Ghana, a positive time-trend was found in the regressions for cassava and for yam (for the whole sample period) and for sorghum (pre-adjustment period). For Burkina Faso, on the other hand, a negative time-trend was found for sorghum (whole sample period) and groundnuts (pre-adjustment period) and a positive time-trend for cotton (whole sample period). These results indicate that structural factors like technology change, environmental degradation, or infrastructure improvement may have had some effects on agricultural production in Ghana and Burkina Faso. Further work will be needed to identify more clearly the factors contributing to these results.

REFERENCES

Askari, H., J.T. Cummings, 1976. *Agricultural Supply Response: A Survey of the Econometric Evidence*. New York: Praeger.

Azam, J.P., T Besley, 1989. *The Supply of Manufactured Goods and Agricultural Development: the Case of Ghana*. Development Centre Papers. Paris, OECD.

Bapna, S.L. 1981. *Aggregate Supply Response of Crops in a Developing Region*. New Delhi: Sultan Chand and Sons.

Bateman, M.J. 1965. "Aggregate and Regional Supply Functions for Ghanaian Cocoa." *Journal of Farm Economics* 47: 384-401.

Behrman, J.R. 1968. *Supply Response in Underdeveloped Agriculture: A Case Study of Four Major Annual Crops in Thailand, 1937-1963*. Amsterdam: North-Holland.

Bevan, D., P. Collier, J.W. Gunning *Controlled Open Economies: A Neoclassical Approach to Structuralism*. Oxford: Clarendon Press.

Beynon, J.G. 1989. "Pricism V. Structuralism in Sub-saharan African Agriculture." *Journal of Agricultural Economics* 40: 323-335.

Bond, M.E. 1983. "Agricultural Response to Prices in Sub-saharan African Countries". In *I.M.F. Staff Papers* 30: 703-726

Chhibber, A. 1989. "The Aggregate Supply Response: A Survey." In: S. Commander (ed.) *Structural Adjustment and Agriculture: Theory and Practice in Africa and Latin America*. London: James Currey.

Fosu, K.Y. 1992. *The Real Exchange Rate and Ghana's Agricultural Exports*. AERC Research Paper 9. Oxford, U.K.: University of Oxford, Centre for the Studies of African Economies.

Fosu, K.Y. 1994. "Structural Adjustment Programme and the Production of Cotton, Coffee and Cocoa in Ghana." Report submitted to CODESRIA. Dakar, Senegal.

Frimpong-Ansah, J.H. 1989. *From Predator to Vampire: the State and the Economy of Ghana*. London: James Currey.

Hall, R.E., J. Johnston and D.M Lilien. 1990. *MicroTSP User's Manual*. Irvine: Quantitative Micro Software. Version 7.0.

Heerink, N., P. Atsma, K.Y. Fosu, 1997. "Farmers' Transport Costs and Agricultural Production in Ghana." In: Asenso-Okyere, W.K., Benneh G. and Tims, W. (Eds.) Sustainable Food Security in West Africa, Kluwer Academic Publishers, 1997, chapter 10.

Jaeger, W. 1992. *The Effects of Economic Policies on African Agriculture*. Discussion Paper No. 147. Washington, D.C.:World Bank.

Peterson, W. 1979. "International Farm Prices and the Social Cost of Cheap Food Policies". *American Journal of Agricultural Economics* 59: 12-21.

Pütz, D. 1992. *Agricultural Supply Response in the Gambia: a Sectoral, Household, and Intrahousehold Analysis*. Kiel: Wissenschaftsverlag Vauk.

Rao, J.M. 1989. "Agricultural Supply Response: a Survey." *Agricultural Economics* 3: 1-22.

Sadoulet, E., A. de Janvry, 1995. *Quantitative Development Policy Analysis*. Baltimore, London: Johns Hopkins University Press.

Savadogo, K. and S. Larivière, 1994. "Ajustement Structurel et Performance Agricole: Quelques leçons de l'expérience d'auto-Ajustement au Burkina Faso." In: F. Heidhues and B. Knerr (eds.) *Food and Agricultural Policies Under Structural Adjustment*. Frankfurt: Peter Lang.

Sedogo, M., H. Michelsen, 1995. "Burkina Faso." In: S.R. Tabor (ed.) *Agricultural Research in an Era of Adjustment: Policies, Institutions, and Progress*. Washington, D.C.: World Bank.

Strycker, J.D. 1990. *Trade, Exchange Rate, and Agricultural Pricing Policies in Ghana*. Washington, D.C.: World Bank.

Thsibaka, T.B. 1986. *The Effects of Trade and Exchange Rate Policies on Agriculture in Zaire*. Research Report 56. Washington, D.C.: International Food Policy Research Institute.

UNDP and World Bank, 1992. *African Development Indicators*. New York: UNDP, and Washington, D.C.: World Bank.

World Bank, 1993. *Ghana 2000 and Beyond: Setting the Stage for Accelerated Growth and Poverty Reduction*. World Bank (Africa Regional Office, West Africa Department).

World Bank, 1995. *African Development Indicators 1994-95*. Washington, D.C.: World Bank.

REGRESSION RESULTS FOR THE EXTENDED NERLOVE MODEL

A. GHANA

TABLE 7A.1 : Regression Results for Harvested Area of Maize: Extended Nerlove Model

Explanatory variables		Period 1966-1982		Period 1966-1989	
		Entire Model	Reduced Model	Entire Model	Reduced Model
P_{t-1}	Maize	9.73	-14.64	7.97	34.85
		(1.36)	(-1.45)	(0.41)	(5.30)
A_{t-1}	Maize	-0.64	-	0.17	-
		(-5.79)		(0.46)	
$P_{ci, t-1}$	Plantain	16.95	-	10.83	-
		(1.81)		(0.50)	
	Cassava	-220.24	-98.60	-86.48	-73.74
		(-14.65)	(-4.51)	(-2.10)	(-3.22)
	Yam	-86.51	-43.95	-46.59	-36.75
		(-9.50)	(-2.39)	(-1.44)	(-1.48)
	Sorghum	-31.31	-	-14.38	-46.83
		(-4.11)		(-0.50)	(-3.71)
	Millet	-7.44	-	-23.69	-26.42
		(-1.28)		(-1.77)	(-2.75)
$P_{ci, t}$	Cocoa	-17.29	-	-2.87	-6.10
		(-6.05)		(-0.53)	(-1.89)
S.D. price	Maize	-7.14	-7.50	-1.54	-
		(-12.76)	(-4.90)	(-1.03)	
S.D. yield	Maize	26.02	20.89	1.78	-
		(11.24)	(4.56)	(0.60)	
Exp. yield	Maize	-47.27	-	2.27	-
		(-3.64)		(0.24)	
D	Dummy	40.59	-	16.60	-
		(6.76)		(1.28)	
adjusted R^2		0.99	0.80	0.64	0.70
LM [probability]		0.89 [0.46]			0.03 [0.97]
Chow F-value [probability]					0.76 [0.64]
Number of observations		16	16	23	23

Note: The numbers in parentheses are t-values; an intercept term was included in all equations

TABLE 7A.2 : Regression Results for Harvested Area of Millet: Extended Nerlove Model

Explanatory variables		Period 1966-1982		Period 1966-1989	
		Entire Model	Reduced Model	Entire Model	Reduced Model
P_{t-1}	Millet	10.84	0.62	-10.62	-7.80
		(1.17)	(0.10)	(-1.68)	(-1.99)
A_{t-1}	Millet	-0.60	-	-0.06	0.30
		(-2.44)		(-0.19)	(1.59)
$P_{ci, t-1}$	Sorghum	3.77	-	3.12	-
		(0.46)		(0.28)	
	Maize	-17.90	-	-7.09	-
		(-1.27)		(-0.75)	
	Yam	34.86	-	0.28	-
		(2.05)		(0.02)	
S.D. price	Millet	-1.04	-	-0.84	-
		(-2.07)		(-1.22)	
S.D. yield	Millet	12.42	-	0.24	-
		(1.77)		(0.08)	
Exp. yield	Millet	-99.77	-67.69	-11.42	-
		(-1.97)	(-2.08)	(-0.65)	
D	Dummy	-2.63	-	0.02	-
		(-0.24)		(0.00)	
adjusted R^2		0.67	0.22	0.10	0.17
LM [probability]		1.19 [0.34]			0.80 [0.46]
Chow F-value [probability]					0.36 [0.79]
Number of observations		16	16	23	23

Note: The numbers in parentheses are t-values; an intercept term was included in all equations

TABLE 7A.3 : Regression Results for Harvested Area of Sorghum: Extended Nerlove Model

Explanatory variables		Period 1966-1982		Period 1966-1989	
		Entire Model	Reduced Model	Entire Model	Reduced Model
P_{t-1}	Sorghum	-16.72	-13.05	-14.87	-7.47
		(-2.67)	(-3.33)	(-1.80)	(-1.40)
A_{t-1}	Sorghum	0.55	0.66	0.28	0.35
		(2.16)	(4.38)	(0.99)	(1.60)
$P_{ci, t-1}$	Millet	-20.05	-17.68	-5.19	-6.92
		(-2.72)	(-4.09)	(-0.99)	(-1.87)
	Maize	11.19	-	2.11	-
		(0.92)		(0.27)	
	Yam	2.14	-	18.70	-
		(0.12)		(1.44)	
S.D. price	Sorghum	1.15	-	2.37	-
		(1.02)		(1.49)	
S.D. yield	Sorghum	5.04	6.46	-1.36	-
		(2.04)	(3.92)	(1.00)	
T	Time-trend	0.69	0.92	-0.14	-
		(0.90)	(3.08)	(-0.33)	
D	Dummy	-4.28	-	2.35	-
		(-0.81)		(0.48)	
adjusted R^2		0.71	0.80	0.22	0.10
LM [probability]		20.26 [0.00]			0.19 [0.83]
Chow F-value [probability]					1.69 [0.21]
Number of observations		16	16	23	23

Note: The numbers in parentheses are t-values; an intercept term was included in all equations

TABLE 7A.4 : Regression Results for Harvested Area of Cassava:
Extended Nerlove Model

Explanatory variables		Period 1966-1982		Period 1966-1989	
		Entire Model	Reduced Model	Entire Model	Reduced Model
P_{t-1}	Cassava	-9.89	-9.67	-12.72	-21.01
		(-0.39)	(-1.18)	(-0.70)	(-2.02)
A_{t-1}	Cassava	-0.26	-	0.12	-
		(-0.40)		(0.54)	
$P_{ci, t-1}$	Plantain	-13.61	-	-13.59	-17.06
		(-0.80)		(-1.30)	(-3.13)
	Maize	-11.04	-	-1.30	-
		(-1.33)		(-0.24)	
	Yam	-5.64	-	4.93	-
		(-0.24)		(0.44)	
$P_{ci, t}$	Cocoa	2.51	-	2.58	-
		(0.79)		(2.13)	
S.D. price	Cassava	-2.02	-	0.71	-
		(-0.24)		(0.12)	
S.D. yield	Cassava	0.57	0.36	0.26	0.26
		(2.31)	(2.17)	(1.69)	(1.81)
T	Time-trend	1.49	-	0.74	1.10
		(1.12)		(1.76)	(5.15)
D	Dummy	-4.97	-	-3.88	-7.61
		(-0.60)		(-1.02)	(-2.55)
adjusted R^2		0.36	0.17	0.65	0.60
LM [probability]		0.33 [0.72]			2.04 [0.16]
Chow F-value [probability]					2.26 [0.11]
Number of observations		16	16	23	23

Note: The numbers in parentheses are t-values; an intercept term was included in all
equations

TABLE 7A.5: Regression Results for Harvested Area of Yam: Extended Nerlove Model

Explanatory variables		Period 1966-1982		Period 1966-1989	
		Entire Model	Reduced Model	Entire Model	Reduced Model
P_{t-1}	Yam	-0.48	-2.34	-14.36	-15.43
		(-1.33)	(-0.78)	(-1.36)	(-1.72)
A_{t-1}	Yam	0.63	0.42	0.32	0.45
		(38.69)	(6.49)	(1.64)	(2.57)
$P_{ci, t-1}$	Plantain	7.80	-	4.81	-
		(27.98)		(1.09)	
	Cassava	-1.16	-	-17.93	-22.39
		(-2.46)		(-1.51)	(-2.34)
	Maize	2.79	-	4.86	-
		(10.05)		(1.05)	
	Sorghum	-4.25	-8.99	-4.51	-
		(-15.01)	(-4.87)	(-0.71)	
	Millet	-1.51	-	-5.97	-7.39
		(-7.71)		(-1.84)	(-1.97)
$P_{ci, t}$	Cocoa	-2.85	-3.19	-1.97	-1.77
		(-32.20)	(-5.30)	(-1.71)	(-1.69)
S.D. price	Yam	-9.13	-7.17	-6.15	-
		(-78.52)	(-5.26)	(-1.58)	
S.D. yield	Yam	0.28	0.24	0.21	0.17
		(116.71)	(13.15)	(3.50)	(2.91)
Exp. yield	Yam	-0.02	-	-0.21	-0.63
		(-0.85)		(-0.69)	(-3.96)
D	Dummy	-0.54	-2.11	1.74	-
		(-2.35)	(-2.12)	(0.57)	
adjusted R^2		1.00	0.98	0.80	0.68
LM [probability]		2.05 [0.24]			1.45 [0.27]
Chow F-value [probability]					1.48 [0.31]
Number of observations		16	16	23	23

Note: The numbers in parentheses are t-values; an intercept term was included in all equations

180

TABLE 7B.1: Regression Results for Harvested Area of Sorghum: Extended Nerlove Model

Explanatory variables		Period 1966-1982		Period 1966-1989	
		Entire Model	Reduced Model	Entire Model	Reduced Model
P_{t-1}	Sorghum/Millet	615.40	-73.36	516.99	-153.48
		(3.86)	(-0.22)	(2.60)	(-1.41)
A_{t-1}	Sorghum	-0.58	-	-0.31	-
		(-3.76)		(-1.85)	
$P_{ci, t-1}$	Maize	111.60	-	19.11	-
		(2.60)		(0.47)	
	Groundnuts	38.93	-	32.39	-
		(2.99)		(1.78)	
	Cotton	45.42	-	40.21	-
		(2.65)		(1.97)	
S.D. price	Sorghum	0.12	-	-0.18	-
		(0.54)		(-0.86)	
S.D. yield	Sorghum	18.8	-	18.95	-
		(5.56)		(4.89)	
T	Time-trend	-1.25	-4.09	-1.81	-3.02
		(-0.94)	(-2.19)	(-1.79)	(-4.51)
D	Dummy	-12.36	-	-10.65	-
		(-1.99)		(-1.60)	
adjusted R^2		0.94	0.42	0.84	0.48
LM [probability]		1.50 [0.27]		3.96 [0.06]	
Chow F-value [probability]				1.02 [0.41]	
Number of observations		16	16	23	23

Note: The numbers in parentheses are t-values; an intercept term was included in all equations

TABLE 7B.2 : Regression Results for Harvested Area of Millet:
Extended Nerlove Model

Explanatory variables		Period 1966-1982		Period 1966-1989	
		Entire Model	Reduced Model	Entire Model	Reduced Model
P_{t-1}	Sorghum/Millet	-1.28	-191.28	-196.44	-206.32
		(-0.00)	(-1.19)	(-1.13)	(-2.48)
A_{t-1}	Millet	-0.24	-	0.05	-
		(-0.84)		(0.18)	
$P_{ci, t-1}$	Groundnuts	35.06	-	-1.30	-
		(0.99)		(-0.05)	
S.D. price	Millet	0.21	-	-0.06	-
		(0.82)		(-0.27)	
S.D. yield	Millet	-6.94	-	4.59	-
		(-0.62)		(0.52)	
Exp. yield	Millet	-70.27	-	-5.74	-
		(-1.88)		(-0.45)	
D	Dummy	-36.11	-	-18.60	-17.50
		(-1.89)		(-1.42)	(-1.96)
adjusted R^2		0.22	0.03	0.16	0.27
LM [probability]		0.21 [0.81]			0.43 [0.66]
Chow F-value [probability]					0.38 [0.77]
Number of observations		16	16	23	23

Note: The numbers in parentheses are t-values; an intercept term was included in all equations

182

Table 7B.3: Regression Results for Harvested Area of Maize:
Extended Nerlove Model

Explanatory variables		Period 1966-1982		Period 1966-1989	
		Entire Model	Reduced Model	Entire Model	Reduced Model
P_{t-1}	Maize	0.94	36.70	8.44	61.73
		(0.03)	(-1.19)	(0.45)	(4.02)
A_{t-1}	Maize	0.07	0.46	0.05	0.43
		(0.32)	(2.17)	(0.33)	(3.37)
$P_{ci, t-1}$	Sorghum	12.75	-	-44.06	-47.32
		(0.14)		(-1.78)	(-2.21)
	Cotton	-6.47	-	-0.99	-
		(-0.60)		(-0.21)	
S.D. price	Maize	-0.39	-	-0.49	-
		(-0.55)		(-0.90)	
S.D. yield	Maize	-0.96	-	-0.99	-6.78
		(-0.28)		(-0.57)	(-4.05)
Exp. yield	Maize	-2.58	-	1.45	-
		(-0.48)		(1.32)	
D	Dummy	-1.29	-	-3.04	-
		(-0.22)		(-1.07)	
adjusted R^2		-0.81	0.41	0.51	0.68
LM [probability]		0.73 [0.51]			0.27 [0.77]
Chow F-value [probability]					0.99 [0.46]
Number of observations		16	16	23	23

Note: The numbers in parentheses are t-values; an intercept term was included in
all equations

Table 7B.4 : Regression Results for Harvested Area of Groundnuts: Extended Nerlove Model

Explanatory variables		Period 1966-1982		Period 1966-1989	
		Entire Model	Reduced Model	Entire Model	Reduced Model
P_{t-1}	Groundnuts	7.45	5.25	5.92	4.42
		(0.85)	(1.08)	(0.83)	(1.25)
A_{t-1}	Groundnuts	0.70	-	0.22	0.47
		(1.16)		(0.65)	(2.46)
$P_{ci, t-1}$	Sorghum/Millet	47.37	-	-19.45	-
		(0.49)		(-0.40)	
S.D. price	Groundnuts	0.02	-	0.02	-
		(0.67)		(0.57)	
S.D. yield	Groundnuts	-1.62	-	1.07	-
		(-0.51)		(0.55)	
Exp. yield	Groundnuts	-2.42	-16.13	-2.59	-
		(-0.15)	(-3.04)	(-0.86)	
D	Dummy	10.01	-	-0.69	-
		(1.28)		(-0.19)	
adjusted R^2		0.14	0.35	0.00	0.21
LM [probability]		0.41 [0.67]			0.05 [0.95]
Chow F-value [probability]					0.45 [0.72]
Number of observations		16	16	23	23

Note: The numbers in parentheses are t-values; an intercept term was included in all equations

TABLE 7B.5: Regression Results for Harvested Area of Cotton: Extended
Nerlove Model

Explanatory variables		Period 1966-1982		Period 1966-1989	
		Entire Model	Reduced Model	Entire Model	Reduced Model
P_{t-1}	Cotton	-2.98	2.88	5.32	4.58
		(-0.51)	(3.07)	(2.07)	(3.25)
A_{t-1}	Cotton	0.03	0.41	0.36	0.38
		(0.05)	(2.24)	(2.18)	(2.80)
$P_{ci, t-1}$	Sorghum/Millet	-27.87	-	17.43	-
		(-0.51)		(0.86)	
S.D. price	Cotton	-0.05	-	-0.02	-
		(-1.32)		(-0.73)	
S.D. yield	Cotton	-0.29	-	0.69	0.47
		(-0.44)		(3.14)	(3.49)
Exp. yield	Cotton	-1.85	-	0.84	0.82
		(-1.06)		(2.20)	(3.62)
D	Dummy	-2.45	-	-0.52	-
		(-1.23)		(-0.47)	
adjusted R^2		0.51	0.61	0.86	0.86
LM [probability]		0.53 [0.60]			2.44 [0.12]
Chow F-value [probability]					2.63 [0.07]
Number of observations		16	16	23	23

Note: The numbers in parentheses are t-values; an intercept term was included
in all equations

Section 4
HEALTH, NUTRITION AND FOOD SECURITY

CHAPTER 8

DETERMINANTS OF HEALTH AND NUTRITIONAL STATUS OF CHILDREN IN GHANA

W.K. Asenso-Okyere, F.A. Asante and *M. Nubé*

Introduction

The interconnections between the health and nutrition of a household are better understood by considering children in the early stages of their life. Children are more vulnerable to the consequences of under-nutrition and/or malnutrition than adults. It is in these formative years that good nutrition gets translated into visible signs on the body and can be measured by using anthropometric methods. Also mortality rates tend to be high for this age group, and the risk of preventable mortality reduces considerably after the age of five. The poverty level of a household may therefore be linked to the nutrition and health status of a child.

Issues of health and nutrition are very important in Ghana due to the low health and nutrition outcomes of a large cross-section of the population, most importantly children. Reutlinger and Selowsky (1978) pointed out that the nutritional status of an infant is the most important policy-induced determinant of the individual's initial physical condition, which in turn determines the effectiveness of further investment in human capital. On the basis of anthropometry about one third of children are malnourished.

A study by Ewusi (1978) in the mid-seventies on the nutritional status of children in six villages in the coastal savannah plains and in the forest zone summarised the incidence of malnutrition in the study area. Using anthropometric measurements based on changes in the upper arm circumference, the survey showed that for most of the villages, more than two-thirds of the children could be considered as being undernourished. He further concluded that on average 70.3 percent of the children were undernourished with 21.5 percent seriously undernourished. Orraca-Tetteh and Watson (1977), in another study of the nutritional situation in Bafi, a village in the Brong-Ahafo region, had a result which was similar to Ewusi (1978).

A nation-wide nutrition survey of 14,000 children conducted jointly by the Nutrition Department of the Ministry of Health and UNICEF in 1986 found that 58.4 percent of pre-school children fell below 80 percent of the U.S. National Centre for Health Statistic (NCHS) weight-for-age standards (World Bank, 1989b). This was roughly twice the level obtained in the first national survey carried out in 1961-62. About 8 percent of children in the country were clinically classified as suffering from marasmus or kwashiorkor. Malnutrition levels were highest in the northern zone (64 percent) and lowest in the coastal zone (48 percent).

The 1988 Demographic and Health Survey (GSS, 1989b) reports levels of malnutrition among 1,841 children between the ages of 3 - 36 months. Chronic malnutrition constitutes 30 percent of the sample reported. The survey further indicated high levels of malnutrition in the north, and the two upper regions which were comparable to the savannah zone. It was also found that children who recently had diarrhoea had high rates of malnutrition. There were pronounced improvements in the rates of malnutrition of children whose mothers had more than a middle school level (10 years of schooling) of education. According to the survey, stunting and underweight occur among almost one in three children aged 3 to 36 months in rural areas, where over two thirds of the Ghanaian population lives. Nearly one in three urban children is stunted and nearly one in three urban children is underweight.

Alderman (1990) using the first round of the Ghana Living Standards Survey (GLSS 1) data found that 31.4 percent of children fell below the weight-for-age standard (median - 2 standard deviation). The GLSS data which is the most recent national survey which groups the nutritional status of children in Ghana by age, gender and agro-ecological zone, revealed that the acute malnutrition for boys was appreciably higher than that for girls in the 6 - 24 month age bracket. The gap closes and reverses in the older age bracket which has less malnutrition. Malnutrition is observed to be severe in the savannah agro-ecological zone. This is consistent with earlier studies which reported higher levels of malnutrition in the northern regions (Levinson, 1988; 1961-62 and 1986 National Nutrition Surveys). Greater Accra has the lowest levels of both acute and chronic malnutrition. The remainder of the coastal agro-ecological zone has appreciably less chronic malnutrition than the forest or savannah regions.

In comparison with the national survey of 1986 the GLSS results indicate some improvement in nutritional conditions. Another indication of the changing pattern of malnutrition in the 1980s comes from data collected by the Catholic Relief Service for children attending maternal child health centres (CRS, 1987). The data showed a slightly improving pattern over the period 1983-1987. However, the data also indicate that even in a 'good' agricultural and nutrition year such as 1980, 35 percent of children were below 80 percent of the weight-for-age standard; in the severe drought of 1983, the figure was 51 percent, and by 1986, after the recovery of agricultural production, it had decreased to 35 percent.

The overall poor health conditions in Ghana are illustrated by the fact that life expectancy at birth in 1990 was only 55 years (UNDP, 1993), which is identical to the average for all low income countries, excluding India and China. In 1993 the Infant Mortality Rate was 83 deaths per 1000 live births, which is higher than the average for low income economies (World Bank, 1993). With respect to health care, between 1987 and 1990, about 76 percent of the Ghanaian population had access to health services (UNDP, 1993). Due to absence of modern facilities many people in the rural areas do get access to health care through herbalists and other unorthodox health care facilities. About 40 percent of all births are attended to by health personnel. In 1984 the number of persons per doctor was 14,890 and 640 per nurse. Through an intensive campaign (Expanded Programme on Immunisation) the proportion of one-year-olds immunised has increased from 34 percent in 1981 to 64 percent between 1989 and 1991. This has drastically reduced the incidence of some childhood killer diseases and has therefore contributed to a reduction in infant mortality.

Infant mortality rate estimated at around 100 deaths per 1000 births in 1973-77, has declined by approximately 22 percent to 77 deaths per 1000 births between 1983-1987. Mortality during childhood has also declined during the period under consideration. The probability of a child dying between birth and age 5 has dropped from 0.187 in 1973-77 to 0.155 in 1983-88 (GSS, 1989b).

The Research Problem and Objective

Past research has indicated the complex relationships between health and nutrition. The relative importance of the various factors which determine health and nutrition is affected by local conditions with respect to climate, food availability, water and sanitation facilities, etc.

The present study aims to define and quantify the factors which determine the health and nutritional status of children in Ghana. The results will be useful for enacting pragmatic policies and designing effective intervention programmes to improve the health and nutrition of people, especially children who tend to be more vulnerable in many instances.

Source of Data and Method of Analysis

The data for this study were based on the first round (1987/88) of the Ghana Living Standards Survey (GLSS I) which was conducted by the Ghana Statistical Service with assistance from the World Bank.

The GLSS I was carried out on a probability sample of 3,200 households from 200 enumeration areas. There were a total of 15,648 individuals in the households selected. A two-stage sampling procedure was used in selecting the sample. In the first stage 200 enumeration areas were selected with probability equal to size after they had been ordered according to agro-ecological zones and then rural/urban/semi-urban localities. Each work load contained 20 households with 4 of them used as replacements.

The GLSS provides data on various aspects of the Ghanaian households' economic and social activities and the interaction between those activities. The data were collected using three questionnaires:

(i) household questionnaire;

(ii) community questionnaire; and

(iii) price questionnaire (on food commodity prices).

Due to the absence of food intake data household expenditures were used in the computation of food availability. This was done by converting food expenditure data to quantities through a price variable and the quantities to calories through a conversion factor obtained from nutrition tables. Prices were instrumented on the basis of location and time, using a deflator. This was necessary as prices were available for a limited number of locations, while for some locations only a few prices were available. A food intake survey technique would have been the best in estimating the quantity of food consumed (Bouis *et al.*, 1992). Food quantity information obtained from expenditure surveys have measurement errors which are often systematically correlated with income, such that the responses of food intakes to increases in income are seriously overstated (Bouis *et al.*, 1992). This may be, firstly, because food quantities are not measured independently of income if total expenditures are the proxy used for income, and secondly, because food transfers in the form of guest and hired worker meals are under-recorded.

Bouis *et al.* have concluded that the use of food expenditure surveys (typically used in economic analysis) has led to severely upward biased estimates, while using food intake survey techniques (developed by nutritionists) gives more accurate estimates of nutrient demand parameters.

The paper uses both bivariate and multivariate analysis in ascertaining the factors that affect the health and nutrition status of children under five in Ghana. Bivariate relationships are investigated through cross-tabulations. Multivariate analysis involved the specification and estimation of a nutrition and health status econometric model.

After this introductory section, the next section presents some bivariate relationships between a composite health and nutrition variable and other variables.

BIVARIATE RELATIONS BETWEEN HEALTH AND NUTRITION
AND THEIR DETERMINANTS

Anthropometry

Table 8.1 gives results on the prevalence of malnutrition in Ghana, by type and by region. The distribution of undernutrition among children, 0-59 months, reveals that 29.7 percent of children are chronically undernourished or stunted (height-for-age z-score below median - 2 standard deviations). This shows that about 1 in 3 children aged 0-59 months are chronically undernourished. About 27 percent of children in Ghana are underweight (weight-for-age z-score below median - 2 standard deviations). This implies that about 1 in 3 children in Ghana are underweight (see also Alderman, 1990).

Acute undernutrition, which is manifested by wasting (weight-for-height z-score that is below median -2 standard deviations of the reference population) can be found with 7 percent of children in Ghana aged 0-59 months. Thus, approximately 1 in 13 Ghanaian children are wasted. The distribution of undernutrition is consistent with results obtained from the Demographic and Health Survey carried out in 1988. The rural areas turn out to be more undernourished (higher percentages of chronically undernourished, underweight and acute undernutrition) than the urban areas. This confirms earlier findings (Boateng *et al.*, 1990) that poverty in Ghana is a rural phenomenon. When chronic and acute malnutrition are considered together, undernutrition is highest in the savannah area followed by the forest and coastal areas.

Food Availability

Table 8.2 provides data on the household per capita kcal availability as obtained from the GLSS. For Ghana as a whole the GLSS indicates an average higher than the figure obtained from national data. For example FAO reports for Ghana per capita kcal availabilities for 1987 and 1988 of 2205 and 2245 per day respectively (FAO, State of Food and Agriculture, 1994).

The data show a very wide range of household per capita kcal availability figures when households are distributed over quintiles on the basis of kcal availability, with an average of 596 kcal/capita/day for the lowest quintile, and 6051 kcal/capita/day for the highest quintile. In a previous study a similar wide range was found by Alderman and Higgins (1992). Apart from data weaknesses, transfers between households may be partially responsible for the wide range in per capita kcal availability. Table 8.2 also provides mean height-for-age and mean weight-for-height z-scores for each quintile. It should be noted that the mean height-for-age or weight-for-height z-scores do not show a correlation with the household per capita food availability. Thus, the data suggest that the GLSS information on household food availability is a poor indicator for actual children's food intake, and the resulting children's nutritional status.

TABLE 8.1: Distribution of Undernutrition Among Children
Aged 0-59 Months in Ghana, 1987/88

	Chronically undernourished	Underweight	Acute undernutrition
Ghana	29.7	27.3	6.7
Locality			
Rural	33.9	30.4	7.1
Urban	20.0	20.0	5.9
Region			
Ashanti	32.3	32.6	9.5
Brong-Ahafo	34.9	30.9	6.3
Central	31.1	29.3	5.8
Eastern	26.0	18.7	4.1
Greater-Accra	20.8	17.3	7.5
Northern	38.9	40.5	11.5
Upper East	28.7	22.3	6.4
Upper West	37.8	37.0	8.4
Volta	28.2	25.7	5.8
Western	27.1	27.5	5.1
Zone			
North	35.8	34.3	9.0
South	28.7	26.0	6.3
Agroecology			
Coastal	23.1	21.8	6.2
Forest	32.6	29.9	6.1
Savannah	33.2	29.8	8.3

Source: Computed from the Ghana Living Standards Survey,
1987/88 data

TABLE 8.2: Household Food Availability and Anthropometry of Underfives,
1987/88

Household per capita kcal quintile	Height-for-age z-score underfives			Weight-for-height z-score underfives		Household kcal per cap per day	
	N	Mean	St Dev.	Mean	St Dev.	N	Mean
1	380	- 1.15	1.27	- 0.45	0.89	634	596
2	359	- 1.17	1.29	- 0.44	0.97	635	1120
3	313	- 1.04	1.47	- 0.62	1.02	633	1658
4	262	- 1.14	1.42	- 0.55	0.94	630	2561
5	192	- 1.32	1.50	0.65	1.00	611	6051
all	1506	- 1.15	1.37	- 0.52	0.96	3143	2370

Source: Computed from GLSS, 1987/88 data

Age of Mother at Birth

In Ghana about 15 percent of all births are to women whose age is below 20 years (Ghana Fertility Survey, 1979/1980). It has been shown by various authors that when a woman gives birth to a child at very young age (teenage-pregnancy) there is an increased risk for a low birth weight baby (ACC/SCN, 1990). There are a number of mechanisms through which the age of the mother at birth could affect birth outcome and also the child's height-for-age z-score in subsequent years. First of all, teenage girls are not yet fully grown up and when they become pregnant, the conditions for fetal growth may be suboptimal, resulting in a low birth weight. Secondly, when the child is born, breastfeeding by the teenage mother may be insufficient and negatively affect the child's growth after birth. And thirdly, various socio-economic factors, related to the age of the mother, and her experience in raising children, may affect the child's nutritional and health conditions, and thus the pattern of growth.

Table 8.3 gives for the underfives in the GLSS-database the relationship between the ages of mothers when the child was born and the height-for-age z-scores of the children at the time of the survey. The table reflects both the effect of the age of the mother during pregnancy and when giving birth, as well as the effect of the age of the mother during child rearing on the anthropometric z-scores of the child. The table shows that children in the age group 0-60 months, whose mother was 17 years or younger when they were born, have significantly lower height-for-age z-scores when compared with children whose mother was 18 years or above.

If the children are divided over two age groups, 0-36 months and 37-60 months, still for both groups height-for-age z-scores are lower for children born when their mother was 17 years or younger in comparison with children born when their mother was 18 years or older. Thus, the data do not allow the differentiation between the effects of the age of the mother when giving birth and the age of the mother during child raising.

Illness, Vaccination and Nutritional Status of Children

Infectious diseases are considered to play a major role in the causation of malnutrition. Illness may affect nutritional status through at least the following mechanisms: reduced food intake (through reduced appetite), reduced absorption or increased intestinal losses of nutrients (e.g. in diarrhoea), increased requirements because of higher metabolic rates (e.g. fever), and reduced requirements as a result of a lower level of physical activity. In many studies it has been shown that there is a negative correlation between infection and anthropometric z-scores (Tomkins and Watson, 1989). In Table 8.4 the relationships between illness and anthropometry of underfives in the GLSS-data set are explored. Information was collected for how many days children were ill over the past 4 weeks. The table indicates for children below 20 months of age a negative correlation between occurrence of illness over the last month and their nutritional status, whether based on the height-for-age or the weight-for-height z-scores. For older children there is still some negative correlation between illness and nutritional status when expressed on the basis of weight-for-height, however not on the basis of height-for-age.

These results give some support to the concept that recent illness has an effect on a short term nutrition indicator such as weight-for-height but much less on a long term indicator such as height-for-age. This issue could to some extent further be explored by analysing the relationships between vaccination, illness occurrence and anthropometry.

TABLE 8.3: Height-for-age Z-scores of Children at Time of Survey and Age of Mother When Giving Birth to the Child, 1987/88

Age of mother when giving birth	Age in months under-fives	Height-for-age Z-score 0-60 months[1]			Height-for-age 0-36 months			Height-for-age 37-60 months			Household per capita expenditure	
	mean	n	mean	std	n	mean	std	n	mean	std	mean	std
Below 17 yrs	29.26	74.00	-1.67	1.42	48.00	-1.37	1.43	26.00	-2.23	1.23	58454	42094
17 yrs	30.45	80.00	-1.47	1.46	49.00	-1.05	1.34	31.00	-2.14	1.40	64047	43215
18-19 yrs	26.23	175.00	-1.08	1.54	123.00	-0.92	1.41	52.00	-1.46	1.70	68038	38850
20-30 yrs	26.72	1096.00	-1.18	1.57	769.00	-0.99	1.60	327.00	-1.64	1.38	66824	56230
31-40 yrs	27.17	512.00	-1.00	1.58	347.00	-0.79	1.52	165.00	-1.43	1.61	64472	57470
41 and above	29.33	150.00	-1.03	1.54	95.00	-0.74	1.42	55.00	-1.53	1.62	51888	36254

Source: Computed from GLSS,1987/88 data

[1] t-test on height-for-age z-scores of children 0-60 months, comparing children born when the mother is 17 years and below, n=154, av.haz=-1.57, or 18 years and above, n=1933, av.haz=-1.11, difference significant at p=0.002 .

TABLE 8.4: Illness and Anthropometry, 1987/88

Number of days ill in last month	N	Average age in months MEAN	HT FOR AGE Z-SCORE						WT For HT Z-SCORE					
			Age Group Children						Age Group Children					
			0-20 months		21-40 months		41-60 months		0-20 months		21-40 months		41-60 months	
			MEAN	STD	MEAN	STD	MEAN	STD	MEAN	STD	MEAN	STD	MEAN	STD
0	1295	28.63	-0.34	1.44	-1.37	1.57	-1.63	1.55	-0.38	1.24	-0.58	0.97	-0.42	0.93
1 - 7	752	28.16	-0.73	1.37	-1.47	1.53	-1.60	1.29	-0.70	1.16	-0.59	0.99	-0.42	0.80
8 -31	225	25.69	-1.05	1.51	-1.36	1.64	-1.65	1.39	-1.02	1.09	-0.75	1.04	-0.69	0.86

Source: Computed from GLSS, 1987/88 data

In Ghana vaccination programmes against diphtheria, whooping cough, tetanus, poliomyelitis and measles are widely implemented. In the GLSS mothers were requested to provide information whether their youngest child was vaccinated against these diseases or not.

Table 8.5 indicates that for the 41-60 months age group approximately 75 percent of the children had been vaccinated. Again, the absolute numbers of children are relatively low as information is only available for the youngest child in the household. With respect to the relationship between vaccination and nutritional status, Table 8.5 indicates that for children in the 20-40 months age group, and even more so for children in the 40-60 months age group, vaccination is associated with better anthropometric z-scores. On the other hand, for the 0-20 months age group there is no difference in nutritional status between vaccinated and non-vaccinated children.

The mechanism through which vaccination is considered to have an effect on nutritional status is a reduction in the occurrence of infectious diseases. However the GLSS-data do not reveal a clear pattern in the occurrence of illness in vaccinated and non-vaccinated children (Table 8.5).

TABLE 8.5: Vaccination and Child Anthropometry , 1987/88

		0-20 months days				21-40 months days					41-60 months days		
Child Vaccinated		haz	whz	ill		haz	whz	ill		haz	whz	ill	
	N	MEAN	MEAN	MEAN	N	MEAN	MEAN	MEAN	N	MEAN	MEAN	MEAN	
Yes	428	-0.56	-0.64	3.6	454	-1.39	-0.53	2.9	306	-1.54	-0.45	2.4	
No	216	-0.44	-0.41	3.2	149	-1.64	-0.86	3.2	103	-2.16	-0.48	2.4	

Source: Computed from GLSS, 1987/88, data

Education of Parents and Child Anthropometry

Mother's education has been found to be an asset to good household nutrition and health. It has been suggested that because education is usually positively correlated with the use of contraception (see, for example, Cochrane and Zachariah, 1983) maternal education tends to be positively correlated with birth interval that, in turn, tends to improve birth outcome. Further, mothers with more education may provide better health care to their children (Grossman, 1972). On the other hand, educated women may increasingly become involved in wage labour, away from home, and child care has to be provided by others (Schultz, 1984).

Table 8.6 shows for the GLSS data set a slightly better average height-for-age z-score for children whose mothers are literate. Clearly also, other factors including income, are likely to play a role in the positive correlation between education of the mother and the z-score of the child.

TABLE 8.6: Literacy of Mother and Child Anthropometry, 1987/88

Mother Literate/ Non-literate	Children's Height-for-age Score			Children's Weight-for-height Score		Household Per Capita Expenses
	N	Mean	Std	Mean	Std	
Mother literate	1459	- 1.23	1.60	- 0.57	1.07	58398
Mother non-literate	616	- 0.97	1.42	- 0.52	1.05	77204

Source: Computed from GLSS, 1987/88 data; household per capita expenses in cedis per year

MULTIVARIATE ANALYSIS

Specification of the Health and Nutrition Model

The health and nutrition of children under five in Ghana will be characterised by an econometric model in this section. The food available in the household which is affected by income and prices is converted into health and nutritional status through a health and nutrition production function. Health and nutrition is measured as one composite variable using anthropometric data, but its short term measure are distinguished from its long term one. In the short term, health and nutrition are presumed to be reflected by the weight-for-height z-score and its long term effect is proxied by the height-for-age z-score of the child.

Food Availability

The food available in the household which is measured in kilocalories is obtained by converting food expenditure information to weight (in kg) using prices. The prices of food commodities included were obtained from the Price and Community Survey conducted alongside the GLSS. Food availability in the household is cast in the framework of a demand function and so it is specified as a function of income (total household expenditure) and composite price for food. Other explanatory variables are the number of non-dependents in the household and a dummy for the location of the household (urban/rural).

Using a mixed logarithmic functional form, the food available in the household is specified as:

$$\text{Log(Foodkh)} = a0 + a1(\text{Durb}) + a2(\text{Ndep}) + \log(\text{Hhexp}) - \log(\text{Pkcal}) \tag{8.1}$$

where Foodkh is total food available to a household per day in kcals, Hhexp is total household expenditure in cedis per day, Pkcal is the price per kcal for the household (Pkcal=total household food expenditure/total household food availability in kcal), Ndep is number of nondependents in household, and Durb is dummy for urban area, with Durb=1 for urban area (residential area of 5000 persons or more), Durb=0 otherwise.

196

Food Price Per Kcal

The aggregate food price per kcal for each household is calculated by dividing total household food expenses by total food available in the household in kcal. The aggregate food price per kcal is hypothesised to be a function of the prices of the major food commodities consumed in Ghana. The equation is specified in a double-log (Cobb-Douglas) formulation and the explanatory variables are deflated by the price of maize to ensure the reduction of inflationary tendencies in data collected over a number of seasons in a period of one year.

The food price equation is as follows:

$$Log(Pkcal) = b0 + log(Pmaize) + b1log(Pmillet/Pmaize) + b2log(Pbread/Pmaize)$$
$$+ b3log(Pcassava/Pmaize) + b4log(Pgarri/Pmaize)$$
$$+ b5log(Pyam/Pmaize) + b6log(Pplantain/Pmaize)$$
$$+ b7log(Pfish/Pmaize) + b8log(Pchicken/Pmaize)$$
$$+ b9log(Pbeef/Pmaize) \tag{8.2}$$

where Pkcal is the price of kcals (cedi/kcal), Pmaize, Pmillet, Pcassava, Pgarri, Pyam, Pplantain, Pfish, Pchicken, and Pbeef are the prices of maize, millet and guinea corn, raw cassava, garri, yam, plantain, fish and fish products, chicken, and beef in cedis per kilogram, respectively.

Individual Food Intake

The individual daily calorie intake (availability) is obtained from the household food availability using a conversion factor:

$$Kcali = (Scale*Foodkh) / Sizeq \tag{8.3}$$

where Kcali is the individual daily food intake estimate in kilocalories, Scale is adult equivalent consumption unit, and Sizeq is household size in adult equivalents.

Short-Term Health and Nutrition

The weight-for-height z-score (whz), which is a measure of short-term health and nutritional status, is hypothesised to depend upon the genetic factors of the parents as denoted by their body mass indexes and heights, income of the household represented by total expenditures, the quality of care available to the child proxied by the state of literacy of the mother, experience of the mother proxied by her age, the health condition of the child as measured by days of illness in the last four weeks and whether the child has been vaccinated against the six childhood diseases, the quality of food served to the child proxied by the availability of light in the community, and the age of the child.

The weight-for-height equation which relates short-term health and nutritional status of children to explanatory variables is given in a mixed quadratic form as follows:

$$Whz = c0 + c1(Htm) + c2(Htf) + c3(Bmim) + c4(Bmif) + c5\log(pcexp)$$
$$+ c6[\log(pcexp)]^2 + c7(Dchlvac) + c8(literacy) + c9(Agemoth)$$
$$c10(Agemnths) + c11(Dlight) + c12(Daysi) \tag{8.4}$$

where Whz is the weight-for-height z-score, Hft is height of the father, Htm is height of the mother, Bmif and Bmim are the body mass index of the father and mother, respectively. Dchlvac is dummy for child vaccination, with Dchlvac=1 for child vaccination, and Dchlvac=0 otherwise. Literacy is dummy for literacy of mother, with Literacy=1 for literate mothers, and Literacy=0 otherwise. Agemoth is age of the mother, Agemnths is age of the child, Dlight is dummy for electricity, with Dlight=1 for electricity, and Dlight=0 otherwise, Daysi is number of days the child had been ill in the last four weeks.

Long-Term Health and Nutrition

Similarly, the long-term health and nutrition variable, height-for-age z-score (haz) is hypothesised to be a function of characteristics of the child, the genetic factors of the parents and some community endowments. The equation is specified in a linear form as follows:

$$Haz = d0 + d1(Hft) + d2(Htm) + d3(Bmim) + d4(Bmif) + d5(Dsex) + d6(Dchlvac)$$
$$+ d7(Schyrs) + d8(Agemnths) + d9(Agemnths)^2 + d10(Agemoth)$$
$$+ d11(Dlight) + d12(daysi) \tag{8.5}$$

where Haz is height-for-age z-score, schyrs is years of schooling of mother and Dsex is gender of the child, with Dsex=1 for female, Dsex=0 otherwise. Other variables are as defined previously.

It should be noted that age of the mother is included in the regression and not age of the mother when giving birth, which is the analysed variable under the section headed 'Age of Mother at Birth' (Table 8.3). For the regression analysis, this has no consequences as the variable age of the mother at birth is a linear combination of the age of the mother and the age of the child.

Estimated Results and Discussion

Using the Statistical Analysis System (SAS) a non-linear ordinary least squares (NOLS) estimation procedure was used in the estimation of the equations (8.1), (8.2) and (8.4) and (8.5). The estimated equations are presented in Tables 8.7 to 8.10.

Food Availability

The food availability equation had an R^2 of 0.81 (Table 8.7). The average share of food expenses as percentage of total expenses is 0.61 (exp[- 0.47 - 0.16*0.33 + 0.015*2.5]).

The urban/rural dummy variable is significant at the 0.0001 level, with a lower share of food expenses as percentage of total expenses in the urban situation. The number of non-dependents has a weak positive effect on the food share.

TABLE 8.7: Dependent Variable: Household Food Availability
in kcal/day

Variable	Coefficient	t-value	Probability
Constant	- 0.47***	-19.46	0.0001
Durb	- 0.16***	- 8.38	0.0001
Ndep	0.015*	1.80	0.072

Sample size = 934   *** significant at < 0.01 percent level
R-square = 0.81   ** significant at < 0.05 level
Adj. R-square = 0.81   * significant at < 0.1 level

Food Price Per Kcal

All the price ratios turned out to be significant at least at the 0.01 level. Of the nine food commodities included in the equation, relative prices of millet, cassava, garri, and yam have a positive effect on the kcal price, while the more expensive foods bread, plantain, fish, chicken and beef affect the relative food price negatively (Table 8.8).

TABLE 8.8: Dependent variable: Relative Food Price per kcal

Variable	Coefficient/elasticity	t-value	Probability
Constant	27.23***	2.58	0.0100
Log(Pmillet/Pmaize)	4.77***	3.55	0.0004
Log(Pbread/Pmaize)	- 9.83***	-3.19	0.0015
Log(Pcassava/Pmaize)	7.55***	3.72	0.0002
Log(Pgarri/Pmaize)	5.98***	2.72	0.0067
Log(Pyam/Pmaize)	1.89***	5.41	0.0001
Log(Pplantain/Pmaize)	- 3.34***	-3.30	0.0010
Log(Pfish/Pmaize)	- 5.43***	-2.90	0.0038
Log(Pchicken/Pmaize)	- 3.51***	-3.86	0.0001
Log(Pbeef/Pmaize)	- 2.23***	-2.60	0.0094

Sample size = 934   *** Significant at <0.01 percent level
R^2 = 0.35   ** Significant at <0.05 percent level
Adjusted R_2 = 0.34   * Significant at <0.1 percent level

Weight-for-Height z-Score

Results for the estimation of the short-term health and nutrition measure (whz) are given in Table 8.9. The coefficients of the body mass index of the mother (bmim) and father (bmif) and the number of days the child had been ill over the last month (daysi) are all significant at less than 0.01 percent level. Thus, genetic and health factors are the most important variables that determine the short-term health and nutritional status of children. The number of days of illness (daysi) has a negative relationship with the whz. Food security is exacerbated by poor health where the child cannot efficiently utilise the available food. This stresses the importance of good health as food security strategy.

TABLE 8.9: Dependent Variable: Weight-for-Height

Variable	Coefficient	t-value	Probability
Constant	-3.834	-0.51	0.6139
Htm	0.003	0.66	0.5077
Htf	0.006	1.33	0.1842
Bmim	0.049***	4.73	0.0001
Bmif	0.046***	3.42	0.0007
Log(pcexp)	0.032	0.02	0.9815
[Log(pcexp)]²	-0.005	-0.08	0.9349
Dchlvac	-0.063	-0.83	0.4074
Literacy	0.041	0.55	0.5849
Agemoth	-0.0002	-0.45	0.6551
Agemnths	0.001	0.51	0.6090
Dlight	0.027	0.32	0.7515
Daysi	-0.021***	-3.22	0.0014

Sample size = 925   *** Significant at < 0.01 percent level
R^2 = 0.06   ** Significant at < 0.05 percent level
Adjusted R^2 = 0.05   * Significant at < 0.1 percent level

The results confirm other findings that literate mothers tend to have a positive impact on the health and nutrition of their children. Availability of electricity in the community also tends to improve the health and nutrition of the children.

Height-for-Age z-Score

The most significant variables that determine the health and nutritional status of children in the long-run are the genetic composition of the parents, and the level of development in the community (proxied by availability of electricity).

The heights of the mother and father and mother's BMI which were positively related to the height-for-age z-score were significant at less than 0.05 level (Table 8.10). Although not significant, years of schooling of the mother had a positive impact and days of illness had a negative impact on the long term health and nutritional status of children.

Finally, in the multivariate analyses of haz and whz as presented in Tables 8.9 and 8.10 the variable on the length of the period of breastfeeding of under-fives is not included. The reason is that the information on the length of breastfeeding was only available for the last child in each household, and including the variable would significantly reduce the sample size. To make a cursory investigation of this effect a separate regression was run for long-term health and nutrition (haz) with length of breastfeeding of the child as one of the explanatory variables (Appendix 8A.1). Results showed a negative and significant relationship with the dependent variables. These startling results suggest that children who continue to be breastfed into their second and third year do not receive or do not consume adequate amounts of supplementary food. Perhaps children who continue to be breastfed receive as supplementary food only porridges of low nutritional value, while children who are completely weaned eat from the nutritionally better family pot. Another possibility could be that children who continue to be breastfed are complacent after breastfeeding and are not too eager to consume supplementary food. The issue of weaning practices in Ghana has already been given attention in various reports, and needs further study (GOG/UNICEF, 1990, p.55; Nubé and Asenso-Okyere, 1996).

TABLE 8.10: Dependent Variable: Height-for-Age

Variable	Coefficient	t-value	Probability
Constant	-12.214***	-7.89	0.0001
Htm	0.041***	5.74	0.0001
Htf	0.028***	4.06	0.0001
Bmim	0.033**	2.27	0.0234
Bmif	0.024	1.29	0.1957
Dsex	0.015	0.17	0.8637
Dchlvac	0.073	0.69	0.4886
Schyrs	0.0005	0.05	0.9609
Agemnths	-0.097***	-9.25	0.0001
$[Agemnths]^2$	0.001***	6.24	0.0001
Agemoth	0.0005	0.88	0.3817
Dlight	0.281**	2.48	0.0132
Daysi	-0.010	-1.09	0.2723

Sample size = 933   *** Significant at < 0.01 percent level
R2   = 0.23   ** Significant at < 0.05 percent level
Adjusted R2 = 0.22   * Significant at < 0.1 percent level

Conclusion and Recommendations

To date most health and nutrition studies undertaken in Ghana have focused on either nutrition and calorie consumption aspects or health and illness aspects of nutrition without linking nutrition and health simultaneously. To bridge this gap this study has tried to find out the factors (publicly and privately) which are the principal determinants of health and nutrition of children under-five in a simultaneous context.

In assessing the relative merits of various public policies to deal with malnutrition among the poor, a better understanding of the household behaviour is the starting point. It is in this context that public policies must be designed, that is, they must take into account the fact that households make the ultimate decisions concerning the family's expenditure on nutrition and health. The following are a few policy implications that can be considered to promote the health and nutrition of Ghanaian children.

(i) The calorie availability can increase if the prices of commodities such as cassava and millet are reduced. While millet and sorghum are the main staple crops in the North, maize, cassava and plantains are the major crops for many Ghanaians in other parts of the country. Cassava is a good source of energy and so its promotion on a large scale to reduce the price can lead to an increase in the caloric availability in many households. Fortunately cassava grows on many Ghanaian soils, but if its cultivation is carried out with the application of fertilizer and improved practices, it is possible to raise its average yield from the present 7 tons per hectare to more than 20 tons per hectare. While there is little likelihood that the government would like to influence directly the price of foodstuffs through subsidies and/or price support, programmes that tend to reduce the cost of transport, or postharvest losses which may be implemented by public agencies, would have significant effects on the food supplies and therefore food availability to the child.

(ii) The size of the household labour force was found to affect food availability positively. Although large family sizes are not recommended, shortage of farm labour has been found to affect farm production negatively, especially during the peak season. Farm labour supply can increase if there is a slowdown in rural-urban migration. Such a slowdown can come about if there is an expansion in social amenities and a rise in rural incomes. The farm labour supply problem can also improve if the efficiency of labour is raised. Extension can play a vital role in this respect.

(iii) For children in the 0-59 months age group both the short-term health and nutrition indicator weight-for-height as well as the long-term health and nutrition indicator height-for-age are negatively correlated with recent illness. This is significant only for the short-term. Improved source of light (electricity) and further expansion of vaccination programmes are likely to contribute to reduced occurrence of illness and a better health and nutrition status of the under-fives in the short- to long-run.

(iv) Prolonged breastfeeding appears to be negatively correlated with the child's long-term nutritional status. Formulation of better weaning foods and more information to women on the importance of adequate supplementary feeding after the age of 6 months may contribute to better child health and nutrition.

(v) There is the need to improve the nutritional status of children as measured by anthropometric indicators (weight-for-height, weight-for-age and height-for-age). Food production policies must be complemented with nutritional and health policies which promote physiological access (utilisation) to food which concern the conversion of food into nutrients that the body needs. On the average 29.7 percent, 27.3 percent and 6.7 percent of children are chronically undernourished, underweight and acutely undernourished, respectively. A national nutrition strategy and related programmes (including a surveillance system) must be put in place in order to address the serious nutritional deficiencies which have been identified.

(vi) The multivariate analysis showed the positive impact of good source of light (electricity) on the nutrition of children. The absence of this publicly provided input may cause food products to become spoiled or contaminated. This may lead to a reduction in the absorption of nutrients, or worse, to gastro-intestinal and other diseases accompained by a drastic reduction in the degree of nutrient absorption, and thus a reduction in the nutritional status. Therefore it is important to emphasise the provision of electricity in the development programme of the country, especially in the rural areas. A rural electrification scheme can be sustained if the people have the capacity to maintain the equipment once it has been put into place. However, rural electrification should be taken as part of an integrated rural development scheme so that benefits from the linkages can be maximised.

(vii) The positive impact of immunisation was made apparent in this study. Results showed that children in the 20-40 months age group and even more so children in the 40-60 months age group had better anthropometric z-scores when they were vaccinated. It is therefore important for the government to step up its campaign to fully immunise most children. The programme can be more successful if both public and private health care centres participate in it and the outreach services are expanded.

ACKNOWLEDGEMENT

The data for the research was based on the Ghana Living Standards Survey which was obtained with the kind permission of the Ghana Statistical Service.

REFERENCES

Alderman, H. and Higgins, P. 1992. Food and Nutritional Adequacy in Ghana. Cornell Food and Nutrition Policy Program. Working Paper 27. U.S.A.

Alderman, H. 1990. Nutritional status in Ghana and its determinants. Social dimensions of adjustment in sub-Saharan Africa. Working Paper No. 3.World Bank, Washington, D.C..

Behrman, J.R. and Deolalikar, A.B. 1988. "Health and nutrition". In Handbook of Development Economics. Vol. 1. Edited by H. Chenery and T.N. Srinivasan. Elsevier Science Publishers.

Boateng, E.O., Ewusi, K., Kanbur, R., Mckay, A. 1990. A Poverty Profile For Ghana, 1987-88. Social Dimensions of Adjustment In Sub-saharan Africa. Working Paper No. 5. World Bank, Washington, D.C.

Boom, G. J. M., M. Nubé, W.K. Asenso-Okyere, 1995, Nutrition, Labour Productivity and Labour Supply of Men and Women in Ghana, Paper submitted for the Conference on "Sustainable Food Security in West Africa", Réseau SADAOC, 13-15 March 1995.

Bos, E., Adolfse, L. and Heerink, N. 1993. Undernutrition: A Constant Threat. Wageningen Economic Papers 1993-4. Faculty of Economics. Wageningen Agricultural University. The Netherlands.

Bouis, H., Haddad, L. and Kennedy, E. 1992. Does it Matter How we Survey Demand for Food? Evidence from Kenya and the Philippines. International Food Policy Research Institute. Washington D.C.

Caldwell, J.C., Reddy, P.H., and Caldwell, P. 1983. The Social Components of Mortality Decline: An Investigation in South India Employing Alternative Methodologies. Population Studies, 37(2).

CRS. 1987. Growth Surveillance System. Annual Report 1986. Catholic Relief Service. Accra. Mimeo.

Ewusi, K. 1978. Planning for Neglected Rural Poor. Accra.

Eyeson, K.K., Ankrah, E.K., 1975, Composition of Foods Commonly used in Ghana, Food Research Institute, Council for Scientific and Industrial Research, Accra-Ghana.

FAO. 1992. Food and Nutrition: Creating a Well-fed World. Food and Agriculture Organisation, Rome.

Ghana Fertility Survey, 1979/80, Ghana Statistical Service, Accra.

Ghana Statistical Service (GSS). 1989a. Ghana Living Standards Survey (GLSS). First Year Report. Sept. 1987-Aug. 1988. Accra.

Ghana Statistical Service (GSS). 1989b. Ghana Demographic and Health Survey(GHDS) 1988. Accra.

GOG/UNICEF. 1990. Children and Women of Ghana: A Situation Analysis. Republic of Ghana and the United Nations Children's Fund.

Grossman, M. 1972. On the Concept of Health Capital and the Demand for Health. *Journal of Political Economy*, 80.

ISSER/SOW, 1993. Background to food Security in Ghana. Institute of Statistical, Social and Economic Research, University of Ghana, Legon, and the Centre for World Food Studies of the Free University, Amsterdam. Working Paper No. WP-93-HFSR-1.

Levinson, F.J. 1988. Combating Malnutrition in Ghana. Mimeo.

Nubé, M. And Asenso-Okyere, W.K. 1996. "Large Differences in Nutritional Status Between Fully Weaned and Partially Breastfed Children Beyond the Age of 12 Months". *European Journal of Clinical Nutrition*, 50.

Orraca-Tetteh, R. and Watson, J.D. 1977. Report of the Re-assessment of the Nutritional Status of the Population of Bafi. Department of Nutrition and Food Science. University of Ghana, Legon.

Reutlinger, S. and Selowsky, M. 1978. "Malnutrition and poverty. Magnitude and Policy Options". World Bank Staff Occasional Paper. No. 28. The Johns Hopkins University Press, Baltimore.

Schultz, T.P. 1984. "Studying the Impact of Household Economic and Community Variables on Child Mortality", *Population and Development Review,* 10.

Shetty, P.S., James, W.P.T., 1994, Body Mass Index, Food and Nutrition Paper 56, FAO, Rome.

Tomkins, A. Watson, F., 1989, Malnutrition and Infection, ACC/SCN, State of the Art Series, Nutrition Policy Discussion paper, N. 5.

UNDP, 1993. Human Development report 1993. United Nations Development Programme.

United Nations. 1990. Assessing the Nutritional Status of Young Children. National Household Survey. Capability Program. United Nations.

World Bank. 1993. World Development Report, 1993. Oxford University Press.

APPENDIX 8A.1

Dependent Variable: Height-for-Age (haz)

Variable	Coefficient	t-value	Probability
Constant	-17.5231***	-9.38	0.0001
Htf	0.0229***	3.10	0.0020
Htm	0.0354***	4.60	0.0001
Bmim	0.0225	1.49	0.1381
Bmif	0.0253	1.27	0.2053
Dsex	0.1198	1.26	0.2080
Dchlvac	0.1717	1.51	0.1326
Schyrs	-0.0047	-0.41	0.6787
Agemnths	-0.1197***	-5.83	0.0001
(Agemnths)^2	0.0015***	4.54	0.0001
Agemoth	-0.0001**	-2.22	0.0267
Dwat	-0.0099	-0.07	0.9435
Dlight	0.0890	0.61	0.5399
Ht	0.1717***	10.28	0.0001
(Ht)^2	-0.0007***	-9.85	0.0001
Daysi	-0.0138	-1.46	0.1449
Mlcbf	-0.0712***	-6.10	0.0001

Sample size = 596   *** Significant at < 0.01 percent level
R^2 = 0.418   **   Significant at < 0.05 percent level
Adjusted R^2 = 0.402   *   Significant at < 0.1 percent level

APPENDIX 8A.2: Dependent variable: Weight-for-Height
(whz)

Variable	Coefficient	t-value	Probability
Constant	-0.0089	-0.001	0.9992
Htm	0.0066	0.995	0.3200
Htf	0.0058	0.906	0.3652
Bmim	0.0383	2.945***	0.0034
Bmif	0.0353	2.054**	0.0405
Log(pcexp)	-0.5168	-0.314	0.7534
[Log(pcexp)]^2	0.0177	0.236	0.8139
Dchlvac	-0.1515	-1.536*	0.1250
Literacy	0.1547	1.578*	0.1151
Agemoth	-0.0006	-1.192	0.2337
Agemnths	0.0187	3.743***	0.0002
Dlight	0.0134	0.123	0.9024
Daysi	-0.0147	-1.800*	0.0723
Mlcbf	-0.0506	-6.235***	0.0001

Sample size = 585   *** Significant at < 0.01 percent level
R^2 = 0.12   **   Significant at < 0.05 percent level
Adjusted R^2 = 0.10   *   Significant at < 0.1 percent level

Source: Computed from GLSS, 1987/88 data

CHAPTER 9

NUTRITION, LABOUR PRODUCTIVITY AND LABOUR
SUPPLY OF MEN AND WOMEN IN GHANA

G.J.M. van den Boom, M. Nubé and W.K. Asenso-Okyere

Introduction

Ghana's population is expected to increase from 14.1 million in 1990 to 18.7 million in 2000 (Bumb *et al.,* 1994). This growth in population along with growth in per capita income and urbanisation will create increased demand for food. Before the launching of the Economic Reform Programme in 1983 Ghana's food production had fallen to an abysmally low level in response to the general deterioration in the economy. Output of food and cash crops declined at a rate of 0.3 percent per year between 1970 and 1980. Cereal production, which exceeded domestic demand by some 200,000 tonnes in 1971-73, registered a deficit of over 300,000 tonnes in 1981-83. Production of starchy staples fell from 7.9 million tonnes to 4.1 million tonnes between 1974 and 1981. As a result of declining food output combined with a rising population, per capita food availability in 1981-83 was 30 percent lower than it was in 1975. Efforts made under the Economic Recovery Programme have improved the food supply situation in the country although food production levels still remain low. Output levels of cocoyam and plantain have not reached their average levels in the early 1970's (Table 9.1).

TABLE 9.1 Production of Major Food Crops and Nutritional Adequacy in Ghana, 1970-1992

	1970/74	1975/79	1980/82	1984/86	1987/89	1990/92
	Thousand metric tonnes					
Maize	452	304	317	509	638	739
Rice	62	69	48	75	84	121
Millet	127	134	129	123	182	107
Sorghum	171	152	151	163	200	212
Cassava	2836	2198	2534	3339	3115	4694
Yam	791	569	454	790	1222	1947
Cocoyam	1210	777	859	1580	1109	1105
Plantain	1809	993	843	1224	1106	1020
	calories available as % of requirements					
Nutrition	87		76			93

Source: PPMED, Ministry of Food and Agriculture, Accra; FAO production yearbooks

Inter-year and intra-year fluctuations form a major problem with the food supply situation in Ghana. With poor storage facilities and limited processing food becomes abundant at harvest time and very scarce during the planting season. Ghana's agriculture which is mainly rain-fed renders itself to the vagaries of the weather. This situation is worsened by price expectations which follow the classical cobweb theorem (a year of low prices give rise to a reduced acreage allocation and therefore lower production in the following year) (Ezekiel, 1938). For example, Bumb *et al.* (1994) have reported that the area planted to maize was a low 280,000 ha. in 1983, and prices were high for that year's production; consequently, 700,000 ha. of land were planted to maize in 1984, and maize prices dropped that year. The adaptive expectations of farmers made them reduce area planted to maize in 1985, and maize prices rose that year. Production therefore exhibits considerable volatility from year to year. Jebuni *et. al.* (1990) have estimated high coefficients of variation for yam (0.21), rice (0.32), maize (0.36) and cassava (0.45). Sometimes shortfalls in domestic production are met with official imports in addition to private imports which are encouraged through a liberal trade regime. These imports tend to improve the food availability situation in the country from time to time and they tend to avoid extreme cases of food insecurity.

Still, problems of under-nutrition and malnutrition persist in Ghana (see for example (Alderman, 1990). The deterioration in the economy that was experienced in the latter 1970's and early 1980's had its toll on nutrition. The daily calorie supply declined from 87 percent of requirements in 1965 to 76 percent in 1985. However, in response to the economic reforms instituted in 1983 which saw improvements in agricultural activities, the per capita per day calorie availability increased to 93 percent of requirements between 1988 and 1990 (UNDP, 1993). Despite improvements in average calories available, nutrition and nutritional status of many Ghanaians is still inadequate to lead a healthy and productive life[1]. For example, classifying the nutritional status of adults using their body-mass-index (BMI = weight/height^2), gives an indication of chronic energy deficiency in adults (Table 9.2).

TABLE 9.2: Nutritional Status of Adults Aged 18 and Above in Ghana, 1987-1989

	men (n=6002)	women (n=7063)
Weight (m)	58.5	54.1
Height (kg)	168.3	157.5
Body-Mass-Index (m/kg^2)	20.6	21.8
	distribution BMI (%)	
Normal range (above 18.5)	81	83
Underweight/ health risk (16 to 18.5)	17	15
Chronic energy deficient (below 16)	2	2

Note: The classification of BMI ranges considered inadequate is taken from Shetty and James (1994) and FAO (1994)

Source: Ghana Statistical Service, GLSS I and GLLS II

---

[1] As this study focuses on nutrition and productivity of adults, the prevalence of under- and malnutrition among children in Ghana will not be discussed, although it is recognized that food shortages have their severest impact on the nutritional status of children (World Bank (1989), Alderman (1990)).

According to commonly used cut-off points it appears that some 2 percent of adults in Ghana can be considered to suffer from chronic energy deficiency. Another 15 percent can be considered to be in a range where underweight and health risks may occur, while many of these will have a reduced capacity for work. Further disaggregation reveals that, in general, rural residents were thinner than their urban counterparts, while also undernutrition seems to prevail more in the savanna zone than the other agro-ecological zones of Ghana.

Labour Productivity and Labour Supply in Ghana

Inadequate income and low productivity are among the main determinants of inadequate nutrition. As on average Ghanaian households earn more than half of their income in agriculture, while for the poorest 10 to 40 percent of the population the figure is even higher at some 70 to 75 percent (Boateng et al., 1990; Seini et al., 1997), low productivity in agriculture is a main policy concern. Most of the research in this area deals with yield improvements through improved technologies and fertilizer use, while the role of human capital and labour productivity receives less attention.

The low agricultural productivity in Ghana is indeed partly due to the low application of chemical fertilizer, unavailability of adequate amounts of organic manure and small average size of farms, preventing economies of scale in management and capital inputs to be fully exploited. Most farms in Ghana range between one and four acres. The fertilizer use levels in Ghana are low as compared with other countries. In 1990 Ghana used less than 5 kg of plant nutrients per hectare of arable land as compared with 12 kg in Nigeria, 53 kg in Zimbabwe, 110 kg in Indonesia and 366 kg in Egypt (Bumb et al., 1994).

The historical importance of the relative neglect of labour productivity as compared to land productivity is that in the past most work in agriculture was done by farm family labour whose opportunity cost was quite low. Agricultural productivity was therefore measured in relation to land which was thought to be scarcer because of the apparent difficulty in getting access to it. However, despite releases of improved varieties for many crops there has not been significant improvements in yields on farmers' fields (Table 9.3). Still, it has been demonstrated in programmes like Global 2000 that it is possible to at least triple maize yields and double cassava yields from present levels if farmers adopt improved practices.

One of the causes which may have prevented sustained yield improvements to appear might be the continuous decline in farm family labour due to increasing schooling of children

TABLE 9.3: Yields of Some Major Food Crops
in Ghana,1988-1992

Crop	1988	1989	1990	1991	1992
Metric tonnes per hectare					
Maize	1.2	1.3	1.2	1.5	1.2
Rice	2.1	0.9	1.7	1.3	1.7
Millet	0.8	0.9	1.7	1.3	1.7
Sorghum	0.8	0.8	0.6	0.9	0.8
Cassava	9.3	8.0	8.4	10.7	10.3
Yam	7.1	5.9	7.4	11.5	10.4
Cocoyam	10.1	6.3	6.2	6.8	6.1

Source: PPMED, Ministry of Food and Agriculture, Accra.

and migration to the urban areas. At the same time, there has been an increase in hired labour as a substitute for unavailable family labour[2]. This has raised the opportunity cost of labour in many of the agricultural areas and led to a growing interest in the role of human capital and labour productivity in development, both inside and outside agriculture.

Assessment of Nutrition-Productivity Relations

While a lot of work has revealed positive effects of increased fertilizer use, improved technologies and education on agricultural productivity in Ghana, little is known about positive effects of improved nutrition. It has been proven to be quite difficult to asses such effects quantitatively as nutrition and labour productivity are mutually related to one another: better nutrition (food consumption and nutritional status) may contribute to higher labour productivity (per hour and total hours) and, conversely, higher productivity contributes to higher incomes and thereby allows improvement of nutrition outcomes (see Strauss, 1993; and Strauss and Thomas, 1995, section 2, for excellent reviews). By implication and due to the simultaneity, a positive correlation between the two does not by itself indicate whether the effect of nutrition on productivity is indeed significant, as nothing can be said about causality as long as the two effects are not separately identified[3].

This simultaneity between nutrition and labour productivity has led to an emphasis on the use of advanced statistical and econometric methods to address the nutrition-productivity issue; indeed in recent years several studies have made an attempt to isolate the effect of nutrition on labour productivity from the reverse effect through correcting the estimation methods for simultaneity bias. Some of these use survey data, like we do in this chapter, and take a household modelling framework as the starting point of the analysis (Strauss, 1986; Sahn and Alderman, 1988; Kyereme and Thorbecke, 1991)[4]. However, for several reasons estimation practices seldom employ all properties of the underlying model. For example estimation of the structural form of a simultaneous equations model requires explicit specification of all relations in the model and identification of all parameters and this may require unavailable data[5].

---

[2] Note that family labour and hired labour need not be homogeneous. Indeed the two are observed to be heterogeneous with low substitutability and higher productivity for hired labour (Deolalikar and Vijverberg, 1991). However, it is realized that this may merely reflect seasonality as hired labour is over-represented in the (productive) peak season. Still part of the difference may also be due to real labour power differences, for example as hired labour may be selected to perform certain work which needs more skills as compared to *average* family labour.

[3] Still, several authors have interpreted a positive correlation between the two as a causal relationship (see e.g. Martorell and Arroyave, 1988). For example, a positive coefficient in a regression of say calorie intake on ability to work may merely reflect the reverse causality by which differences in ability to work cause income differences, which in turn cause differences in food demand and calorie intake.

[4] Studies employing experimental data are not discussed in this paper, although they have contributed a lot on a better understanding of the nutrition-productivity links, see Korjenek (1992) for example for an overview.

[5] The possibility of specification errors in model equations which are not of primary concern is also sometimes used as an argument for not estimating the full underlying model (Higgens and Alderman, 1993, p 6). However, we consider this a less valid argument as explicit recognition of specification errors seems preferable over implicit errors hidden in reduced form or limited information estimation methods.

Moreover, estimation of a system of nonlinear equations requires advanced econometrics, while the full range of estimation methods for such a system has not been available in widely used econometric software packages until recently[6].

For all these reasons, most of the work in this area has estimated single equations, either in their reduced form and allowing the use of Ordinary-Least-Squares techniques (presuming that all right hand side variables are exogenous) or in their structural form and employing limited information Instrumental-Variable methods such as 2-Stage-Least-Squares or Limited-Information-Maximum-Likelihood to correct for simultaneity bias (presuming that the instruments are reasonably correlated with the right hand side endogenous variable, but exogenous and uncorrelated with the error term). Reduced form estimation has the disadvantage that a priori knowledge regarding the underlying model is not taken into account, while Instrumental-Variable methods may be prone to the choice of functional forms and instruments in the first stage of the estimation procedure (Bound et al., 1993). The latter is especially problematic when instruments are poorly correlated with the instrumented variables as is usually the case for nutrition and productivity variables.

In this chapter we propose a structural form to investigate the mutual nutrition-productivity interactions. The relations considered of main relevance for these interactions are cast within a utility maximization framework and estimated with data from a household survey in Ghana (Ghana Statistical Service, 1989). A Full-Information-Maximum-Likelihood method developed in Amemiya (1985) is used to estimate the parameters of the implied nonlinear simultaneous equations systems, while the specification departs from a re-parameterisation of a utility function developed in Hausman (1980).

MUTUAL INTERDEPENDENCE BETWEEN NUTRITION AND LABOUR EARNINGS

The Efficiency Wage Relation and the Nutritional Status Production Function

Household food consumption and labour allocation decisions are known to have important consequences for the availability of calories and the nutritional status of the household members. The most widely used starting point to analyse such decisions is to specify some objective of the household (usually a utility function), which it maximises subject to a budget constraint and a time constraint. Such an approach may lead to the estimation of a demand and labour supply system which takes prices p, labour productivity w and hence full income $M = \mu + wT$ (where μ is non-labour income and T is labour endowment in hours) to be exogenous to the household's decision. The estimates may be employed for subsequent assessment of food consumption, calorie intake and nutritional status of the household members under alternative assumptions for the exogenous variables.

However, if nutrition has an effect on the efficiency of labour and hence on the labour productivity, then (full) income is affected by the household's food consumption and labour allocation decision and the usual assumption that it can be considered to be exogenous income no longer holds and needs relaxation.

An appropriate way to do so is to let labour endowment refer to hours of labour power T_E

---

[6] For example the Full Information Maximum Likelihood estimator employed in this chapter has been programmed using the SAS/ETS econometric software and this FIML option was not available before version 6.0, September, 1993.

(or efficient hours) rather than to clock hours T. The household's food consumption and labour supply decision then takes the renumeration w_E of an efficient hour of labour to be exogenous to its own behaviour, whilst labour power endowment T_E and renumeration w of a clock hour are endogenous[7], viz. dependent on food consumption c, time share ℓ not devoted to work and nutritional status N. Such a relationship may also depend on several individual, household and community characteristics z[8].

$$T_E = f_E(c,\ell,N \mid z) \tag{9.1}$$

Figure 9.1 illustrates possible relationship between food consumption, leisure, nutritional status and labour power.

A relation like (9.1) is frequently referred to as an efficiency-wage relation, as it implies that labour earnings are based on efficiency units of labour or hours of labour power, rather than on clock hours (cf. Stiglitz, 1976; Keyzer, 1993). Most applications have employed market wages and salaries per clock hour as a proxy for labour productivity of workers and employees and used these to estimate a relation like (9.1) (Sahn and Alderman, 1988; Haddad and Bouis, 1991). However, the relation is equally applicable in situations of self-employment. In fact, a lion's share of the sample studied in this chapter concerns persons who are self-employed or family workers and the term wage is to be interpreted as the remuneration for an hour of labour and not as the market wage rate.

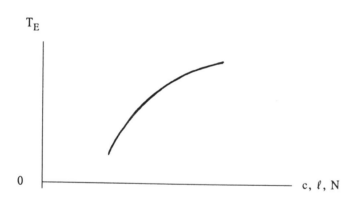

Figure 9.1:   Labour Power as a Function of Food Consumption,
             Leisure and Nutritional Status

---

[7] The majority of persons in the sample are self-employed and it seems reasonable to assume that their labour power rather than hours worked per se determine labour earnings. For wage earners employed outside the household, such an assumption presumes that the employer is aware of the labour power of his employees and rewards effective labour hours rather than clock hours.

[8] For example, gender, age, type of work, household composition, place of residence, availability of infrastructure and utilities may have an impact on the efficiency-wage relation as well as on the nutritional status production function and the utility function specified below.

Using the efficiency wage relation, the value of labour endowment can now be expressed either as the product of labour earnings w per clock hour and labour endowment T of clock hours or as the product of labour earnings w_E per efficient hour and labour endowment T_E of efficient hours.

As argued above, the nutritional status N of household members is itself a function of food consumption c and the time share ℓ spent on care, leisure and home tasks[9], while as before the form of such a relationship further depends on certain individual, household and community characteristics z.

$$N = f_N(c, \ell \mid z) \tag{9.2}$$

Figure 9.2 illustrates the relation between nutritional status, food consumption and labour supply.

In economics such a relationship appears in the household production theory as a relation with (food) consumption and leisure time as inputs into the production of a certain nutritional or health status (e.g. Rosenzweig and Schultz, 1983; Cebu Study Team, 1992). We shall use this interpretation as it is flexible in its specification and in its choice of variables for estimation.

Note, however, that from a more nutritional or biological point of view the relation may also be looked at as an energy balance, relating calorie intake (c) to body weight (N) while controlling for the level of activity level as reflected in the labour allocation between work and other activities ℓ (e.g. Spurr, 1984; Svedberg, 1991). Although the latter interpretation may be more appealing as it has the advantage of incorporating a priori nutritional knowledge into the analysis, it has also some drawbacks. One problem may be the instability of the coefficients

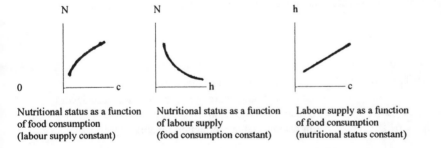

Nutritional status as a function of food consumption (labour supply constant)	Nutritional status as a function of labour supply (food consumption constant)	Labour supply as a function of food consumption (nutritional status constant)

Figure 9.2: Relationship Between Nutritional Status (N), Food Consumption (C) and Labour Supply h = 1 − ℓ

of the balance, as individuals may adapt their basal metabolic rate. Some authors have argued that such adaptation would enable a person to maintain his or her body weight at the same activity level under a reduction of up to 30 percent in calorie intake, implying that at caloric intakes above some low threshold level nutrition would have no further effect on productivity

---

[9] It is convenient to let the time share ℓ spent on care, leisure and home tasks be the complement of the more directly productive time share, say h and to refer to the latter as labour supply.

(Sukhatme and Morgen, 1982). However, the evidence for this hypothesis has been challenged in subsequent literature as it is based on small samples (e.g. Strauss, 1993), while others have found an adaptation of less than 10 percent and also the threshold level, when it exists, might well be high enough to be relevant for large parts of the population of developing countries (Martorell and Arroyave, 1988; Scrimshaw and Young, 1989; Payne and Lipton, 1994). Thus, more than through metabolic adaptations, it would seem that people cope with energy deficiencies, either by accepting a temporary or even lasting reduction of body weight or by adapting their behaviour through a reduction of energy demanding activities (Waterlow, 1985).

A further point in estimating equation (9.2) as an energy balance is that detailed information is required on activities and calorie intake at the individual level and the survey data used in this chapter are not particularly suited for this. For example activity levels relate to the type of work, which is defined in broad categories only, while the derivation of calorie intakes is problematic, as only figures for the value of food consumed at the household level are reported[10]. Given such data problems, which are common to most surveys, we do not pursue the estimation of an energy balance per se but instead interpret equation (9.2) in the more general context of a nutritional status production function.

Behavioral Model with Nutrition-Productivity Relations

In order to explore the efficiency-wage relation (9.1) and the nutritional status production function (9.2) in their mutual interrelations, the two are incorporated into a behavioral model of a household. At given prices p, labour earnings w_E per efficient hour, non-labour income μ and characteristics z and taking into account the relationships (9.1) and (9.2) between nutrition and labour productivity, households are assumed to choose consumption c and to allocate a share $h = 1 - \ell$ of its labour power to earn income and a share ℓ to other activities (the latter including leisure, but also care and home tasks) so as to maximise some utility function u.

$$\max_{c,\,\ell} \quad u(c,\ell \mid z) \quad : \text{utility function}$$

s.t. $\qquad\qquad\qquad\qquad\qquad\qquad\qquad\qquad\qquad\qquad$ (9.3)

$$pc + w_E T_E \ell \le \mu + w_E T_E \quad : \text{budget constraint (full income)}$$
$$h = 1 - \ell \qquad\qquad\qquad : \text{time constraint; share labour power devoted to work}$$
$$T_E = f_E\,(c,\ell,N \mid z) \quad : \text{efficiency-wage relation}$$
$$N = f_N\,(c,\ell \mid z) \qquad : \text{nutritional status production function}$$

The remainder of the chapter is concerned with an empirical elaboration of the model (9.5). Its specification and estimation will put us in a position to discuss the significance of nutrition-productivity interrelations in Ghana and to simulate the consequences of certain changes in variables largely beyond the influence of the household, but possibly affected by policy measures.

---

[10] It is well known that there are many pitfalls in the derivation of calorie intakes from food demand, including possible upward biases of calorie-income elasticities and unobserved redistributions (Bouis and Haddad (1992), Minhas (1991)). A main problem with the GLSS data used here is the absence of direct information of quantities consumed. This might be overcome by using separate information on food prices prevailing at the year of the survey. Although such data are partly available, their use appeared to lead to an unacceptable range of calorie intakes. Per capita calories per day would be practically zero in many households, whilst in many others intake would exceed 10 thousand kilo-calories (see also Alderman and Higgens (1992), Table 9.1).

Ghana Living Standards Survey

The data employed to estimate the nutrition-productivity relationships outlined above are taken from the Ghana Living Standards Survey. The survey was conducted by the Statistical Service of Ghana with the assistance of the World Bank. Three rounds of the survey have been completed, covering the periods 1987/88 (GLSS I), 1988/89 (GLSS II) and 1990/91 (GLSS III). A two-stage sampling procedure was used; in the first stage, the enumeration areas were selected with probability proportional to size after they had been arranged by local council, ecological zone and rural, semi-urban and urban localities, and in the second stage 20 households were selected to make each work load (4 of them were kept as reserve). The data were collected at three levels: household, individual and community levels. Household data include income, expenditure, housing, household enterprises and assets. Data on individual household members cover demographic characteristics, education, health, employment and time use, migration and anthropometry. For the local community data were collected on social services availability (education and health), communication, transportation, food prices, and general economic and social characteristics.

This chapter employs data of the first round of the GLSS (Ghana Statistical Service, 1989). The data contain an extensive overview of amongst others the expenditure patterns and income sources of more than 3000 households and also of the individual characteristics like age, sex, weight, height, education, employment and labour earnings of most of the household members. In Table 9.4 we have put together both individual and household level information considered relevant for the nutrition and labour productivity relations.

The figures concern a sub-sample of 643 men and 504 women between 18 and 65 years of age, with a body mass index below 22.5, reporting to have worked up to 50 hours a week on average during the last year and to have earned up to 300,000 cedis per year and living in households with an adult equivalent expenditure level below 100,000 cedis per year[11]. The table includes a short description of the selected variables along with their mean and standard deviation. Some salient aspects of the data are the following.

Men tend to allocate 22 percent more of their time to work. However, since the role of women in housekeeping, child care and reproduction is not accounted for, the reverse is often true in terms of work load; the ratio of male over female labour in home production is generally quite small in many African countries, and after correction for this men tend to have 15 to 40 percent more leisure time (McGuire and Popkin, 1990, Table 9.1). Earnings per hour worked are less different, male labour being 8 percent more productive on average[12].

---

[11] The cut-off points are used to confine the analysis to persons for which the nutrition-productivity relation is most relevant. Maximum labour earnings considered are well above the average income of a household, while the cut-off point for the expenditure level is somewhere near the fourth quintile, meaning that only observations in the upper quintile are not in the sub-sample.

[12] Larger labour supplies and higher productivity for men are also observed in other countries. For example, in a sample of agricultural households in Peru, men work 20 percent more than women, while male labour is estimated to be twice as productive (Jacoby, 1992). As compared to the 8 percent productivity difference in the Ghana sample, the latter results suggest considerable differences between the two countries.

TABLE 9.4: Summary Statistics of Nutrition and Labour Productivity Data for Men and Women in Ghana

Variable	Description	Men N=643 Mean	Men N=643 Std Dev	Women N=504 Mean	Women N=504 Std Dev
H	Time allocated to work (share of 2500)	0.60	(0.24)	0.49	(0.23)
W	Hourly labour earnings (cedis)	55.92	(32.85)	51.68	(31.08)
WH	Labour earnings (1000 cedis)	83.02	(59.12)	60.22	(45.28)
C	Food consumption (per adult e.g. 1000 cedis)	49.89	(22.43)	48.11	(21.63)
μ	Consumption less earnings (C-WH, 1000 cedis)	-33.13	(61.59)	-12.10	(49.79)
N	Body weight (kg)	57.07	(6.58)	49.34	(5.65)
HT	Height (m)	1.69	(0.07)	1.58	(0.06)
BMI	Body mass index (weight/ squared height)	19.93	(1.59)	19.80	(1.60)
NMAX	Weight corresponding to bmi=22.5	64.39	(5.05)	56.06	(4.49)
INACT	Dummy for inactive due to illness	0.09	(0.28)	0.09	(0.29)
COA	Dummy coastal zone	0.34	(0.47)	0.34	(0.48)
FOR	Dummy forest zone	0.40	(0.49)	0.44	(0.50)
GAC	Dummy Greater Accra region	0.14	(0.35)	0.12	(0.33)
URB	Dummy urban area	0.33	(0.47)	0.33	(0.47)
SIZE	Household size	5.94	(3.24)	6.26	(3.33)
KID	Dummy for presence children under 5	0.63	(0.48)	0.71	(0.45)
AGE	Age	36.88	(11.48)	36.13	(11.76)
SELF	Dummy for self-employed/family worker	0.59	(0.49)	0.87	(0.33)
LIT	Dummy variable for literacy	0.59	(0.49)	0.30	(0.46)
PRI	Years primary education	3.97	(2.75)	2.53	(2.88)
SEC	Years of post-primary education	2.95	(3.15)	1.40	(2.18)
FARM	Dummy job in agriculture	0.59	(0.49)	0.54	(0.50)
INDU	Dummy job in industry	0.13	(0.33)	0.13	(0.33)
SERV	Dummy job in service sector	0.28	(0.45)	0.33	(0.47)
LAND	Acres farmed (acres per adult)	1.17	(2.14)	1.10	(1.62)
MANU	Dummy for manual work	0.78	(0.41)	0.69	(0.46)
WATER	Dummy for treated water	0.27	(0.44)	0.25	(0.43)
LIGHT	Dummy for electricity	0.24	(0.43)	0.20	(0.40)
DEFL	General deflator (September 1987=1)	1.19	(0.13)	1.18	(0.13)

Notes:

- Labour earnings include both wage earnings outside the household and earnings from self-employment inside the household, the latter accounting for the lion's share. Witness for example the fact that 59 percent of the men and 87 percent of the women in the sub-sample are self-employed or family workers.

- Maximum labour supply is set to 2500 clock hours per year (50 weeks, 50 hours per week), while the maximum BMI for which nutritional status may have an impact on labour power is set to 22.5, which is well above cut-off points used to define energy adequacy, see Table 9.2. Besides, both the summary statistics and the estimates appear to be robust against alternative cut-off points.

Source: Computed from Ghana Living Standards Survey 1987/88, Ghana Statistical Service

Men and women live in households with comparable levels of food consumption around an average of slightly less than 50,000 cedis per adult equivalent, i.e. some 135 cedis per day[13]; at prices between 40 and 60 cedis per thousand Kcal (40 cedis for calories from maize and cassava; 60 cedis for calories from rice and bread, for example), this average represents 2300 to 3400 Kcal per day (Alderman and Higgens, 1992). As cheaper calories contribute about two-thirds of the total, the average calorie price is only slightly more than 45 cedis, leading to an average daily food consumption close to 3000 Kcal for men and 2700 Kcal for women.

Somewhat surprisingly, the average Body-Mass-Index of men and women is practically the same. Thus, the apparent difference in BMI between men and women in Table 9.2 (20.6 for men as compared to 21.8 for women) disappears when a cut-off point of 22.5 is used; this suggests that a classification of inadequate BMI ranges need not be gender specific (Shetty and James, 1994, p.19).

With respect to the remaining variables, it stands out that few women earn an income outside of the household (only 13 percent as compared to 41 percent of the men). Also, education is quite divergent between men and women and, for example, only 30 percent of the women is literate against almost 60 percent of the men.

Model Specification

Coming to the specification of our model we start with the utility function and propose the following form.

$$u(c,\ell) = -\,(\beta+\gamma\ell)\,\exp[-\gamma(\alpha+\ell+\gamma c)\,/\,(\beta+\gamma\ell)]$$
where (9.4)
$$\beta+\gamma\ell > 0 \text{ and } \alpha+\ell+\gamma c \le 0$$

This specification is inspired by the utility function developed and employed in Hausman (1980). One may note that (9.4) is slightly different from Hausman's seminal model where hours worked, rather than the share of leisure in total time appears as the second argument in the utility function. In effect Hausman's budget constraint $pc \le m$ refers to cash income $m = \mu + wTh$, whereas our budget constraint $pc+wT\ell \le M$ employs full income $M = \mu + wT$ (and thus, since $h + \ell = 1$, $M = m + wT\ell$). In fact (9.4) can be seen as a re-parameterisation of the Hausman utility function[14]. Maximisation of the utility function (9.4) subject to the budget constraint $pc + wT\ell \le M$ yields the following consumer demand system[15].

---

[13] At the prevailing exchange rate of 150 cedis for a US dollar, this implies a food consumption of less than a dollar per adult equivalent per day.

[14] Letting $\hat{h} = hT$ denote hours worked, Hausman proposes the utility function $\hat{u}(c,\hat{h}) = (\hat{\gamma}\hat{h}-\beta)\,\exp[-\hat{\gamma}(\hat{\alpha}-\hat{h}+\hat{\gamma}c)/(\beta-\hat{\gamma})\hat{h}]\,/\,\hat{\gamma}^2$, requiring $\beta-\hat{\gamma}\hat{h} > 0$. Employing the complementarity between labour supply and leisure, that is $\hat{h} = (1-\ell)T$, and after the re-parametrisation, $\hat{\alpha} = (\alpha+1)T$, $\beta = (\beta+\hat{\gamma})T$ and $\hat{\gamma} = \gamma$ one can obtain (4) as $u(c,l) = (\hat{\gamma}\,/\,T)^2\,\hat{u}(c,(1-\ell)T)$. Conversely, and employing the same re-parametrisation, the original utility function $\hat{u}$ can be derived from our utility function u through $\hat{u}(c,\hat{h}) = (T\,/\,\gamma)^2\,u(c,(T-\hat{h})\,/\,T)$.

[15] This demand system can be obtained either from the Kuhn-Tucker conditions of maximization of $u(c,\ell)$ subject to $pc+wT\ell \le M = \mu + wT$ or from differentiation of the corresponding indirect utility function. It may be verified that this dual function is given by $v(p,wT,M) = \exp(\gamma wT\,/\,p)\,(\alpha\gamma-\beta+\beta\gamma wT\,/\,p+\gamma^2 M\,/\,p)$.

$$\ell = -\alpha - \beta \, wT / p - \gamma \, M / p$$

$$c = M / p + \alpha \, wT / p + \beta \, (wT / p)^2 + \gamma \, (wT / p)(M / p)$$

(9.5)

Several remarks can be made with respect to this model. First, it follows immediately from the time constraint that the labour supply time share h equals:

$$h = 1 + \alpha + \beta \, wT / p + \gamma \, M / p = 1 + \alpha + (\beta + \gamma) \, wT / p + \gamma \mu / p$$
(9.5′)

A second remark concerns the role of the two inequalities $\beta + \gamma \ell > 0$ and $\alpha + \ell + \gamma c \leq 0$. It appears that these are the very conditions for the utility function to be proper and to yield a demand system when maximised on a budget set[16].

Note further that the level of utility is negative everywhere. This does not pose problems, because after all utility levels can be mapped into any region by monotonic transformations, without affecting the consumers' decisions. Also, without loss of generality we may presume that β is positive, while α and γ are likely to be negative; β must be positive in order to make zero leisure $\ell = 0$ feasible; $-\alpha$ is the committed 'leisure' time share devoted to non-income earning activities and thus the presumption that α is negative equals the assumption that everybody needs some leisure time, whatever his or her circumstances; finally, $-\gamma$ is the marginal increase of the demand for leisure from an increase of full income and thus the presumption that γ is negative is just the assumption that leisure is a normal and not an inferior commodity. Then from the monotonicity of the utility function it follows that the minimum utility level is reached at $\ell = 0$ and $c = -\alpha / \gamma$ and equals $u(-\alpha / \gamma, 0) = -\beta$. The demand for leisure is bounded above by $\beta / -\gamma$ and the level of utility climbs to zero as consumption c reaches infinity and leisure ℓ comes close to its upper bound.

As we have argued the household's consumption and labour allocation decision cannot take full income to be exogenous whenever labour power depends on such decisions and a labour power function or efficiency wage relation enters the behavioral system. Unfortunately both labour power itself and labour earnings from an extra hour of labour power supplied are seldom observed and the GLSS is no exception in this case. However, we do have information on clock hours worked as well as on the labour earnings from a clock hour and it appears that we can rewrite the efficiency wage relation so as to make it estimable. Recalling that the value of labour endowment can either be written in terms of these observable variables, i.e. w times T, or in terms of the unobservable efficiency wage and labour power, i.e. w_E times T_E, so that we can write $wT = w_E T_E$. In order to capture nonlinearities in the labour power or efficiency wage relation (9.1) the following semi-logarithmic/quadratic form is employed in the estimation[17].

$$w = w_E T_E / T = \alpha + \beta \log(c) + \gamma \log(h) + \delta_1 N / N_{max} + \delta_2 (N / N_{max})^2$$
(9.6)

One may note that in this specification the nutritional status is assumed to affect labour productivity insofar as body weight N deviates from some normative body weight N_{max}, corresponding to a Body-Mass-Index of 22.5. Thus, the term N/N_{max} is proportional to the Body-Mass-Index, which to some extent reflects differences in a person's capacity to work due

---

[16] To be precise the inequalities can be shown to be both necessary and sufficient for the utility function u to be increasing in both ℓ and c and strictly quasi-concave in (c,ℓ).

[17] Both for the efficiency wage relation and for the nutritional status production function several variants of the forms finally chosen have also been used, but gave similar or less satisfactory results.

to (chronic) energy deficiencies (Spurr, 1984; Shetty and James, 1994).

Finally for the nutritional status production function (9.2) we employ the following semi-logarithmic form.

$$N = N_{max} + \alpha + \beta \log(c \, / \, h) \tag{9.7}$$

To complete the specification of the model, we have to specify how the parameters of the nutrition and labour productivity functions (9.5), (9.6) and (9.7) depend on certain individual, household and community characteristics z. With respect to the constant terms, we allow these to vary with agro-ecological zone, with locality and with household size. This reflects that the nutrition and productivity relations might differ between, for example, a large household living in the rural savannah, and a small household living in Accra (the households being otherwise similar). Also, the constant terms incorporate a linear and quadratic effects for the age, representing life cycle effects in both labour supply, nutritional status and productivity[18] and, finally, whether or not people are self-employed is added to have a potential effect. Thus the parameter α in the above equations all include the following terms (see Table 9.4 for a legend of the variable names).

$$\alpha_1 \, FOR + \alpha_2 \, COA + \alpha_3 \, URB + \alpha_4 \, AGE + \alpha_5 \, AGE^2 + \alpha_6 \, SELF + \alpha_7 \log(SIZE) \tag{9.8}$$

Apart from these differences by zone (savannah, forest, coastal zone), by locality (urban, rural), by household size, by age and by type of employment, some other factors may be expected to play an important role in the nutrition-productivity relations. For example labour supply is known to be restrained by the care to be devoted to children and we add this effect to the constant term. Thus constant term α of the labour supply function (9.5') is specified as in (9.8) and further includes an effect α_8 KID.

For the efficiency wage relation (9.6) we allow for additional effects of education (literacy, years of primary and secondary education) and an additional sectoral effect (agriculture, industry or services).

$$\alpha_8 \, LIT + \alpha_9 \log(1+PRI) + \alpha_{10} \log(1+SEC) + \alpha_{11} \, INDU + \alpha_{12} \, SERV \tag{9.9}$$

For the nutritional status function (9.7), additional effects are introduced to reflect the way individuals adapt their nutritional status at varying consumption levels and labour efforts. For example, nutritional status may be affected by illness, possibly reducing the appetite to consume, by the availability of treated water and electricity, increasing the efficiency of the relationship, by the type of work he or she does, as laborious work requires extra consumption to prevent nutritional hardship and finally by genetic differences. Such considerations lead to the following specification of the term β in the nutritional status production function (9.7):

$$\beta_0 + \beta_1 \, INACT + \beta_2 \, WATER + \beta_3 \, LIGHT + \beta_4 \, MANU + \beta_5 \, HT \tag{9.10}$$

---

[18] For the efficiency wage relation, the age effect represents years of experience.

Error Specification

To estimate the above system of equations an error term is added to the labour supply equation (9.5), the efficiency wage relation (9.6) and the nutritional status production function (9.7). The errors are assumed to be normally distributed with expectation zero and a covariance matrix which is the same for all observations. Formally, let ϵ_{im} and ϵ_{jn} refer to the error terms of the i-th and j-th equation (i,j = 1,2,3) and the m-th and n-th observations (m,n = 1,2,...,M, with M = 643 for men and M=504 for women). Then, with $\epsilon_m = (\epsilon_{1m}, \epsilon_{2m}, \epsilon_{3m})'$ and $\Sigma = [\sigma_{ij}]$ we have the following error specification[19].

$$E\{\epsilon_{im}\} = 0$$

$$E\{\epsilon_{im}, \epsilon_{jn}\} = \Delta_{mn}\, \sigma_{ij} \tag{9.11}$$

$$\epsilon_m \in N(0, \Sigma)$$

Such assumptions regarding the error terms are commonly employed and in the current analysis they mean that differences in the nutrition-productivity relations amongst men and differences amongst women other than those captured in the analysis are symmetrically distributed around an estimated mean, while the likelihood of a difference to occur is considered to be independent of the specific person.

Parameter Estimates and Statistical Properties

The labour supply equation, the efficiency wage relation and the nutritional status production function have been estimated as a nonlinear simultaneous system of equations employing the Full Information Maximum Likelihood method (FIML) as discussed extensively in Amemiya (1985, chapter 8). Let $y_m = (h_m, w_m, N_m)'$, x_m and θ denote the endogenous variables, the exogenous variables and the parameters of the model, respectively, and let $f(y_m, x_m, \theta) = (f_1, f_2, f_3)'$ denote the functional specifications of the labour supply, labour productivity and nutritional status production equations. Then, we can write the model to be estimated in short.

$$y_m = f(y_m, x_m, \theta) + \epsilon_m \quad , m = 1,2,...,M \tag{9.12}$$

The FIML estimator of θ stems from maximisation of the likelihood of observing y_m and x_m under the normality assumption (9.11). The logarithm of this likelihood can be obtained by taking the logarithm of the density function of the multivariate normal distribution and using the respective definitions of the 3×3 covariance and Jacobian matrices Σ and J_m.

$$\Sigma(\theta) = 1 / M \ \sum_{m=1}^{M} (y_m - f(y_m, x_m, \theta)) (y_m - f(y_m, x_m, \theta))'$$

$$J_m(\theta) = \partial (y_m - f(y_m, x_m, \theta)) / \partial y_m \quad m = 1,2,...,M \tag{9.13}$$

---

[19] The symbols E, N and Δ_{mn} denote mathematical Expectation, Normal distribution and the Kronecker delta ($\Delta_{mn} = 1$ if m = n and $\Delta_{mn} = 0$ if m ≠ n).

This leads to the following log-likelihood as a function of the parameters θ.

$$L(\theta) = -0.5 \, M \, (3 + 3 \log(2\pi) + \log|\Sigma(\theta)|) \; + \; \sum_{m=1}^{M} \log |J_m(\theta)| \qquad (9.14)$$

As compared to other estimation methods, FIML has the advantage that there is no need for choosing instrumental variables. Also, under the presumed multivariate normal distribution of errors, the method is known for its efficiency, that is it generates unbiased minimum variance parameter estimates[20]. An alternative estimation method would be the nonlinear 3-Stage-Least-Squares method (3SLS). Computationally, 3SLS is less expensive and more readily applicable in most existing software (see also footnote 6).

However, unlike the FIML method, 3SLS employs instrument variables and as standards for choosing instruments do not exist, estimates may be sensitive to the particular choice made (see section headed 'Assessment of Nutrition-Productivity Relations'). In addition, the 3SLS method is less efficient under normality. If, however, normality is violated, 3SLS still yields unbiased estimates, while FIML may then produce poor results. Because of this sensitivity of the FIML method for normality, we have tested this assumption and it appeared that the hypothesis that the residuals are normally distributed is not rejected by the data[21]. Thus, FIML estimation seems an appropriate method to estimate our model.

Table 9.5 gives point estimates and asymptotic t-scores of the parameters, together with the R^2-measure for the goodness of fit by equation and the value of the maximised likelihood.

The t-scores indicate that most of the effects are significant, including those which relate to the mutual endogeneity between nutrition and labour earnings. Also, the R^2 are fairly high as this type of cross-section survey data usually exhibit considerable measurement errors, while the specification of the nutrition-productivity relations is bound to be incomplete due to uncaptured (and unobservable) effects such as genetic differences. It is further noteworthy that the estimates appeared not to be sensitive for alternative choices for full employment and for maximum Body-Mass-Index[22].

It appears that several important parameters differ between men and women. Still, the fit of the data on the model is approximately equal for men and women, although the wage equation is somewhat better for the males. This may be due to the fact that more men receive a market wage rate for which the measurement error is probably smaller than for a self-employment wage rate.

Significance of the Nutrition-productivity Relations: Partial Effects

According to our estimates the proposed interrelations between nutrition and labour productivity are significant in Ghana and different between men and women. Looking at the equations separately and without taking the endogeneity into consideration, the estimates give rise to the following observations. Note carefully, however, that the observations are partial as they

---

[20] Note that unbiasedness property presumes that there is no bias due to sample selection. Indeed, such a selectivity bias is often encountered in estimating wage and labour supply equations, viz. if only data on market wages and market employment are used in the estimation (e.g Haddad and Bouis, 1991). Fortunately, we have wage data for both employed labourers *and* for self-employed/family workers and thus need not correct for sample selection.

[21] The Shapiro-Wilk statistic for the residuals varies between 0.95 and 0.98 and exceeds the critical value of 0.92, p=0.01 (see Royston, 1982).

[22] Both parameter values and their significance are hardly affected with changes of plus and minus 20 percent.

TABLE 9.5: Parameter Estimates for the Nutrition-Productivity Model
(Nonlinear Full-Information-Maximum-Likelihood)

Equation		Men		Women	
		\multicolumn{4}{c}{Parameter estimates}			

Labour supply $\quad h = 1 + \alpha + \beta\, wT/p + \gamma\, M/p$		Men		Women	
	Constant	-0.263	(3.59)	-0.235	(2.75)
	FOR	-0.007	(0.41)	0.033	(1.84)
	COA	0.002	(0.13)	0.040	(1.94)
	URB	0.033	(1.96)	0.013	(0.75)
	AGE_2	0.003	(0.76)	0.006	(1.47)
	AGE^2	-0.032E-3	(0.63)	-0.085E-3	(1.62)
	SELF	-0.041	(2.31)	-0.062	(2.24)
	log(SIZE)	-0.024	(1.94)	-0.059	(3.51)
	KID	0.015	(0.97)	-0.028	(1.45)
	log(1+LAND)	0.034	(2.92)	0.003	(0.23)
β	wT/p	2.236E-6	(11.01)	2.366E-6	(7.44)
γ	M/p	-4.078E-6	(20.26)	-4.243E-6	(14.62)
	Adjusted R^2	0.58		0.57	

Hourly earnings $\quad w = \alpha + \beta\, \log(c) + \gamma\, \log(h) + \delta_1\, N/N_{max} + \delta_2\, (N/N_{max})^2$		Men		Women	
α	DEFL	21.647	(2.25)	-12.744	(1.21)
	URB*GAC	-0.835	(0.18)	-16.833	(3.12)
	FOR	3.174	(1.04)	-4.202	(1.09)
	COA	-4.848	(1.42)	6.062	(1.55)
	URB	-3.277	(1.07)	2.545	(0.71)
	AGE_2	2.244	(3.30)	1.198	(1.36)
	AGE^2	-0.024	(2.71)	-0.011	(1.00)
	SELF	-6.264	(1.86)	-11.384	(2.49)
	log(SIZE)	6.606	(3.59)	8.904	(3.57)
	INDU	6.850	(1.74)	-3.382	(0.67)
	SERV	9.217	(2.55)	6.362	(1.64)
	LIT	-5.292	(1.12)	-7.044	(1.30)
	log(1+PRI)	-1.997	(0.89)	1.392	(0.59)
	log(1+SEC)	8.593	(3.18)	9.583	(2.55)
β	log(c)	34.326	(107.17)	24.475	(85.82)
γ	log(h)	-11.105	(2.25)	-16.328	(16.06)
δ_1	N/N_{max}	-9.928E+2	(17.10)	-5.801E+2	(8.82)
δ_2	$(N/N_{max})^2$	5,897E+2	(10.36)	3.315E+2	(5.96)
	Adjusted R^2	0.32		0.26	

Nutritional status $\quad N = N_{max} + \alpha + \beta\, \log(c/h)$		Men		Women	
α	Constant	-11.455	(5.59)	-5.681	(2.58)
	FOR	-0.416	(0.88)	0.526	(1.12)
	COA	-0.093	(0.18)	0.053	(0.11)
	URB	-0.265	(0.50)	-0.952	(2.01)
	AGE_2	0.134	(1.30)	-0.074	(0.68)
	AGE^2	-0.002	(1.73)	0.000	(0.22)
	SELF	1.240	(2.59)	-0.110	(0.18)
	log(SIZE)	0.812	(3.07)	0.649	(1.79)
β	Constant	4.582	(2.55)	6.454	(3.11)
	INACT	-0.361	(1.46)	0.277	(0.82)
	WATER	0.065	(0.35)	0.814	(3.02)
	LIGHT	0.403	(2.21)	-0.034	(0.11)
	MANU	-0.145	(0.75)	-0.247	(1.12)
	HT	-2.500	(2.39)	-4.059	(3.13)
	Adjusted R^2	0.54		0.53	

Log-likelihood of system	5150 (N=643)		3934 (N=504)	

Notes:
- Absolute value of asymptotic t-score in parentheses; values of 1.65 and above are significant at p=0.05.

- See Table 9.4 for legend and description of variables.

concern only immediate effects and neglect indirect effects due to the simultaneity of the estimated equations. Determination of the full effects of changes in certain exogenous variables can only be done through simulation and to that we shall turn in the next sub-section.

As compared to women living in households of 6 persons, those in households with only 3 persons are estimated to reallocate 5.9 percent of their time away from housekeeping; however, men tend to decrease their labour supply with only 2.4 percent as household size increases from 3 to 6 persons. Also, the correlation between household size and labour productivity (as measured by earnings per hour), is stronger for women, who are estimated to gain about 35 percent more from a doubling of household size, probably reflecting the presence of adolescents in larger households who can relieve women's work load.

Life-cycle effects in labour supply are not very strong and, if anything can be concluded from our estimates, there may be some life-cycle effects for women rather than for men. The reverse holds for the age effect in hourly labour earnings, which seems nonexistent for women, while for men experience seems to pay and hourly earnings are estimated to reach a maximum at the age of 47 years. Persons who are self-employed or work as a family labourers, are estimated to allocate less time to work, viz. some 4 and 6 percent less for home employed men and women, respectively. Also they earn less per hour, i.e. on average some 11 and 21 percent less for men and women respectively. This may reflect the lack of job opportunities, although part of the difference may also be attributed to selectivity if the more productive and able persons are overrepresented in those with a job[23]. Note however that this selectivity problem has been taken care of insofar as differences are due to observed differences in education levels and nutritional status.

With respect to education, literacy and primary education do not seem to contribute to labour productivity, while post-primary education does have a significant impact. For example, four years of middle school lead to an estimated increase of 25 to 30 percent of hourly labour earnings. This suggests that an expansion and improvement of middle and secondary education in Ghana is of some importance. The insignificance of primary education is usually attributed to its low quality and the stagnating enrolments. On its turn the lack of perspectives for post-primary education has been argued to be one of the main impediments for improvements in primary education (Glewwe, 1991; Lavy, 1992).

A final remark related to the variables considered exogenous concerns the effects of treated water and electricity in the nutritional status production function. The estimates suggest that the presence of treated water is significant for women, whereas electricity has an effect on men. Electricity and treated water are not correlated with the nutritional status of women and men respectively. A possible explanation for this might be that women's work load is relieved when fetching and home treatment of water become easier, while men's workload might be more connected with activities related to lightening and energy supply.

Coming to the effects of the endogenous variables, Table 9.6 provides labour supply and wage rate elasticities evaluated at the sample mean. With regard to labour supply, an increased wage rate (i.e. an increase of hourly labour earnings) has two effects. One is the direct price effect, which is positive as it becomes more attractive to work and the opportunity cost of leisure time rises. Due to this effect a 1 percent higher wage rate is estimated to cause an increase of labour supply of about 0.5 to 0.6 percent. However, this effect is wiped out through the income effect of a wage increase. Actually, the income effect is estimated to exceed the price effect

---

[23] Note also that part of the difference might also stem from an underreporting bias, as underreporting of home employment and home earned labour income may be more severe.

TABLE 9. 6:   Elasticities in the Nutrition-Productivity Relations

	Men	Women
Labour Supply h		
Wage	-0.43	-0.49
Price effect	(0.52)	(0.63)
Income effect	(-0.95)	(-1.12)
Income	-0.23	-0.10
Hourly Earnings w		
Food consumption	0.61	0.47
Labour supply	-0.20	-0.32
Body weight	0.75	0.06
Body-mass-index	0.82	0.06
Nutritional Status N		
Food consumption	0.0057	0.0019
Labour supply	-0.0057	-0.0019

considerably so that eventually a 1 percent wage increase leads to a decrease of labour supply of 0.4 to 0.5 percent. As expected, the income elasticity is also negative and a 1 per increase of nonlabour income is estimated to lead to a 0.23 and 0.10 percent decrease of labour supply of men and women respectively.

Hourly labour earnings of both men and women are estimated to respond positively to food consumption, negatively to labour supply and positively to nutritional status. With an elasticity of less than 0.1, the latter effect is quite small for women, while for men a 1 percent increase of weight or body-mass-index is estimated to add about 0.8 percent to their productivity (both figures at the sample mean). This is in line with figures obtained elsewhere; Deolalikar, (1988) for example estimates an elasticity of wage rates with respect to weight-for-height between 0.3 and 0.7 for farmers in South India.

A 1 percent increase of food consumption is estimated to affect male wage rates by 0.61 percent and female rates by 0.47 percent. These elasticities are quite close to the food consumption elasticity of 0.55 found for agricultural productivity in rural Sierra Leone (Strauss, 1986). However, they are much bigger than estimated elasticities in Asian countries. For example, Sahn and Deolalikar, (1988) found a food consumption elasticity of 0.21 for men and a neglectible elasticity for women in Sri Lanka, while Deolalikar (1988) and Haddad and Bouis (1991) did not find any evidence for productivity effects of calorie intake for male dominated samples in South India and the Philippines, respectively[24].

An increased labour supply affects wage rates reversely and on average a 1 percent increase decreases productivity by 0.2 to 0.32 percent for men and women respectively. The estimated effects of food consumption and labour supply on a person's nutritional status are generally small; for example, starting from average food consumption and labour supply, a change with 50 percent in either of the two is estimated to affect body weight by only 150 to 250 grammes for men and 75 to 100 grammes for women. This points to the fact that adjustments of the body-mass-index are a matter of long term energy deficiencies rather than variation of

---

[24] The statistical evidence from the latter studies, however, seems rather weak as wage equations show a poor fit with R^2's of less than 0.10.

food consumption and hours worked in the previous year. Note, however our earlier concerns that the GLSS data are not particularly suited for estimating these effects.

Simulating Income Changes, Middle School Enrolment and Job Growth

The effects described above disregard the endogeneity of some of the right hand side variables by which labour supply, hourly labour earnings and nutritional status are determined simultaneously. Here, we take account of the endogeneity and estimate the full effects of the exogenous variables, whose value is primarily beyond the influence of the household's food consumption and time allocation decisions. For this purpose the estimated relations are employed to simulate nutrition and productivity outcomes under selected assumptions regarding the exogenous variables. Recall from equation 9.12 that $y = (h, w, N)'$ and θ denotes the endogenous variables and the parameters, respectively. Then, at (arbitrary) values x_s for the exogenous variables and letting $\hat{\theta}$ be the estimates of parameters θ, a simulation $\hat{y}_s$ of the endogenous variables is defined as the value of labour supply, hourly labour earnings and nutritional status which satisfies the estimated equations, viz.

$$\hat{y}_s = f(\hat{y}_s, x_s, \hat{\theta}) \tag{9.15}$$

The results of selected simulations are summarised in Table 9.7[25]. The base run uses actual values of t exogenous variables and comparison with the sample averages of Table 9.4 suggests that the model is able to give reasonable predictions of the nutrition-productivity outcomes. The other simulations illustrate what might happen as a result of a change in certain circumstances. As expected the nutritional status is not sensitive with respect to the simulations performed. In contrast, food consumption, labour supply and hourly earnings fluctuate considerably.

A redistribution of incme μ so that for every worker the difference between his or her labour earnings and his or her food consumption is equalised (viz. 33,130 and 12,100 cedis for men and women, respectively) leads to a considerable increase of labour supply. Those who earn too little in the base run to finance their own food consumption plus some additional fixed sum are forced to make longer hours; men have to work an extra 20 percent at a comparable productivity, whereas women have to add as much as 50 percent to income earning activities and see their productivity decrease with some 8 percent. The extra efforts allow workers to increase their food consumption with about 40 percent.

If, in another simulation, workers are given an extra 25,000 cedis (or in prevailing cases they keep an extra 25,000 cedis for their own food consumption) then both males and females can reduce their workload by some 17-18 percent and see their productivity rise with 10-12 percent so that total labour earnings decrease with some 6-9 percent, i.e. with some 5 to 6 thousand cedis. The extra income of 25,000 cedis leads to an increased food consumption of 20,000 cedis, an effect similar to the increased efforts in the previous simulation.

Conversely, an income decrease of about half the value of the initial food consumption leads to a considerable loss of productivity (16 percent for men and 12 percent for women). To prevent earnings to fall and to limit the decline of their own food consumption for workers increase their labour supply (20 percent for men and 27 percent for women). Women seem

---

[25] The solution for the three endogenous variables is obtained through Newton's method, which is an iterative scheme adjusting the values of y in a next iteration with the inner product of the inverse Jacobian matrix J_s (defined in equation 9.13) and the values of y in the current iteration (Ortega and Rheinboldt, 1970, chapter 7).

TABLE 9.7: Validation and Simulation Results

| | | Changes income μ: | | | | |
		Redistri-bution	Overall increase	Overall decrease	Middle school	Job growth
Men	Base	$\mu = -33$	$\mu = \mu + 25$	$\mu = \mu - 25$	SEC=4	SELF=1
Labour supply h	0.59	+ 20	– 18	+ 20	+ 2	+ 3
Hourly earnings w	59.10	– 1	+ 12	– 16	+ 14	+ 6
Labour earnings whT	80.47	+ 22	– 6	+ 4	+ 15	+ 8
Food consumption μ+whT	47.34	+ 37	+ 42	– 46	+ 26	+ 14
Body-Mass-Index N / N$_{max}$	19.92	+ 0.1	+ 0.2	– 0.4	+ 0.3	0.0
Women	Base	$\mu = -12$	$\mu = \mu + 25$	$\mu = \mu - 25$	SEC=4	SELF=1
Labour supply h	0.48	+ 50	– 17	+ 27	– 6	+ 6
Hourly earnings w	57.01	– 8	+ 10	– 12	+ 20	+ 20
Labour earnings whT	64.48	+ 35	– 9	+ 11	+ 11	+ 27
Food consumption μ+whT	52.38	+ 43	+ 37	– 35	+ 13	+ 33
Body-Mass-Index N / N$_{max}$	19.79	+ 0.1	+ 0.3	– 0.2	+ 0.1	+ 0.2

Notes:
- Figures in the table are averages for the respective samples.

- Base run is used for validation and reproduces h, w and N from parameter estimates and actual values of exogenous variables; simulations produce h, w and N under selected hypothetical assumptions regarding the exogenous variables; simulation figures give average increase in percentage of the base run.

more successful and can increase their labour earnings with 11 percent, compensating for some 13 percent of the initial 48 percent decline in food consumption; for men the initial decline in food consumption is slightly bigger (53 percent) and they can compensate for only some 7 percent of it.

The next simulation illustrates the effects of middle school enrolment and reveals that labour productivity would increase with an estimated 14 and 20 percent, with an additional but smaller effect of 2 and -6 percent on labour supply for men and women, respectively. Labour earnings increase by 15 and 11 percent, allowing male and female workers to increase their food consumption with 26 and 13 percent respectively.

A final simulation is concerned with a situation in which the negative effects of being self-employed or being a family worker vanish, e.g. as a result of labour market developments and job growth. This is especially beneficial for women, who are enabled to earn 20 percent more per hour while labour supply rises with 6 percent. For men, who are more frequently not dependent on self-employment or family work, the effect is smaller and, according to our estimates, men can add an average of 8 percent to their labour earnings (6 percent higher wage rate and 3 percent higher labour supply).

Summary and Conclusion

Improvements of living standards of the poor through accumulation of human capital are often argued to be both an objective *and* a means for an equitable development (UNDP, 1993; Behrman, 1993). However, much of the empirical research trying to substantiate this claim is not conclusive as it remains difficult to isolate the effects of human capital on income from the reverse effect. Still, gradually emerging evidence suggests that such effects may be significant.

In this chapter we have focused on the relationships between nutrition, labour productivity and labour supply in Ghana within a nonlinear simultaneous equations system derived within a utility maximisation framework. A Hausman-type utility function has been employed, while for the wage equation and nutritional status production function logarithmic/quadratic specifications were used. Nonlinear Full-Information-Maximum-Likelihood was used to estimate the simultaneous system of equations. Both the use of the utility function and the use of the estimation method to account for simultaneity is new in the nutrition-productivity literature.

The estimates reveal that hourly labour earnings of both men and women in Ghana respond positively to food consumption, negatively to labour supply and positively to nutritional status. With an elasticity of less than 0.1, the latter effect is quite small for women, while for men a 1 percent increase of weight or body-mass-index is estimated to add some 0.8 percent to their productivity. The elasticity of labour productivity with respect to food consumption is estimated to be 0.61 and 0.47 for men and women, respectively.

With regard to labour supply, the direct price effect of increased hourly labour earnings, by which it becomes more attractive to work and opportunity costs of other activities rise, is wiped out through the full income effect, by which one can afford to spend less time on income earning activities. Eventually, at the sample average, a 1 percent increase of hourly earnings leads to a decrease of labour supply of 0.4 to 0.5 percent, although this effect diminishes proportionally as initial hourly earnings are lower and initial labour supply is higher. As expected, the income elasticity is also negative and a 1 per increase of nonlabour income is estimated to lead to a 0.23 and 0.10 percent decrease of labour supply of men and women respectively.

Nutritional status seems not very sensitive to food consumption and labour supply decisions and the estimated effects are generally small; for example, starting from average food consumption and labour supply, a change with 50 percent in either of the two is estimated to affect body weight by only 150 to 250 grammes for men and 75 to 100 grammes for women. This points to the fact that adjustments of nutritional status are primarily a matter of long term energy deficiencies rather than variation of food consumption and hours worked in the short or medium term.

Nutrition-productivity outcomes have been validated and simulated for a few selected assumptions regarding some of the exogenous variables, which are largely beyond the household's influence, but may be of considerable relevance for policy purposes. For example simulating a situation in which the negative effects of being self-employed or being a family worker vanish, shows that labour market developments and job growth may be particularly beneficial for women, who are enabled to earn 20 percent more per hour while their labour supply rises with 6 percent. For men, who are more frequently not dependent on self-employment or family work, the effect is smaller and, according to our estimates, men can add an average of 8 percent to their labour earnings. Further knowledge of the relations between nutrition, labour productivity and labour supply may help to shed more light on the effectiveness of certain development policies to ameliorate living standards in Ghana.

ACKNOWLEDGEMENT

The data for the research was based on the Ghana Living Standards Survey which was obtained with the kind permission of the Ghana Statistical Service.

REFERENCES

Alderman, H., 1990. Nutritional Status in Ghana and its Determinants. Working Paper 3, Social Dimensions of Adjustment in Sub-Saharan Africa, World Bank, Washington.

Alderman, H. and P. Higgens, 1992. Food and Nutritional Adequacy in Ghana. Working Paper 27, Cornell Food and Nutrition Policy Program.

Amemiya, T., 1985 Advanced Econometrics. Basil Blackwell, Oxford.

Behrman, J.R., 1993. The Economic Rationale for Investing in Nutrition in Developing Countries. *World Development* 21, 1749-71.

Bouis, H.E. and L.J. Haddad, 1992. Are Estimates of Calorie-Income Elasticities too High? A Recalibration of the Plausible Range. *Journal of Development Economics* 39, 333-64.

Boateng, E.O., K. Ewusi, R. Kanbur and A. McKay, 1990. A Poverty Profile for Ghana 1987-88. Working Paper 5, Social Dimensions of Adjustment in Sub-Saharan Africa, World Bank, Washington.

Bound, J., D. Jaeger and R. Baker, 1993. The Cure Can be Worse Than the Disease: a Cautionary Tale Regarding Instrumental Variables. University of Michigan.

Bumb, B.L., J.F. Teboh, J.K. Attah and W.K. Asenso-Okyere, 1994. Ghana Policy Environment and Fertilizer Sector Development. IFDC, Muscle Shoals, USA and ISSER, University of Ghana, Legon.

Cebu Study Team, 1992. A Child Health Production Function Estimated from Longitudinal Data. *Journal of Development Economics* 38, 323-51.

Deolalikar, A.B. and W.P.M. Vijverberg, 1987. A Test of Heterogeneity of Family and Hired Labour in Asian Agriculture. Oxford Bulletin of Economics and Statistics 49, 291-305.

Deolalikar, A.B., 1988. Nutrition and Labor Productivity in Agriculture: Estimates for Rural South India. *Review of Economics and Statistics* 70, 406-13.

Ezekiel, M., 1938. The Cobweb Theorem. Quarterly Journal of Economics.

Ghana Statistical Service, 1989. Ghana Living Standards Survey: First year report, September 1987 - August 1988. Accra.

Glewwe, P., 1991. Schooling, Skills and the Returns to Government Investment in Education: an Exploration Using Data from Ghana. Working Paper 76, Living Standards Measurement Study, World Bank, Washington.

Haddad, L.J. and H.E. Bouis, 1991. The Impact of Nutritional Status on Agricultural Productivity: Wage Evidence from the Philippines. Oxford Bulletin of Economics and Statistics 53, 45-68.

Hausman, J., 1980. The Effect of Wages, Taxes, and Fixed Costs on Women's Labor Force Participation. Journal of Public Economics, 161-94.

Higgens P.A. and H. Alderman, 1993. Labor and Women's Nutrition: a Study of Energy Expenditure, Fertility, and Nutritional Status in Ghana. Policy Research Working Paper 1009, World Bank, Washington.

Jacoby, H.G. 1992. Productivity of Men and Women and the Sexual Division of Labor in Peasant Agriculture of the Peruvian Sierra. *Journal of Development Economics 37*, 265-87.

Jebuni, C.D., Asuming-brempong and K.Y. Fosu, 1990. Ghana Economic Recovery Programme and Agriculture, USAID, Accra.

Korjenek, P.A., 1992. The Relationship Between Consumption and Worker Productivity: Nutrition and Economic Approaches. Soc. Sci. Med. 35, 1103-13.

Kyereme, S.S. and E. Thorbecke, 1991. Factors Affecting Food Poverty in Ghana. *Journal of Development Studies 28*, 39-52.

Keyzer, M.A., 1993. Welfare Assessment of the Efficiency Wage Argument. Staff Working Paper 92-07, Centre for World Food Studies, Vrije Universiteit, Amsterdam.

Lavy, V., 1992. Investment in Human Capital: Schooling Supply Constraints in Rural Ghana. Working Paper 93, Living Standards Measurement Study, World Bank, Washington.

Martorell, R. and G. Arroyave, 1988. Malnutrition, Work Output and Energy Needs. In: K.j. Collins and D.f. Roberts Eds., Capacity for Work in the Tropics, Cambridge: Cambridge University Press.

Minhas, B.S., 1991. On Estimating the Inadequacy of Energy Intakes: Revealed Food Consumption Behaviour Versus Nutritional Norms (Nutritional Status of Indian People in 1983). *Journal of Development Studies 28*, 1-38.

Mcguire, J. And B.M. Popkin, 1990. Beating the Zero Sum Game: Women and Nutrition in the Third World. Discussion Paper 6, Undp Acc/subcommission on Nutrition.

Ministry of Food and Agriculture, 1990. Ghana Medium-term Agricultural Development Programme (Mtadp): An Agenda for Sustained Agricultural Growth and Development (1991-2000), Ppmed, Accra.

Ortega, J.M. and W.C. Rheinboldt, 1970. Iterative Solutions of Nonlinear Equations in Several Variables. New York: Academic Press.

Payne, P. and M. Lipton, 1994. How Third World Rural Households Adapt to Dietary Energy Stress. Ifpri, Washington.

Rosenzweig, M.R. and T.P. Schultz, 1983. Estimating a Household Production Function: Hetrogeneity, the Demand for Health Inputs and Their Effects on Birth Weight. *Journal of Political Economy 91*, 723-46.

Royston, J.P., 1982. An Extension of Shapiro and Wilk's W-test for Normality to Large Samples. *Applied Statistics* 31, 115-24.

Sahn, D.E. and H. Alderman, 1988. The Effects of Human Capital on Wages, and the Determinants of Labor Supply in a Developing Country. *Journal of Development Economics 29*, 157-83.

Scrimshaw, N.S. and V.R. Young, 1989. Adaption to Low Protein and Energy Intakes. Human Organisation 48, 20-30.

Seini, W.A., V.K. Nyanteng and G.J.M. Van Den Boom, 1997. Income and Expenditure Profiles and Poverty in Ghana. In: W.K. Asenso-Okyere, G. Benneh and W. Tims (eds.) Sustainable Food Security in West Africa, Kluwer Academic Publishers, Chapter 4.

Shetty, P.S. and W.P.T. James, 1994. Body Mass Index: a Measure of Chronic Energy Deficiency in Adults. Food and Nutrition Paper 56, Rome: Food and Agriculture Organization.

Spurr, G.B., 1984. Physical Activity, Nutritional Status, and Physical Work Capacity in Relation to Agricultural Productivity. In: E. Pollitt and P. Amante Eds., Energy Intake and Activity, New York: Alan R. Liss.

Stiglitz J.E., 1976. The Efficiency Wage Hypothesis, Surplus Labour, and the Distribution of Income in Ldc's. Oxford Economic Papers 28, 185-207.

Sukhatme, P.V. and S. Margen, 1982. Autoregulatory Homeostatic Nature of Energy Balance. *American Journal of Clinical Nutrition 35*, 355-65.

Strauss, J., 1986. Does Better Nutrition Raise Farm Productivity? *Journal of Political Economy 94*, 297-320.

Strauss, J. , 1993. The Impact of Improved Nutrition on Labor Productivity and Human-resources Development: An Economic Perspective. In: P. Pinstrup-Andersen ed., the Political Economy of Food and Nutrition Policies, Baltimore: John Hopkins University Press.

Strauss, J. And D. Thomas, 1993. Human Resources: Empirical Modeling of Household and Family Decisions. Forthcoming in T.N. Srinivasan and J.R. Behrman eds., Handbook of Development Economics Volume 3, Amsterdam: North-holland.

Svedberg, P. 1991. Poverty and Undernutrition in Sub-saharan Africa: Theory, Evidence, Policy. Monograph Series 19, Institute for International Economic Studies, Stockholm: Stockholm University.

UNDP, 1993. Human Development Report 1993. United Nations Development Programme, New York.

Waterlow, J.C., 1985. What Do We Mean by Adaptation? In: K. Blaxter and J.C. Waterlow eds., Nutritional Adaption in Man, London: Libbe.

World Bank, 1989. Ghana Population and Nutrition Sector Review. Report No. 7597-GH. The World Bank, Washington, D.C.

Section 5
TRADE AND FOOD SECURITY

CHAPTER 10

FARMERS' TRANSPORT COSTS AND AGRICULTURAL PRODUCTION IN GHANA

N. Heerink, P. Atsma and *K. Y. Fosu*

Introduction

Transport costs play an important role in agricultural development. High transport costs that result from inadequate transport infrastructure are to a large extent responsible for the large discrepancies that frequently exist between food prices paid by consumers in urban areas and the prices received by farmers producing these crops in low-income countries, particularly in Africa (see e.g. Ahmed and Rustagi, 1987). Reduction of transport costs may considerably increase the levels of agricultural production and food consumption.

Transport costs faced by African farmers do not only consist of direct or actual costs like fuel, vehicle hire, vehicle maintenance, and so on, but also of the labour and leisure time foregone. Agriculture in large parts of Africa is mainly a "footpath economy", where farmers spend much of their time headloading commodities from fields to home, from villages to markets. Human porterage poses massive constraints on rural labour supply, severely limits production potential at the farm level, and adds significantly to production and marketing costs (World Bank, 1993).

Studies examining the role of transport costs in economic development usually concentrate on "modern" transport networks and their effects on trade flows, profit margins of traders, and discrepancies between production and consumption prices. Studies on agricultural production sometimes include indices of transport infrastructure among the explanatory variables (see e.g. the review in Ahmed and Donovan, 1992, Ch. 5). But little attention has thus far been paid to rural household transport and the effects of transport costs faced by farmhouseholds on agricultural production decisions (see also Bryceson and Howe 1993, Fosu *et al.* 1995a and 1995b, Hine 1993, Hine *et al.*, 1983a and 1983b).

In analysing the effect of farmers' transport costs on production, it is important to make a distinction between the monetary costs and the costs in terms of time spent on transport activities. Costs incurred on transporting output and inputs are an important element of farmers' incomes. Transport time, however, does not directly affect monetary income. By reducing leisure time and posing serious health risks, it does affect farmers' incomes and welfare indirectly. In addition, agricultural production is likely to be affected favourably by reduction of transport time and health improvement.

The purpose of this chapter is to present an empirical analysis of the effects of farmers' transport costs on agricultural production in Ghana. Rural regions in Ghana have a low feeder road density (only 89 m/km²). Most feeder roads are poorly maintained and unusable during the rainy season. Human porterage therefore plays a very important role in the transport of agricultural produce at the village level (World Bank, 1993, p. 35). Information on costs and time involved in transporting agricultural output has been collected in a nationwide Agricultural Economic Survey that was conducted in Ghana in 1987. This information is used to carry out an empirical analysis of the impact of transport costs and time on agricultural production.

The chapter is structured as follows: The next section presents a classification of transport costs involved in agricultural production and input use. In the section headed 'Production Theory and Transport Costs', a procedure for modifying neo-classical theory of agricultural production to include the effects of changes in transport costs incurred by farmhouseholds is outlined. The resulting model is estimated empirically on the basis of the aforementioned survey data for Ghana. Results for cassava and maize, the two major crops in Ghana, are presented in the subheading 'Empirical Results for Ghana'. Conclusions are then presented.

A Classification of Transport Costs

As in other economic sectors, a production process in agriculture can be viewed as the conversion of inputs into an output. The consumption of an output and the production of inputs usually take place by different economic agents at different geographical locations. Transport of inputs and outputs is therefore of direct importance to agricultural production.

A schematic representation of transport activities with regard to agricultural inputs and outputs is given in Figure 10.1.[1] External inputs, i.e. inputs not produced or provided by the agricultural households themselves, have to be transported from the place where they are produced (or imported), to the different sales outlets spread throughout the country (line 1). Purchase of external inputs by the agricultural households takes place at domestic input markets. Usually, the inputs are transported by the farmer from the input market to the farmhouse (line 2). Once they are needed, the inputs (whether they are bought or home-produced) are transported from home to the different fields (line 3). When the crop is full-grown, the harvest is transported in the opposite direction, i.e. from the fields to the farmhouse (line 4). Part of the harvest may be consumed by the farmhousehold. The surplus is transported to output markets (line 5). Finally, the output sold is transported to the domestic consumer or exported (line 6). Output may be sold (and inputs bought) at the farm gate, in which case the market and the farm coincide and lines 2 and 5 are absent.

The market prices that farmers pay for inputs and receive for their outputs depend on the transportation costs made by traders of inputs and outputs (lines 1 and 6). For the individual farmer, however, these market prices may be considered as given. Most of the farmers tend not to sell directly to consumers, and the proportion of farmers who export is much smaller than the proportion of those who do not in Ghana. In deciding how much to produce and sell and the amount of various inputs to use, a farmhousehold will consider the transport costs which the household has to bear itself. The analysis presented therefore concentrates on transport costs related to lines 2 to 5. Changes in these transport costs will directly affect agricultural production costs, and hence production decisions of farm households.

Production Theory and Transport Costs

The effects on agricultural production of transportation costs incurred by farmers can be analysed by incorporating transport costs into standard neo-classical theory of agricultural production. This section examines how this can be done. The Cobb-Douglas production function will be used as an example to illustrate the analysis.

The basic unit of analysis is the agricultural household. Agricultural households are

---

[1] In reality, the marketing system of agricultural inputs and output may be more complex than Figure 10.1 suggests. In particular, different levels of wholesale or retail may exist in a marketing chain. The input and output markets in this figure represent only the first level in the marketing channel.

concerned with questions such as how much labour to devote to the cultivation of each crop, whether or not to use purchased inputs, which crops to grow in which fields, and so on. Only one goal, that of short term profit maximisation, is usually explored. Other important assumptions include perfect competition in the markets for farm inputs and outputs, and unlimited working capital for the purchase of variable inputs.

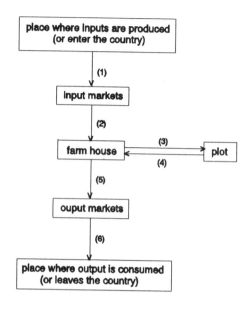

Figure 10.1: Transport Activities in Agricultural Production

The *agricultural production function* is determined by the local conditions such as soil type and climate and by the techniques used. In theory, the function includes quantities of all resources which may influence production. Inputs are rates of resource use and output is a rate of production per unit of time. The period of time to which the production function applies must be short enough to avoid changes in techniques which would alter the shape of the function, yet long enough to include the whole production process. In the remainder of this chapter we will consider annual crops only, because the two (most important) food commodities considered in the empirical part of this chapter are annual crops. The relevant time period for annual crops is a single cropping season. The agricultural production function can be written as:

$$Q_0 = f(X_1 \dots X_n, Z_1 \dots Z_m) \tag{10.1}$$

where

Q_0	= The given output in a given time period
$X_1 .. X_n$	= Variable inputs 1 to n
$Z_1 \dots Z_m$	= Fixed inputs 1 to m

Variable inputs are those which change with the volume of output over the period of time to which the production function applies (e.g. fertilizer, seeds, pesticides, etc.). Fixed inputs are inputs which cannot be changed by the agricultural households during the period to which the production function applies (e.g. land, bullocks, tools, machinery, and buildings). To make economic sense the response of output to increasing levels of variable input(s) must be rising but at a decreasing rate:

$$\frac{\partial Q_0}{\partial X_i} > 0 \; ; \quad \frac{\partial^2 Q_0}{\partial X_i^2} < 0 \quad \text{for } i = 1,...,n \tag{10.2}$$

The *demand for variable inputs* by agricultural households depends on the prices of inputs and the output price. For any variable input in farm production, the optimum level of its use occurs when the extra return just equals the extra costs per unit. Transport activities affect the costs of obtaining inputs and selling output, and hence influence decisions with regard to production level and input use. Not only input and output prices matter in farmers' production decisions, but also the transport costs that farmers incur in obtaining inputs, collecting output, and selling surpluses. With respect to inputs, the costs involved in transporting an additional unit of input (lines 2 and 3 of Figure 10.1) should be added to the market price of input paid by the farmer. The costs of transporting an additional unit of output (lines 3 and 4) should be subtracted from the market price of output. The transport costs from the plots to the home (line 3) applies to the whole production quantity. The transport costs from and to the market applies to the marketed surplus only. Hence, the level of variable input use is optimal for input i (i = 1,...,n) when the following condition is satisfied:

$$\frac{\partial Q_0}{\partial X_i} \; P_{q_0}^* = P_{x_i}^* \tag{10.3}$$

where

$$P_{x_i}^* = P_{x_i} + P_{t_{2i}} + P_{t_{3i}} \tag{10.4}$$

$$P_{q_0}^* = P_{q_0} - (P_{t_4} + P_{t_5}) \tag{10.5}$$

and

$P_{q_0}^*$ = Price per unit of a given output, adjusted for farmers' transport cost

$P_{x_i}^*$ = Price per unit of variable input i, adjusted for farmers' transport cost

P_{q_0} = Market price per unit of a given output

P_{x_i} = Market price per unit of variable input i

$P_{t_{2i}}$ = Transport cost from input market to farm house of (an additional unit of) variable input i (line 2)

$P_{t_{3i}}$ = Transport cost from farm house to plot of (an additional unit of) variable input i (line 3)

P_{t_4} = Transport cost of (an additional unit of) output from plot to farm house (line 4)

P_{t_5} = Transport cost of (an additional unit of) marketed surplus from farm house to output market (line 5)

The quantity of variable inputs used in the production process thus depends on the adjusted output price and the adjusted prices of variable inputs. The *supply curve* can be derived through substitution of the demand functions of variable inputs into the production function. This gives

$$S_0 = h(P_{q_0}^*, P_{x_1}^* ... P_{x_n}^*, Z_1 ... Z_m)$$
(10.6)

In agricultural economics much use is made of the Cobb-Douglas production function, because it is simple, and easy to manipulate and to interpret. It can be written as:

$$Q_0 = \alpha X_1^{\beta_1} ... X_n^{\beta_n} Z_1^{\gamma_1} ... Z_m^{\gamma_m}$$
(10.7)

$$\log Q_0 = \log \alpha + \sum_{i=1}^{n} \beta_i \log X_i + \sum_{j=1}^{m} \gamma_j \log Z_j$$
(10.8)

where

α, β_i, γ_j = Unknown coefficients

Application of condition (10.2) gives:

$$\alpha \beta_k X_k^{\beta_k - 1} \prod_{\substack{i=1 \\ i \neq k}}^{n} X_i^{\beta_i} \prod_{j=1}^{m} Z_j^{\gamma_j} = \frac{P_{x_k}^*}{P_{q_0}^*}$$
(10.9)

or:

$$\log(\alpha \beta_k) + (\beta_k - 1) \log X_k + \sum_{\substack{i=1 \\ i \neq k}}^{n} \beta_i \log X_i + \sum_{j=1}^{m} \gamma_j \log Z_j = \log P_{x_k}^* - \log P_{q_0}^*$$

Rearranging this equation gives the demand function for input k:

$$\log X_k = c_0 + \sum_{\substack{i=1 \\ j \neq k}}^{n} c_{1i} \log X_i + \sum_{j=1}^{m} c_{2j} \log Z_j + c_3 \log P_{q_0}^* - c_3 \log P_{x_k}^* \qquad (10.11)$$

where $c_0 . c_3$ are coefficients equal to:

$$c_0 = \frac{1}{-\beta_k + 1} \log(\alpha \, \beta_k)$$

$$c_{1i} = \frac{\beta_i}{-\beta_k + 1}$$

$$c_{2j} = \frac{\gamma_j}{-\beta_k + 1}$$

$$c_3 = \frac{1}{-\beta_k + 1}$$

The effect of transport costs can be separated from the unadjusted price effect as follows:

$$\log X_k = c_0 + \sum_{\substack{i=1 \\ j \neq k}}^{n} c_{1i} \log X_i + \sum_{j=1}^{m} c_{2j} \log Z_j + c_3 \log P_{q_0} - c_3 \log P_{x_k}$$

$$+ c_3 \log\left(\frac{P_{q_0} - (P_{t_4} + P_{t_5})}{P_{q_0}}\right) - c_3 \log\left(\frac{P_{x_k} + P_{t_{2k}} + P_{t_{3k}}}{P_{x_k}}\right) \qquad (10.12)$$

Substitution of (10.11) or (10.12) into equation (10.8) gives the supply equation as a function of adjusted output price and the adjusted price of input k. By repeating this procedure for all inputs, a system of n equations can be obtained from which a reduced form system can be derived containing only (adjusted) prices and fixed input quantities as right hand variables. For the case of one input (i=1), which is the case used in the empirical analysis in the next section, the supply equation equals:

$$\log Q_0 = \log \alpha + \beta_1 c_0 + \sum_{j=1}^{m} (\beta_1 c_{2j} + \gamma_j) \log Z_j + \beta_1 c_3 \log P_{q_0}^* - \beta_1 c_3 \log P_{x_1}^* \quad (10.13)$$

or

$$\log Q_0 = \log \alpha + \beta_1 c_0 + \sum_{j=1}^{m} (\beta_1 c_{2j} + \gamma_j) \log Z_j + \beta_1 c_3 \log P_{q_0} - \beta_1 c_3 \log P_{x_1}$$

$$+ \beta_1 c_3 \log\left(\frac{P_{q_0} - (P_{t_4} + P_{t_5})}{P_{q_0}}\right) - \beta_1 c_3 \log\left(\frac{P_{x_1} + P_{t_{2_1}} + P_{t_{3_1}}}{P_{x_1}}\right) \qquad (10.14)$$

The supply elasticities with respect to prices and transport costs are as follows:

$$\eta_{q_0} = \beta_1 c_3 \frac{P_{q_0}}{P_{q_0} - (P_{t_4} + P_{t_5})}$$

$$\eta_{x_1} = -\beta_1 c_3 \frac{P_{x_1}}{P_{x_1} + P_{t_{2_1}} + P_{t_{3_1}}}$$

$$\eta_{t_{45}} = -\beta_1 c_3 \frac{P_{t_4} + P_{t_5}}{P_{q_0} - (P_{t_4} + P_{t_5})} = -\eta_{q_0} \frac{P_{t_4} + P_{t_5}}{P_{q_0}} \qquad (10.15)$$

$$\eta_{t_{23}} = -\beta_1 c_3 \frac{P_{t_{2_1}} + P_{t_{3_1}}}{P_{x_1} + P_{t_{2_1}} + P_{t_{3_1}}} = \eta_{x_1} \frac{P_{t_{2_1}} + P_{t_{3_1}}}{P_{x_1}}$$

where

η_{q_0} = Price elasticity of output

η_{x_1} = Elasticity of output with respect to variable input price

$\eta_{t_{45}}$ = Elasticity of output with respect to transport cost of output

$\eta_{t_{23}}$ = Elasticity of output with respect to transport cost of variable input

Information on market prices and transport costs made by farmers can be used to calculate the relevant prices. The costs involved in transporting output and input(s) on foot can be approximated by multiplying the time involved in transportation by the prevailing market wage rate. In this case, the market wage rate serves as a shadow price of time owned by the farm household (see e.g. Singh et al., 1986).

It should be noted that the coefficients of the market price variables and the variables representing the transport cost effect are the same in equation (10.14). For a profit maximising farm household, the impact of an increase in transport cost (per unit output) is similar to the impact of an output price decline of the same (absolute) size. This may not be the case for a decrease in the time spent on transportation, even when the market wage rate is used as the shadow price of transport time. A reduction of the time spent on transportation does not have a direct effect on profits. It does, however, increase the time available for agricultural production and it may also improve farmers' health and hence agricultural productivity. In the case of a staple food crop, it may even happen that crop production responds inversely to income increases resulting from an increase in market price or a decrease in transport costs (negative supply response), but positively to a decrease in the time spent on transporting the output. In empirical applications, it is therefore desirable to distinguish a separate variable that measures the effect of time involved in transportation on agricultural output.

EMPIRICAL RESULTS FOR GHANA

Description of the Data Set Used

The effect of farmers' transport costs on agricultural production was tested empirically using the results of a farm household survey which was conducted in Ghana in 1987/1988. This survey includes questions concerning the costs and time involved in transporting output. Unfortunately, no questions were asked on transport costs involved in buying external inputs.

The Agricultural Economic Survey of Ghana was conducted between late 1987 and early 1988 by the Ghana Cocoa Board. Random probability sampling was used to select and interview 701 agricultural households from all regions in the country. The 701 farmers are engaged in 3119 different crop growing activities, implying an average of 4.45 crops per household. Cassava and maize are the two crops that are grown most frequently (see Table 10.1). These two crops are therefore chosen for our analysis. The first crop is the major starchy staple in Ghana, whereas the second is the major cereal in Ghana.

Besides questions on crop production, prices, land and labour availability, and input use, the questionnaire also contains questions on transport of output from farm to the farmer's home and from home to market. With regard to transport from home to market, respondents were asked to state the man-hours involved if carried on foot or by cart, and to state the cost involved if carried by truck or tractor. The question on transport from farm to home only asked how long it takes in man-hours to carry out the transport. Although rural household transport in Ghana, like elsewhere in Africa, is mainly a task of women (see e.g. Bryceson and Howe, 1993), no questions were asked on the gender of the person carrying out the transport. Table 10.2 summarises the transport costs for cassava and maize for the households that are included in the regression analysis. Only households selling (part of) the output on a market are selected (see below). The average time needed to transport cassava from the field to the farmer's home is

TABLE 10.1: Distribution of Crop Growing Activities in Ghana, 1987-88 (701 Households)

Annual Crops	% of Total Activities	Perennial Crops	% of Total Activities
Root crops	36.5	Tree crops	25.8
Cassava	17.6	Plantains	11.0
Cocoyam	7.1	Cocoa hybrid	5.8
Yam	6.9	Oil palm	3.8
Groundnut/peanut	4.6	Cocoa traditional	12.7
		Oranges	0.9
Cereals	27.2	Coconut	0.6
Maize	16.2	Bananas	0.5
Sorghum/guinea corn	5.2	Other	0.4
Millet	3.6		
Rice		Other crops	2.0
		Sugar cane	0.8
Vegetables	8.5	Pine apple	1.2
Beans/peas	3.9	Tobacco	0.1
Pepper	1.7	Other	0.9
Tomato	1.4		
Okra	0.9		
Garden egg	0.3		
Other	0.3		

Source: Calculated from Agricultural Economic Survey of Ghana

more than 2 hours for 100 kg of cassava, and more than 5 hours for 100 kg of maize. Average production of cassava is 2891 kg for the cassava-producing households in the sample (see Table 10.3). This means that these households need about 66 hours to carry the entire cassava production to the farmhouse. Average maize production is 611 kg for the maize-producing households in the sample (see Table 10.6), implying that farm households need about 36 hours on average to carry home the total production of maize. One can infer from Table 10.2 that about 40 percent of the farmers carry their produce from the home to the market on foot or by cart, while one third of the cassava farmers and one half of the maize farmers use a tractor or truck. The costs involved are about 50 percent higher for maize than for cassava. The average selling price per kg of maize is more than three times that of cassava. As a consequence, total farmers' transport costs constitute about 11 percent of the selling price for cassava and around 5.5 percent of the selling price for maize.

Regression Results for Cassava

The sample used for the regressions explaining cassava production consists of 165 households. These households:

— plant and harvest cassava,

— sell (part of) the produced quantity,

— use cassava cuttings, measured in bundles, as an input (most households which grow cassava use cuttings as the only variable input),

— provide information on transport costs of cassava to a market.

TABLE 10.2:   Average Costs (Cedis Per 100 Kg) of Transporting Output, Ghana 1987-88

	Farm to home	Home to market
Cassava (Average price per 100 kg: 1539 cedis)		
Man-hours	2.27 (n=165)	2.89 (n=67)
Man-hours x wage rate (cedis)	82.90 (n=165)	110.90 (n=67)
Costs truck or tractor (cedis)	-	119.50 (n=56)
No costs	-	0.00 (n=42)
Average costs (cedis)	82.90 (n=165)	85.60 (n=165)
Maize (Average price per 100 kg: 4859 cedis)		
Man-hours	5.18 (n=136)	1.83 (n=55)
Man-hours x wage rate (cedis)	158.80 (n=136)	50.30 (n=55)
Costs truck or tractor (cedis)	-	176.00 (n=71)
No costs	0.00 (n=3)	0.00 (n=13)
Average costs (cedis)	155.40 (n=139)	109.80 (n=139)

Source: Calculated from Agricultural Economics Survey of Ghana

n   denotes the number of farm households in sample making this type of transport costs

Households with missing values for one or more of the relevant variables were excluded. In addition, 16 households were excluded because of unrealistic values for one or more of the relevant variables.[2]

In order to estimate the supply response of cassava production to farmers' transport cost, the following regression equations are used:

$$\log(QC_i) = c_{1c} + c_{2c}\log(PRC_i) + c_{3c}\log(PTC_i) + c_{4c}\log(LC_i) + c_{5c}\log(AC_i) + c_{6c}D1 + u_{1i}$$
$$\text{for } i = 1,...,165$$

$$\log(MC_i) = c_{7c} + c_{8c}\log(PRC_i) + c_{9c}\log(PTC_i) + c_{10c}\log(LC_i) + c_{11c}\log(AC_i) + c_{12c}D1 + u_{2i}$$
$$\text{for } i = 1,...,165$$

QC_i	=	Quantity of cassava produced by household i (in kg)
MC_i	=	Quantity of cassava marketed by household i (in kg)
PRC_i	=	Ratio of farmgate price of cassava (PQ) to input price (PX); defined as ratio of average selling price of cassava (cedis/kg) minus costs involved in transporting cassava from home to market by truck or tractor (cedis/kg) to the average price paid for cassava cuttings (cedis/bundle) for household i
PTC_i	=	Ratio of farmgate price corrected for transport time ($PQ - P_cT$) to uncorrected farmgate price (PQ_c); defined as average selling price of cassava minus costs involved in transporting cassava from home to market by truck or tractor minus transport costs (man-hours multiplied by wage rate[3]) from farm to home per kg cassava produced, minus transport costs (man-hours multiplied by wage rate) of cassava carried on foot or by cart from home to market (cedis/kg), divided by average selling price of cassava minus costs involved in transporting cassava from home to market by truck or tractor for household i.
Lc_i	=	Amount of labour spent on harvesting (uprooting and cutting) cassava by household i (in man-hours)
Ac_i	=	Harvested area of cassava for household i (in acres)
$D1$	=	Dummy variable; equals one for households in coastal zone, zero otherwise
$c_{1c}...c_{12c}$	=	Unknown coefficients
u_{1i}, u_{2i}	=	Random disturbance terms with standard properties

Transport costs, whether they are made from the field to the farmhouse or from the farmhouse to the market, affect total production of a crop as well as the marketed quantity of that crop. We do not want to go into farmhousehold decisions with respect to home consumption and marketing of crop production. Both the produced quantity and the marketed surplus of cassava are therefore used as dependent variables in the model.

Because the effect of price changes may differ from the effect of changes in transport time, the farmgate price is separated from output transport time in the way suggested in the previous section. The coefficient of the farmgate price, however, should in principle be equal to the coefficient of the input price. By dividing the farmgate price by the input price, this restriction is imposed on the model. The coefficient of the resulting price ratio, $\log(PRC_i)$, may be either positive or negative. In the latter case, the supply curve is 'backward-sloping' (see e.g. Levi and Havinden, 1982). The coefficient of the transport time variable, $\log(PTC_i)$, is expected to be

---

[2] These 'outliers' concern households with a yield of more than 11,200 kg per acre (the maximum yield achieved in some isolated cases in Ghana; see Ministry of Agriculture, 1991: Table 4.1.8), with a cassava price of more than 89,358 cedis per kg (three times the average rural wholesale price in 1987), with a price paid for seedlings that exceeds 1,000 cedis per kg, or with average transport costs that exceed the price received for cassava.

[3] Average wage rate of the administrative region to which the household belongs; computed from Ghana Living Standards Survey 1987-1988.

positive.

The labour variable, LC_i, measures the man-hours involved in harvesting cassava. The questionnaire also includes questions on the time involved in land preparation and crop care, but these questions do not make a distinction between crops. No questions are asked on wages paid for hired labour, nor is there any information available on shadow wages for family labour at the farm level. For that reason, labour is modelled as a (semi-)fixed input instead of a variable input. The second (semi-)fixed input in the model is the land area used for cassava production, AC_i. Positive signs are expected for both variables.

Finally, a dummy variable for households living in the coastal zone has been added to the regression equation. No weather data are provided by the survey. For that reason, the impact of agro-climatic factors is approximated by distinguishing the three major agro-ecological zones in Ghana, i.e. the coastal zone, the forest zone, and the savannah zone. Since cassava is not grown in the savannah zone, only one dummy variable (for the coastal zone) is added to the model.

TABLE 10.3: Means And Standard Deviations of Variables
used in Regressions for Cassava

		Mean	Standard deviation
QC:	Cassava production (in kg)	2891	3148
MC:	Cassava marketed (in kg)	1577	2310
PRC:	Ratio of farmgate price to input price	0.096	0.102
PTC:	Ratio of farmgate price corrected for transport time to farmgate price	0.906	0.133
LC:	Labour spent on harvesting (man-hours)	37.5	40.9
AC:	Harvested area (in acres)	2.67	2.16
D1:	Dummy for coastal zone	0.42	0.49

Source: Calculated from Agricultural Economic Survey of Ghana

The average cassava production of the households in the sample is almost 3,000 kg. More than half of this production is sold at a market. Furthermore, Table 10.3 indicates that about 37.5 hours per year is spent on the average on harvesting cassava, while the average harvested area of cassava farmers producing for the market equals 2.67 acres. About 42 percent of the cassava farmers in the sample live in the coastal zone. The regression results are presented in Table 10.4. The two price ratios are significant at the 5 percent testing level. The ratio of output to input price has a negative sign, while the variable measuring the effect of transport time has a positive sign. The amount of labour spent on harvesting cassava has a significant positive effect on production and marketed quantity, but the effect of harvested area does not differ significantly from zero.[4] Finally, the dummy variable for households in the coastal region is at the edge of significance. The estimated coefficient indicates that cassava production in the

---

[4] Similar results are obtained for the variable AC (harvested area of cassava) in two additional regressions of which the results are not shown here: (1) a production function of cassava with cassava cuttings, labour, and harvested area as explanatory variables, and (2) a supply function with the transport time variable excluded. A possible explanation for this unexpected result is the presence of a high correlation between log(AC) and log(LC). The correlation coefficient of these two variables equals 0.61, which is by far the highest correlation coefficient of any pair of explanatory variables in the equation.

coastal zone is on average about ten percent higher than the forest zone when other factors are controlled for. The R-squared is of a satisfactory level for both equations, taking into account that the regressions are based on a micro-level, cross-section data set. The values of the F-statistic indicate that the variables included in the regression equations have a significant joint influence on the dependent variables.

TABLE 10.4: Regression Results for Cassava Supply Function

EXPLANATORY VARIABLES	DEPENDENT VARIABLES	
	log(QC)	log(MC)
log(PRC)	-0.35	-0.47
	(-5.03)	(-5.25)
log(PTC)	1.61	1.47
	(5.82)	(4.09)
log(LC)	0.48	0.40
	(5.80)	(3.77)
log(AC)	0.06	0.05
	(0.51)	(0.33)
D1	0.09	0.12
	(1.55)	(1.54)
R^2	0.43	0.31
F-value	23.70	14.11

Note: The numbers in parentheses are t-values. An intercept term was included in the regressions.

From these results, the elasticities of production (marketed surplus) with respect to farmgate price, input price, and transport time can be obtained as follows:

$$
\begin{aligned}
\eta_{qp_c} &= c_{2c} + c_{3c}\frac{PT_c}{PQ_c - PT_c} \\
\eta_{mp_c} &= c_{8c} + c_{9c}\frac{PT_c}{PQ_c - PT_c} \\
\eta_{qx_c} &= -c_{2c} \\
\eta_{mx_c} &= -c_{8c} \\
\eta_{qt_c} &= -c_{3c}\frac{PT_c}{PQ_c - PT_c} = -\eta_{qp_c} - \eta_{qx_c} \\
\eta_{mt_c} &= -c_{9c}\frac{PT_c}{PQ_c - PT_c} = -\eta_{mp_c} - \eta_{mx_c}
\end{aligned}
\qquad (10.16)
$$

where

η_{qp_c} = Elasticity of cassava production with respect to farmgate price
η_{mp_c} = Elasticity of marketed quantity of cassava with respect to farmgate price
η_{qr_c} = Elasticity of cassava production with respect to variable input price
η_{mr_c} = Elasticity of marketed quantity of cassava with respect to variable input price
η_{qt_c} = Elasticity of cassava production with respect to transport time cost
η_{mt_c} = Elasticity of marketed quantity of cassava with respect to transport time cost
PQ_c = Farmgate price of cassava ; defined as average selling price of cassava minus costs involved in transporting cassava from home to market by truck or tractor (in cedis per kg)
PT_c = Costs (man–hours multiplied by wage rate) of transporting cassava from home to market by foot or cart (in cedis per kg)

The resulting elasticities, calculated at sample means, are presented in Table 10.5. Both production and marketed quantity respond negatively to an increase in the farmgate price of cassava and positively to an increase in input price. A decline in farmers' transport cost, however, has a positive effect on cassava production. The estimated elasticities indicate that a decline of transport time by seven percent will increase cassava production (and marketed quantity) by one percent.

TABLE 10. 5:   Estimated Elasticities for Production and Marketed
Quantity of Cassava

Elasticities	Production	Marketed quantity
Farmgate price	-0.20	-0.33
Input price	0.35	0.47
Transport time cost	-0.15	-0.14

In interpreting these results, it should be taken into account that the average transport time cost is only about one tenth of the average farmgate price. It may therefore be interesting to consider the response of cassava supply to absolute price changes, in addition to the response to percentage changes, as well. The estimated elasticities indicate that when the average farm-gate price increases by 0.50 cedis (from 14.98 to 15.48 cedis) per kg, average cassava production would decline from 2,891 kg to 2,872 kg per farm household. When, on the other hand, average transport costs decline by 0.50 cedis per kg (which is about 40 percent of the transport time cost), average cassava production will increase to around 3,060 kg per household.

Regression Results for Maize

The sample used for the maize regressions contains 139 households. These households:
— plant and harvest maize,

246

— sell (part of) the produced quantity,

— use maize seed as an input (most households that grow maize use seed as the only variable input),

— provide information on transport costs of maize to a market.

Households with missing values for one or more of the relevant variables were excluded. In addition, 10 households were excluded because of unrealistic values for one or more of the relevant variables.[5]

The regression equations used to estimate the supply response of maize production to farmers' transport cost are similar to the ones used for cassava:

$$\log(QM_i) = c_{1m} + c_{2m}\log(PRM_i) + c_{3m}\log(PTM_i) + c_{4m}\log(LM_i) + c_{5m}\log(AM_i) + c_{6m}D1 + c_{7m}D2 + v_{1i} \qquad \text{for } i = 1,..,165$$

$$\log(MM_i) = c_{8m} + c_{9m}\log(PRM_i) + c_{10m}\log(PTM_i) + c_{11m}\log(LM_i) + c_{12m}\log(AM_i) + c_{13m}D1 + c_{14}D2 + v_{2i} \qquad \text{for } i = 1,...,165$$

where

Q_{mi}	=	Quantity of maize produced by household i (in kg)
M_{mi}	=	Quantity of maize marketed by household i (in kg)
PR_{mi}	=	Ratio of farmgate price of maize (PQ_m) to input price (PX_m); defined as ratio of average selling price of maize (cedis/kg) minus costs involved in transporting maize from home to market by truck or tractor (cedis/kg) to the average price paid for maize seed (cedis/kg) for household i
PTM_i	=	Ratio of farmgate price corrected for transport time ($PQ_m - PT_m$) to uncorrected farmgate price (PQ_m); defined as average selling price of maize minus costs involved in transporting maize from home to market by truck or tractor minus transport costs (man-hours multiplied by wage rate[6]) from farm to home per kg maize produced, minus transport costs (man-hours multiplied by wage rate) of maize carried on foot or by cart from home to market (cedis/kg), divided by average selling price of maize minus costs involved in transporting maize from home to market by truck or tractor for household i.
LM_i	=	Amount of labour spent on cutting (or picking) and shelling maize by household i (in man-hours)
AM_i	=	Harvested area of maize for household i (in acres)
$D1$	=	Dummy variable; equals one for households in coastal zone, zero otherwise
$D2$	=	Dummy variable; equals one for households in forest zone, zero otherwise
$c_{1c}..c_{14}$	=	Unknown coefficients
v_{1i}, v_{2i}	=	Random disturbance terms with standard properties.

---

[5] These 'outliers' concern households with a yield of more than 2,000 kg per acre (the maximum yield achieved in some isolated cases in Ghana; see Ministry of Agriculture 1991: Table 4.1.8), with a maize price of more than 157.749 cedis per kg (three times the average rural wholesale price in 1987), with a price paid for maize seed exceeding 1,000 cedis per kg, or with average transport costs that exceed the price received for maize.

[6] Average wage rate of the administrative region to which the household belongs; computed from Ghana Living Standards Survey 1987-1988.

The definitions of the two price ratios are consistent with the specifications used for the cassava equations. The labour variable differs slightly from the one used for the cassava regressions. The question on labour operations in the survey provides information on the man-hours involved in harvesting as well as shelling of maize. The labour input variable measures the time input for both these activities. Furthermore, an extra dummy variable which equals one for households living in the forest zone is added to the equations, because maize is grown in all three agro-ecological zones.

Average maize production of the households in the sample is more than 600 kg (Table 10.6). About 56 percent of this production is sold at a market. The farmgate price of maize is about 18 percent higher than the price (per kg) which farmers paid for obtaining maize seed. Furthermore, the table indicates that more than 50 hours per year is spent on the average on harvesting and shelling maize. The average harvested area of maize farmers who produce for the market equals 2.24 acres, which is 0.43 acres less than the harvested area of cassava. About 31 percent of the maize farmers in the sample live in the coastal zone, while about 34 percent live in the forest zone.

The regression results are presented in Table 10.7. The ratio of output to input price does not have a significant effect on production and marketed quantity of maize. The transport time variable, however, has a significant positive effect in both equations. This result again demonstrates the importance of the time farmers need to transport their produce in explaining agricultural output in Ghana. Harvested area also has a significant positive effect on production and marketed quantity. Labour input has a significant impact on the marketed quantity of maize only.[7] Finally, the results for the dummy variables indicate that maize production in the savannah region is about 15 percent higher than it is in the coastal region, and about 10 percent higher than maize production in the forest region (when other factors remain fixed). The values of the R-squared and the F-statistic are comparable to those obtained for the cassava equations.

The corresponding elasticities are presented in Table 10.8. The elasticity of production (and marketed quantity) with respect to farmgate price is positive, but small. As in the case of cassava, the time involved in transporting output is negatively related to production and marketed quantity. The estimated elasticities are smaller than the elasticities that were obtained for cassava.

As in the case of cassava, it should be remembered in interpreting these results that the average farmgate price is more than 25 times the average transport time cost. When the average farmgate price of maize increases by 0.50 cedis (from 47.70 to 48.20 cedis) per kg, the estimation results indicate that the average maize production will increase by 0.2 kg, on a total average household production of 611 kg. only. The same absolute decline of transport time cost per kg (from 1.75 to 1.25 cedis per kg) would increase average maize production from 611 to 618 kg per household.

---

[7] The low significance of the estimated coefficient for labour input in the equation for maize production may be caused by the high correlation between harvested area, log(AM), and labour input, log(LM). The correlation coefficient beween these two variables equals 0.67, which is the highest correlation coefficient of all pairs of explanatory variables in the equation.

TABLE 10. 6: Means and Standard Deviations of Variables
Used in Regressions for Maize

		Mean	Standard Deviation
QM:	Maize production (in kg)	611	611
MM:	Maize marketed (in kg)	344	457
PRM:	Ratio of farmgate price to input price	1.18	0.94
PTM:	Ratio of farmgate price corrected for transport time to farmgate price	0.95	0.12
LM:	Labour spent on harvesting and shelling (man-hours)	51.2	57.4
AM:	Harvested area (in acres)	2.24	1.89
D1:	Dummy for coastal zone	0.31	0.46
D2:	Dummy for forest zone	0.34	0.47

Source: Calculated from Agricultural Economic Survey of Ghana

TABLE 10. 7: Regression Results for Maize Supply Function

	DEPENDENT VARIABLES	
EXPLANATORY VARIABLES	log(QM)	log(MM)
log(PRM)	-0.01 (-0.15)	0.01 (0.06)
log(PTM)	1.10 (5.17)	1.13 (4.49)
log(LM)	0.09 (1.26)	0.18 (2.00)
log(AM)	0.61 (5.56)	0.55 (4.25)
D1	-0.15 (-2.53)	-0.16 (-2.42)
D2	-0.10 (-1.66)	-0.11 (-1.51)
R^2	0.43	0.37
F-value	16.32	12.84

Note: The numbers in parentheses below the parameter estimates are t-values. An intercept term was included in the regressions.

TABLE 10.8: Estimated Elasticities for Production and Marketed Quantity of Maize

Elasticities	Production	Marketed quantity
Farmgate price	0.03	0.05
Input price	0.01	-0.01
Transport time cost	-0.04	-0.04

Conclusion

Costs made by farm households in transporting agricultural ouput and inputs may have an important effect on agricultural production decisions, particularly in sub-Saharan Africa. Four different types of farmers' transport costs may be distinguished: Costs of transporting inputs to the farmhouse, costs of transporting inputs from the home to the field, costs of transporting output from the field to the farmhouse, and costs of transporting marketed surplus from the farmhouse to the output market. By modifying standard neo-classical production theory, the effects of these transport costs on agricultural production can be separated from market price effects.

Data from the Agricultural Economic Survey conducted in Ghana in 1987 show that farm households which sell part of their production at a market spent on the average more than 2 hours to transport 100 kg of cassava from the field to the farmhouse and more than 5 hours to do the same for maize. About 40 percent of the farmers carry their produce from the home to the market on foot or by cart. The average time involved in this transport is almost 3 hours per 100 kg cassava and almost 2 hours per 100 kg maize. Total farmers' transport costs are estimated to constitute about 11 percent of the selling price for cassava and 5.5 percent of the selling price for maize.

Results from a regression analysis based on these survey data show that farmers' transport time has a significant negative effect on the production and marketed quantity of cassava as well as maize. The estimated coefficients indicate that a seven percent reduction of farmers' transport time will increase cassava output by one percent. For maize, a reduction of 25 percent of transport time is needed to increase output by one percent. Furthermore, the results indicate that cassava has a negative supply response (price elasticity of supply equals -0.20), while maize has a small positive supply response to price changes (price elasticity of maize supply equals 0.04). When looking at the effects of absolute price changes, the results indicate that an output price increase of 0.50 cedis per kg (in 1987 prices) would decrease average cassava production by 19 kg (on a total production of 2891 kg) and increase average maize production by 0.2 kg (on a total production of 611 kg) only. A similar decline of 0.50 cedis of transport time cost per kg, on the other hand, would increase average cassava production by 170 kg and average maize production by 7 kg per household.

These results suggest that improvements in local transport infrastructure that reduce farmers' transport costs should complement price incentive policies in Ghana. Getting prices right only is not the best policy to promote agricultural production and food security. In the case of cassava, higher (relative) farmgate prices are even likely to result in lower production, while reductions in the time needed to transport cassava on foot or by cart are likely to stimulate cassava production in a significant way.

REFERENCES

Ahmed, R. and C. Donovan, 1992. *Issues of Infrastructure Development: A Synthesis of the Literature*. Washington, D.C.: IFPRI.

Ahmed, R. and N. Rustagi, 1987. Marketing and Price Incentives in African and Asian Countries: A Comparison. In: D. Elz (ed.) *Agricultural Marketing Strategy and Pricing Policy*. Washington, D.C.: World Bank.

Bryceson, D.F. and J. Howe, 1993. Rural Household Transport in Africa: Reducing the Burden on Women? *World Development* 21, pp. 1715-1728.

Fosu, K.Y., N. Heerink and E. Probee, 1995a. Extended Annotated Bibliography on Public Goods and Food Security in Ghana. Working Paper SADAOC/PGS Project.

Fosu, K.Y., N. Heerink, K.E. Ilboudo, M. Kuiper and A. Kuyvenhoven (1995b) Public Goods and Services and Food Security: Theory and Modelling Approaches. SADAOC/PGS Paper Presented at Réseau SADAOC Seminar, Accra, Ghana, 13-15 March 1995.

Ghana Cocoa Board, 1988. 1987 Agricultural Economic Survey of Ghana, Accra.

Hine, J.L., 1993. Transport and Marketing Priorities to Improve Food Security in Ghana and the Rest of Africa. In: H.-U. Thimm and H.Hahn (eds.) *Regional Fod Security and Rural Infrastructure*. Münster: LIT Verlag.

Hine, J.L., J.D. Riverson and E.A. Kwakye, 1983a. Accessibility and Agricultural Development in the Ashanti Region of Ghana. TRRL Report SR 791. Crowthorne: Transport and Road Research Laboratory.

Hine, J.L., J.D. Riverson and E.A. Kwakye 1983b. Accessibility, Transport Costs and Food Marketing in the Ashanti Region of Ghana. TRRL Report SR 809. Crowthorne: Transport and Road Research Laboratory.

Levi, J. and M. Havinden, 1982. *Economics of African Agriculture*. London, Longman.

Ministry of Agriculture, 1991. *Agriculture in Ghana: Facts and Figures*. Accra, Presbyterian Press.

Singh, I., L. Squire, and J. Strauss (eds.) 1986. *Agricultural Household Models: Extensions, Applications, and Policy*. Baltimore and London: Johns Hopkins University Press.

World Bank, 1993. *Ghana 2000 and Beyond: Setting the Stage for Accelerated Growth and Poverty Reduction*. World Bank (Africa Regional Office, West Africa Department).

CHAPTER 11

DISTORTED BEEF MARKETS AND REGIONAL LIVESTOCK TRADE IN WEST AFRICA: EXPERIENCE FROM THE CENTRAL WEST AFRICAN CORRIDOR

P. Q. van Ufford and *A. K. Bos*

Introduction

Within the agricultural sector of Sahelian countries the production of livestock is an important sub-sector. With the continuing integration of pastoralists in society, the importance of cattle sales as a source of income has increased. The resulting cash incomes enable these pastoralists to pay for cereals, school fees and taxes; the provision of efficient marketing channels contributes to the stabilisation of their incomes. The study of the regional cattle trade will likewise contribute to the debate on regional integration issues. Most research on regional integration to date has concentrated on the prospects for a regionally integrated cereal market, according to the Mindelo (1986) and Lomé (1989) conferences. However, the livestock sector represents a second potential source for market integration. One of the pertinent questions within this debate is: can the West African region achieve a higher level of self-provisioning with regard to beef production? During many years the national herd in Sahelian countries produced a surplus above national consumption. Traditionally, this surplus was used for export to the coastal countries like Cote d'Ivoire and Ghana, thereby competing successfully with beef imports from the world market.

Since the mid-1980s however, the regional cattle trade has been losing its competitiveness as a result of two factors that seriously distorted the markets of meat and livestock:

(a) the overvalued exchange rate of the CFA franc, the monetary unit used by the West and Central African countries of the francophone zone; and

(b) the heavy subsidies on beef exports from the European Community in the form of restitutions to beef exporters, which led to the 'dumping' of European beef on the West African market.

Paralleled with the 50% devaluation of the CFA franc of January 1994 the distortive effect of beef imports from the European Union (EU) diminished as a result of successive reductions of the export restitution rate by the EU. Therefore, it is important to find out to what extent the resulting reduction of market distortions will restore the competitiveness of cattle trade by Sahelian countries. The aim of this chapter is to consider the trade of cattle and meat in the central West African corridor in the light of the recent monetary adjustment combined with the reduced meat subsidies by the EU. We will concentrate on two main questions:

(a) To what extent can the inefficiency which is often attributed to the cattle trade between the countries of the West African central corridor, be considered as a constraint for its competitiveness?

(b) What are the short-term effects of the reduced market distortions? The emphasis will be on trade volumes, transaction costs and prices. The selection of 1993 and 1994 as years of reference will enable us to assess the direct impact of the devaluation of the CFA franc in combination with the reduction in EU subsidies.

LIVESTOCK TRADE IN THE CENTRAL WEST AFRICAN CORRIDOR[1]

Characteristics of the Trade System and Alternative Supply on Coastal Markets

In this section we will describe the functioning of the livestock trade and also present some quantitative data on cattle and meat flows in the central West African corridor. Marketing costs on two regional trade routes as well as the cost structure of imported, extra-regional meat will be examined.

The central West African corridor comprises five countries: Mali, Burkina Faso, Cote d'Ivoire, Ghana and Togo[2]. Livestock trade in this region has been an important economic activity already before the days of independence. Its history is marked by an erratic pattern of trade flows, but on the whole the number of animals traded has steadily grown until 1975. Information and figures on livestock trade before independence are inadequate. According to different authors (Hopkins, 1973; Arhin, 1979), networks of Mossi and Hausa traders brought cattle from the present Burkina Faso to markets in Ghana and Togo as early as the first part of the twentieth century. Cattle were bartered for various products like kola nuts, textiles and kauri shells. Together with the flourishing of notably the Ghanaian economy, the cattle trade of the region experienced significant growth. As a result, in the 1960s Ghana alone accounted for almost 50% of cattle exports from Burkina Faso (Josser, 1990).

For several reasons, including the collapse of the Ghanaian economy and the simultaneous growth of the Cote d'Ivoire economy, this trend was reversed from about 1970 onwards. Ever since, Cote d'Ivoire has been the principal customer for Burkinabe cattle.

In the central corridor the livestock trade consists of a three-stage pattern (Herman, 1983). In the first stage, several small and dispersed collection markets serve as assembly points for petty traders, brokers and cattle holders selling some of their animals. In our case these collection markets are situated in the Seno, Oudalan and Soum provinces in northern Burkina Faso. During the second stage cattle is conveyed to wholesale markets; the largest wholesale/export market in Burkina Faso is Pouytenga, situated in the middle of the country. Other wholesale markets are Fada N'Gourma, Kaya and Bobo-Dioulasso. The wholesale traders operating on these markets possess considerable financial resources which enable them to carry out export transactions requiring large amounts of private capital, in the absence of credit facilities. The third and final stage involves the transportation, usually by truck or by train, to consumer markets. In order of importance the main consumer markets in the central corridor are Abidjan (Cote d'Ivoire), Bouake (Cote d'Ivoire), Kumasi (Ghana) and Accra (Ghana) as well as Lome (Togo) and Ibadan/Lagos (Nigeria). In the 1990-1993 period Abidjan alone accounted for almost 80%-90% of all central corridor exports (MARA/SSA/DSAP, 1993, 1994). At the destination markets, cattle is usually sold to butchers through a fixed broker. In most cases, both

---

[1] Most data presented in this section were obtained during field work in Burkina Faso, Ghana and Cote d'Ivoire in 1993.

[2] This paper will concentrate mainly on Burkina Faso, Ghana and Cote d'Ivoire. As a consequence, Mali and Togo will largely be left out of the discussion.

butchers and brokers belong to the same ethnic group or country of origin as the trader (for example: Mossi traders sell to Mossi butchers). The broker not only provides the traders with food but could be an effective means of fostering regional trade which could be enhanced by removing some of the existing structural obstacles. Before continuing our analysis of the effects of the CFA franc devaluation and the reduction of EU subsidies, we will elaborate on the efficiency of regional livestock trade.

The impediments to regional trade that were identified included: high taxes as well as elaborate formalities with respect to exports; lack of a coherent regional trade policy; scarce credit facilities; a malfunctioning information system and a frequent incidence of illicit taxes (CILSS & CEBV, 1992). With respect to the removal of these barriers, some results have been achieved during recent years. Several taxes which hampered trade relations between Burkina Faso and Cote d'Ivoire have been lifted. Since the end of 1992, practically all import or export taxes relating to cattle have been removed in these two countries. Furthermore, the establishment of so-called 'societes de convoyage', which can be engaged to take care of all formalities en route, has achieved a reduction in the payments of illicit taxes on the Pouytenga – Abidjan itinerary.[3]

Finally, steps have been taken to set up a regional market information system and national 'cadres de concertation', consultative bodies in which traders and officials collaborate. Concerning trade between Burkina Faso and Ghana, no specific improvements were noted in 1993. Official Ghana statistics do not mention any cattle imports from Burkina Faso or Mali. An unrealistically high import tax as well as various formalities still restrain traders from using the official channels. Nevertheless, in 1993 an estimated 13,000 animals were imported illegally from Burkina Faso. An analytical breakdown of the cost structure of cattle trade could be used to get some idea of the various transaction costs and hence the efficiency of the livestock marketing channel. Tables 11.1 and 11.2 summarise all costs relating to a transaction from Pouytenga to Abidjan and from Pouytenga to Accra, respectively, in April 1993[4].

In both cases, the purchase price is the most important cost item, constituting some 80% of all costs, leaving transaction costs at roughly 20%. The second important cost item, in Tables 11.1 and 11.2, is truck hire. This appears to be higher in Ghana but it should be noted that transport handling costs and illicit taxes were included in the 'truck hire/walking' item. A remarkable difference is found in the profit margin (14% for Cote d'Ivoire and 4% for Ghana). This probably results from the very character of the transaction. In this example, the cattle exported to Cote d'Ivoire had already been fattened in Burkina Faso. Thus, it had a very competitive position on the market in Abidjan. This was not the case with the Pouytenga–Accra transaction where cattle had to compete with local Ghanaian cattle (the fattening of the cattle for the Abidjan destination explains the large difference in weight and purchase price compared to Accra). Furthermore, the profit margin in Ghana is influenced by the exchange rate between the cedi and the CFA franc. Compared with the reproduced results of the various marketing studies (cf. Table 11.3) the Ghana profit margin of 4% is at the very low end of the range for this item. The amount spent on illicit taxes on the Pouytenga– Abidjan itinerary works out at 1.2% of total costs. This relatively low rate is confirmed by figures on the itineraries in Table 11.3. So, illicit taxes do not seem to be as large a problem as some authors thought them to be (cf. Holtzman & Kulibaba, 1992; NOVIB, 1993).

---

[3] Usually, a trader pays 40,000 CFA francs to hire the services of a 'societe de convoyage'. The representative of the latter travels together with the cattle transport and negotiates at every roadblock encountered. Being adequately equipped with ministerial letters, etc., he is in a good position to negotiate a smooth passage. This will save the trader up to 60,000 CFA francs per journey.

[4] Note that the average weight of a cow is approximately 400 kilos for Table 11.1 and 250 kilos for Table 11.2.

TABLE 11. 1: Cattle Trade Costs on the Pouytenga-
Abidjan Itinerary (in CFA Francs)

Cost specification	Total	Cost/head	% Total
Purchase (32 cows)	2,800,000	87,500	81.8
Truck hire	300,000	9,375	8.8
Transport handling	90,000	2,813	2.6
Illicit taxes	40,000	1,250	1.2
Burkina Faso customs	97,130	3,035	2.8
Broker	32,000	1,000	0.9
Other	65,700	2,053	1.9
Transaction costs	624,830	19,526	18.2
Total costs	3,424,830	107,026	100%
Sales revenue	3,840,000	120,000	
Profit margin	415,170	12,974	12.1%

TABLE 11.2: Cattle Trade Costs on the Pouytenga -
Accra Itinerary (in CFA Francs[5])

Cost specification	Total	Cost/head	% Total
Purchase (32 cows)	1,600,000	50,500	80.0
Truck hire/walking	255,000	8,000	13.0
Burkina Faso customs	97,130	3,035	5.0
Broker	8,000	250	0.4
Other	32,000	1,000	1.6
Transaction costs	392,130	12,285	20.0
Total costs	1,992,130	62,285	100%
Sales revenue	2,080,000	65,000	
Profit margin	87,870	2,715	4.4%

We consider Tables 11.1 and 11.2, based on our case study, as quite representative for the costs structure in general. This becomes clear when other West African examples of cattle trade costs are taken into account. In this respect, Table 11.3 provides a regional overview.

The analysis of transaction cost clarifies that: (a) illicit taxes and custom payments form a minor share of total costs; (b) the non-official trade channel to Cote d'Ivoire; (c) a breakdown into separate cost items proves that transaction costs of long distance, cross-border cattle trade are rather modest and leave a large share of the wholesale price to the cattle owner.

In line with our case study represented in Tables 11.1 and 11.2, Table 11.3 leads us to the conclusion that regional cattle trade is not as inefficient as it was thought to be.

---

[5] Note that this table is based on an illegal transaction (which avoids paying import taxes in Ghana). As a result some factors are estimates. The 'truck hire/walking' item includes illicit taxes and transport handling costs.

TABLE 11.3: Cattle Marketing Costs (as a Percentage of Total Costs)of Various Regional Itineraries

	Pouyt.-Abidjan (BF-RCI)	Segou-Abidjan (Mali-RCI)	Dougab.-Abidjan (Mali-RCI)	Pouyt.-Abidjan (BF-RCI)	Zinder-Lagos (Niger-Nigeria)
Purchase price	82	82	76	78	82
Transp. & handling	11	11	15	15	11
Illicit taxes	1	3	3	1	1
Customs	3	2	2	3	5
Brokers	1	1	2	1	2
Various costs	2	1	2	2	—
Total costs	100	100	100	100	100
Transaction costs	18	18	24	22	18
Profit margin	12	7	12	9	26

Source: Holtzman & Kulibaba (1992), Kulibaba (1991), Ancey (1991), van Helden & Quarles van Ufford (1994).

BF = Burkina Faso
RCI = Republique Cote d'Ivoire

The other conclusion to be drawn from the various cost analyses is that cattle producers receive a large share of the consumer price. Price determining factors at that 'end of the line' are therefore decisive for the price obtained by cattle farmers. Distortion of the markets in the main consuming areas will lead to a depressing effect on producer prices in Sahelian countries. In other words, profitability in cattle production depends to a large extent on the factors that influence the functioning of beef markets in cities like Abidjan and Accra. One of these factors is the competition by frozen European beef.

THE SHORT-TERM IMPACT OF REDUCED MARKET DISTORTIONS

Impact on the Cost Structure of Imported European Beef

In order to get some insight into the elements that are responsible for the price level of imported beef, we have broken down the cost structure of imported beef before and after the devaluation, thereby also incorporating the realised decrease in the EU subsidy. Table 11.4 was put together on the basis of various sources and is meant to give some indication of the changes in costs rather than to provide exact data. Moreover, some elements have changed since early 1994, such as the purchase price in Europe, which has gone down.

Table 11.4 shows that, in contrast to the expected doubling of beef prices as a consequence of the CFA devaluation, retail prices of imported beef in Cote d'Ivoire went up by some 50% only; the CIF West Africa price, though, did double indeed. This moderate price increase was mainly due to a decrease in import taxes and import levies. This cost specification illustrates that the EU subsidy, although reduced, is still a highly distortive element. Particularly if beef prices in Europe would fall without a change in the restitution level, this would affect the wholesale price in West Africa to such an extent that regional trade flows would suffer again from competing frozen beef imports. This unfavourable effect on the regional cattle trade flows would be reinforced if, for some reason, West African importing countries would diminish or eliminate existing import levies and taxes, as has been the case in Cote d' Ivoire in 1994.

The price increase of imported beef was accompanied by a large drop in all extra-regional beef imports (cf. Table 11.5), as these were replaced by increased imports from Sahelian countries. At the same time, the price of local fresh meat went up by approximately 15% - 20%, to fluctuate between 850 FCFA and 900 FCFA (CEBV, 1994).

TABLE 11.4: Cost Structure of Imported European Beef,
Before and After Devaluation
(In CFA Franc/Kilo)

Cost specification	1993	1994
Purchase Price (Europe)	608	1,216
EU Subsidy[6])	541	983
FOB Europe	67	233
Cost of Freight	67	133
Transit/Sanitary Inspection	40	40
Profit Margin Exporter	17	30
CIF West Africa	191	436
Import Taxes	67	90
General Costs	70	70
Profit Margin Importer	20	36
Import Levy	100	50
Wholesale Price (West Africa)	448	682
General Costs	40	50
Profit Margin Retailer	40	70
Retail Price	528	802

Source: CEBV (1994) and Rolland (1995)

In other words, the relative price change has resulted in a trade substitution towards fresh beef, having a more favourable price/quality relationship. Also interesting is that such a substantial trade diversion occurred as a result of a relative price change of some 30%. This demonstrates the importance of the cross-price elasticity of consumer demand. If this is also the case in the opposite direction, it could mean that a relative increase in the price of beef from Sahelian origin, or a price decrease for European beef, could easily shift consumer demand towards the latter.

The potential threat from EU exports, having still a high export subsidy element, may weaken in the coming years. In this respect it should be taken into account that the recent GATT agreement obliges the EU to diminish the subsidised export volume of beef by 21%,

---

[6] This figure represents data for October 1993 and March 1994. After the first reduction by 15% in June 1993 (for West Africa only), three successive reductions of restitutions of around 5% each (for all destinations) were decided upon by the EU (July 1993, November 1993, January 1994). The total reduction amounts to 28% (NOVIB, 1994).

leaving a maximum of 817,000 tons eligible for export restitution at the end of the transitional period (the year 2000) (van Berkum, 1994). Like now a very minor part may reach West African countries. As to the export restitution — which is still a potential market distortion — the ECAM (European Community Agricultural Model) scenario foresees a drop in the intervention price, so that in 1996 the EU export price will be equal to the world market price for beef (van Leeuwen, 1992).

Regarding the quantity aspect it is noted that the same ECAM scenario foresees that the EU would even temporarily become a net importer of beef. Whatever the realistic value of the various scenarios based on the GATT agreement and EU budget problems, it is rather improbable that the heavy market distortions of the early nineties will reappear.

Impact on Cattle Flows and on Volumes of Imported Beef

Table 11.5 summarises Burkina Faso's cattle exports towards the coastal countries of the central corridor as well as the beef imports for the same group of coastal countries.

The figures of live-cattle imports (in numbers) from Burkina Faso are not comparable with the figures of frozen beef (in tons) from the European Union. Neither are their qualities comparable: consumers consider the former highly superior to the latter. Just for the sake of comparing these two trade flows, we estimate that 1 ton of frozen beef is equivalent to 7 live adult heads of cattle from the Sahel.

From the above data we conclude that the sudden 62% drop in beef imports from the EU is compensated for by an equally sudden boost in the imports of cattle from Burkina Faso.

During the first two months following the devaluation of the CFA franc a steep increase in Burkina cattle exports to Cote d'Ivoire was noted. Over the year 1994, total exports to Cote d'Ivoire went up by 16%. Beef imports in Cote d'Ivoire, though, have largely disappeared. In Ghana, the devaluation of the CFA franc has evidently led to a large increase in cattle imports. In the first quarter of 1994 alone, Burkina Faso's cattle exports towards Ghana by far exceeded the total number of such exports in all of 1993.

TABLE 11.5: Coastal countries: Imports of cattle from
Burkina Faso and imports of EU beef

	Burkina Faso's Cattle Exports (in numbers)		European beef Imports (in tons)	
	1993	1994	1993	1994
Importing countries				
Cote d' Ivoire	90,000	105,000	12,200	1,900
Ghana	13,000	50,000	29,700	11,850
Togo & Benin	2,000	15,000	7,800	3,600
Total	105,000	170,000	49,700	17,350

Sources: LPIU (1995), Roland (1995), MARA/SSA/DSAP (1993, 1994), Eurostat Statistics (1994, 1995).

Total exports to Ghana for the year 1994 almost quadrupled (cf. Table 11.5). The explanation is that the CFA franc devaluation caused a de facto appreciation of the Ghanaian currency, the cedi. This has led to a renewed interest from Ghanaian traders in Sahelian cattle, backed by a growing demand originating from the large consumption centres of Accra and Kumasi[7]. As a result of the continuing depreciation of the cedi, the increase in import taxes on frozen beef, and the lower cedi price of Sahelian livestock, beef imports from the EU into Ghana have become less attractive. Consequently, the beef import figure in Ghana has been reduced by nearly two thirds in one year.

For cattle producing countries in the Sahel, and for Burkina Faso in particular, the sudden upward move in cattle exports is economically very important, also because the volume effect is accompanied by an upswing in prices. However, if increased sales are not accompanied by an adequate increase in Burkina Faso's cattle stock, such a high trade-flow figure will prove unsustainable in the long run. Reliable figures on cattle stock development over the years are missing to judge if Burkina Faso will be able to maintain a certain cattle export surplus (Rolland, 1995[8]).

It is very important to investigate this issue. For the year 1994 we assume that, as a result of the large devaluation, a contraction in the effective internal demand has caused a reduction in the number of cattle slaughtered in Burkina Faso itself. Such consumption reduction could have partly caused the large increase in cattle exports. However, on the medium-term policies must ensure that the growth of cattle stock and of annual off-take will keep ahead of the growth rate of the national beef consumption.

Impact on Cattle Prices in Burkina Faso

In Burkina Faso the export market prices of so-called 'exportable bulls' (i.e. the most common bull to be exported) have risen considerably. Compared to the years 1991-1993, during which period the price level was quite stable, the average increase has been around 50% (cf. Figure 11.1). This upward shift results from the increased cattle demand from abroad.

The resulting higher revenues for cattle producers are partly offset by the increase in the production costs of cattle rearing. In order to quantify the profitability of cattle rearing one has to know details of the cost structure. The cost items for imported goods (such as veterinary products and cattle feed) have suddenly doubled in price after the devaluation; local cost components, including labour, have had a lower price rise (the general price rise at the national level in Burkina Faso over the first six months of 1994 was approximately 30%). More accurate information is needed before one could judge the extent to which cattle production in Burkina Faso has become more profitable since 1994, when compared to the time before the devaluation.

---

[7] Note that all data concern officially registered exports. The Ghanaian government has not yet abolished the high import tax on cattle and traders continue smuggling the animals into the country. Although most of the exports in the direction of Ghana are registered as such in Burkina Faso, the exact number is probably higher, as can be expected from a trade which has a partly unofficial character.

[8] Referring to the latest studies on forecasts of supply and demand in coastal and Sahelian countries, a growing deficit in the former countries is foreseen for the years between 1995 and 2000 and doubts are expressed as to the Sahelian countries' capability to ensuring supplies for the coastal markets. In this respect, Rolland (1995) notes that'.. given the shortcomings on the information side, none of these countries is in a position to make a serious estimate of available export supply..'(p.43).

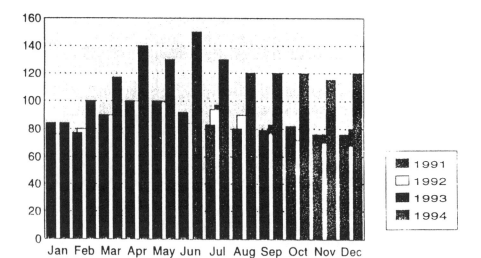

Sources: Ministry of Agriculture (Burkina Faso), field surveys

Figure 11.1: Average Price 'Exportable Bull' on Pouytenga Market
(In CFA Francs), 1991-1994

Conclusion and Major Questions

The first question we dealt with in this chapter concerned the limitations to the development of the regional cattle trade in West Africa, since regional marketing channels were criticised for their inefficiency. Based on detailed studies on transaction costs of regional cattle trade we concluded that this trade does not seem to be as inefficient as it was thought to be. Since the purchase price of cattle makes up such a high percentage of the ultimate consumer price, we also concluded that the main factor for lowering costs and improving the comparative advantage of Sahelian cattle production for export should be found in improving the efficiency at the production level.

The second concern of this chapter was the serious distortion of the beef markets in West Africa by two main causes: the overvaluation of the CFA franc and the dumping of EU beef. These distortions had resulted in a loss of the original competitiveness of the Sahelian cattle producing countries, Burkina Faso and Mali in particular. In early 1994, these trade distortions were considerably diminished through a full correction of the overvaluation of the CFA franc and a reduction of the high subsidy on EU frozen beef exports. We investigated the extent to which these sudden changes have affected import flows and prices of Sahelian cattle and of EU frozen beef on the short term. We arrived at two conclusions:

The Improved Competitiveness of the Regional Cattle Trade

The 'repair' of the two distortions has, on the short term, completely changed the shares of cattle imports both for a CFA country (Cote d'Ivoire) and for a non-CFA country (Ghana). The traditional cattle exporting countries of the Sahel zone have benefitted from the increased trade flow and higher prices. The 'boost' in cattle exports from Sahelian countries proves that these countries have regained the competitiveness which they had before 1975, when they were nearly the sole supplies of beef in the coastal countries of West Africa. Since it is likely that the national beef deficit of the coastal countries will grow (Rolland, 1995), this raises the question if higher trade flows from Sahelian exporting countries can be sustained in the long run. The increased trade flows require a continuous growth of the cattle stock, i.e. the national herds of Sahelian countries.

This question of the supply response over time by the various types of pastoralists in the region is seriously in need of research.

The Consumer Demand's Sensitivity for Relative Beef Prices

The market shares of the two sources of beef, Sahelian cattle and frozen beef imports from the European Union, appear to be quite sensitive to relative price changes at the West African beef markets. With a moderate price rise in 1994, the beef imports from the EU practically had to give away their market share to the competing trade channel of Sahelian cattle. A high cross-price elasticity in the demand for beef may imply that the reversed change in the price relationship in favour of EU beef (lower import price, or a rise in Sahelian cattle prices) could bring back an increased market share for the latter. Consequently, two major questions have to be raised:

(a) How will the import prices of (EU) frozen beef develop in the coming years?

(b) How will the national surplus and production costs of Sahelian cattle develop over the next few years?

Thus, both our conclusions on measured short-term results lead to important questions with respect to the development of the cattle sector in Sahelian countries.

REFERENCES

Ancey, V., 1991. Image de la production et des dchanges de betail dans le couloir central Mali - Burkina Faso-Cote d'Ivoire. Paris, Club du Sahel.

Arhin, K., 1979. West-African Traders in Ghana in the Nineteenth and Twentieth Centuries. London, Longman Inc.

Berkum, S. van, 1994. Gevolgen van het GATT-akkoord voor de EU-land-bouw. Den Haag, Landbouw-Economisch Instituut (LEI-DLO).

CEBV, 1994. Impact et accompannement du changement de parite du franc CFA dans le secteur de l'elevage dans la sous region. Ouagadougou, CEBV.

CILSS & CEBV, 1992. Seminaire regional sur la commercialisation du betail et de la viande dans les pays du Sahel et de la cote: rapport final. Nouakchott, CILSS.

Helden, W. van & P. Quarles van Ufford, 1994. Faire du Bizness: Une etude sur la commercialisation du betail et de la viande en Afrique de l'ouest. Universite d'Amsterdam, Reseau de Recherche SADAOC Working paper WP 94 FTFS-1.

Herman, L., 1983. The livestock and meat marketing system in Upper Volta: an evaluation of economic efficiency. Ann Arbor, CRED, University of Michigan.

Holtzman, J.S. & N.P. Kulibaba, 1992. Livestock marketing and trade in the central corridor of West Africa. Washington, USAID.

Hopkins, A.G., 1973. An economic history of West-Africa. London, Longman group Ltd.

Josserand, H.P., 1990. Systemes ouest-africains de production et d'echanges en produits d'elevage: aide-memoire synthetique et premiers elements d'analyse regionale. Paris, CILLS/Club du Sahel.

Kulibaba, N.P., 1991. Livestock and meat transport in the Niger-Nigeria corridor. Bethesda, Maryland, Abt. Associates Inc.

Leeuwen, M.G.A. van, 1992. Rundvleesmarkt en rundvleesbeleid in de EEG-een modelmatige verkenning. Den Haag, Landbouw-Economisch Instituut.

LPIU, 1995. Analysis of meat and animal products imports: occasional report No.1, Accra, Livestock Planning Intelligence Unit, Ministry of Food and Agriculture.

MARA/SSA/DSAP, 1993, 1994. Bulletin Annuel Statistique de l'Elevage: 1993 et 1994. Ouagadougou, Direction des Statistiques Agro-Pastorales/Ministere de l'Agriculture et des Ressources.

NOVIB, 1993. De gesubsidieerde export van EG rundvlees naar West-Afrika: Hoe de Europese vleesberg de veehouderij in de Sahel ondermijnt. Den Haag, NOVIB.

NOVIB, 1994. 4th Report on the European campaign to stop the dumping of EU beef in West Africa. Den Haag, NOVIB.

Quarles van Ufford P., 1994. De franc CFA gedevalueerd: een studie naar de gevolgen van de devaluatie van de franc CFA voor de veehandel in de 'couloir central' van West-Afrika. Universiteit van Amsterdam, ISG, unpublished report.

Rolland, J.P., 1995. Foreign trade policy and regional trade in meat and livestock: report of the November mission (November 1-26, 1994). Paris, Solagral, Ministere de la Cooperation.

SOLAGRAL/IRAM/CEBV/Ministere de la Cooperation, 1993. Etude politiques de commerce exterieur des produits d'origine animale en Afrique de l'ouest et du centre. Paris, Solagral/Ministere de la Cooperation.

Staatz, J., 1980. Livestock and Meat Marketing in West Africa: Volume III, Ivory Coast and Mali. Ann Arbor, CRED, University of Michigan.

CHAPTER 12

FRAMEWORK TO ASSESS THE PERFORMANCE OF FOOD COMMODITY MARKETING SYSTEMS IN DEVELOPING COUNTRIES WITH AN APPLICATION TO THE MAIZE MARKET IN BENIN

C. Lutz, and *A. van Tilburg*

Ō/3

Q/3

Introduction

A properly functioning market for agricultural products is generally perceived as the best organisational structure to achieve more efficient decisions of producers, consumers and traders (cf. Bardhan, 1990). However, efficient trade does not automatically follow from spatial differentiation between production and consumption and a network of connecting markets. A mix of institutional regulations coordinates the exchange process. In this respect, the question is not to choose between the two extremes — market or government planned system — but to develop the right institutional environment that facilitates exchange and market performance. Getting prices right is a major objective in agricultural policies (cf. Timmer, 1986). Nevertheless, it is not easy to determine the right mix of market freedom and government intervention to realise the optimum outcome (cf. Killick, 1989). "The challenge for the 1990s is to balance the enabling macro-economic environment with sound sectoral policies and the alleviation of micro-economic constraints" (Lele, 1992, p.576).

Problem of Definition

The subject of the chapter is to present a framework to analyse the performance of food commodity marketing systems operating between producers and consumers and to apply them to the maize market in Benin. With the introduction of structural adjustment programmes in the eighties and nineties in many developing countries, the performance of food commodity marketing systems became more critical. `More market and less government' has been the slogan.

Initially, methods to analyse the performance of food markets originated mainly from the Industrial Organisation theory (e.g. Harris, 1979). More recent studies also applied the theories of Transaction Costs Economics (e.g. Williamson, 1985) and Institutional Economics (e.g. North 1990; Eggertson, 1990).

Aims

The aim of this chapter is to present a framework to assess the performance of food commodity marketing systems by:

(a) discussing concepts relevant for performance assessment of food commodity marketing systems;

(b) pressing a framework to evaluate the performance of a food commodity marketing system; and

(c) applying the framework to maize marketing in Benin.

The chapter is organised as follows. Aim (a) is covered by discussing concepts from industrial organisation theory, marketing and institutional economics (next section). Aim (b), the framework to evalaute the performance of a food commodity marketing system is presented under the subheading ' A Framework of Analysis for a Food Commodity Market'. In the section headed ' An Application of the Framework: the Functioning of the Maize Market in Benin', a part of this framework is applied to the maize market in Benin; aspects of structure and conduct are discussed and the functioning of spatial arbitrage is assessed.

APPROACHES TO ASSESS THE PERFORMANCE OF MARKETING SYSTEMS

Industrial Organisation Theory

Industrial organisation (IO) theory deals with the functioning of markets (Tirole, 1988), or with resource allocation through market systems (Scherer and Ross, 1990), or with the structure of firms and markets and their interactions (Carlton and Perlof , 1994).

Calton and Perlof (1994) distinguish two major approaches to the study of industrial organisation. The first approach is the structure-conduct-performance analysis. An industry's performance (the success of an industry in producing benefits for consumers) depends on the conduct (policies) of firms which, in turn, depends on the market structure (factors that determine the competitiveness). The second approach (price theory) uses economic incentives to explain market phenomena. Specific applications of price theory such as transaction cost analysis and game theory are considered to be helpful in explaining the structure, conduct and performance of markets.

Competition is assumed to be a main force that helps to solve economic performance problems. Several arguments in favour of perfect competition have been put forward, e.g. political arguments and efficiency arguments, but there are also qualifications and doubts (Scherer and Ross, 1990). Political arguments in favour of perfect competition are decentralisation of power, impersonal interplay of competitive market forces (the invisible 'hand') and freedom of opportunity to choose particular economic activities. Efficiency arguments in favour of competition regard the assumed profit maximising behaviour of firms resulting in efficiency of resource allocation.

The problem with perfect competition is that it is an ' ideal type' that in its pure form hardly exists: for example, marketing and information problems are supposed not to exist in a perfectly competitive market.

Initially, a hypothetical market operating under perfect competition was taken as a preferred standard to measure market performance (Bain, 1959). The standard was changed from perfect competition to 'workable competition', when it was realised that perfect competition can hardly be found in its pure form in the real world (e.g. Sosnick, 1958). Later, workable competition was replaced by contestability (Baumol et al., 1988). In a perfectly contestable market, entry is free and exit is costless. Entrants and incumbents compete on equal terms, and entry is not impeded by fear of retaliatory price alterations. Entry by new competitors and expansion by smaller existing (fringe) rivals may threaten the position of a dominant firm in the market. The dominant firm must then abandon its attempt to maximise short-run profits and instead reduce its price to a level at which new entry and the expansion of fringe rivals are discouraged.

Market structure analysis or structure-conduct-performance (SCP) analysis (e.g. Cubbin, 1988; Mario and Mueller, 1983; Scherer and Ross, 1990) is an approach that has often been used to evaluate (food) commodity marketing systems. The main hypothesis is that market structure and market conduct affect market performance. The structure of a market is related to the degree of buyer and seller concentration, market transparency and entry and exit conditions. The market structure concerning a particular product may vary from perfect competition to dual monopoly. Market conduct is related to marketing policy and collusion of firms. Market performance characteristics may include level and seasonality of prices and level and distribution of profits among market participants. In a dynamic context, feedback relationships between performance, structure and conduct are assumed. In traditional SCP analysis, an actual market is characterised by scores on the dimensions of perfect, workable or contestable competition. Dependent on the scores, the competition is called 'near' or 'highly' perfect, workable, contestable, or imperfect. A common hypothesis is that the lower the degree of competition in a market (e.g. oligopoly, monopoly), the higher product prices and profits are.

Marketing

Core concepts of marketing (Kotler, 1992) regard: a managerial process guiding the supply of products or services which satisfy a need of customers who are willing to engage in exchange through a marketing network.

Approaches in Marketing Theory

In the course of time, different approaches were employed to study the role of marketing:

— The commodity approach focuses the analysis on the flow of products from producers to consumers in the marketing channel. This approach suits the subject of this chapter: The performance of food commodity marketing systems.

— In the functional approach, the main commercial services offered by traders and marketing institutions are analysed (Kohls and Downey, 1972): exchange functions (buying and selling), physical functions (transport, storage, processing) and facilitating functions (standardisation, financing, risk-bearing and market intelligence).

— The institutional approach analyses the structure, role and performance of marketing institutions such as marketing boards, marketing cooperatives, auctions and future markets.

— The marketing management approach (planning implementation and control of marketing programmes) assumes that at least one party takes the initiative to achieve desired responses from other parties. The instruments are product policy, price policy, distribution policy and communications policy (Kotler, 1992). The marketing management approach is particularly relevant in markets where products can be differentiated and buying behaviour of customers can be influenced by suppliers (Meulenberg ,1986).

Marketing channels are sets of interdependent organisations involved in the process of making a product or service available for use or consumption (Stern and El-Ansary, 1992). The fundamental activity in marketing channels is the transaction, i.e. the act of exchange between economic agents (Achrol, Reve and Stern, 1983). Conventional marketing channel networks (CMC) are distinguished from vertical marketing systems (VMS).

In conventional marketing channels (CMC), subsequent sages in the assembly and distribution of commodities are connected by a market. The coordination among channel members is primarily achieved through bargaining and negotiation. The actors tend to be preoccupied with volume, costs and investments in a single stage of the marketing process (Stern and El-Ansary, 1992).

Vertical marketing systems (VMS) are developed to achieve control over the costs and quality of the functions performed by various subsequent channel members. A VMS is a network of interconnected units or firms, in which market exchange transactions are substituted by the entrepreneur-coordinator (Coase, 1937). Coordination among channel members is achieved through the use of comprehensive plans and programmes. Mechanisms to achieve coordination and cooperation among channel members may be based on the use of power through rewards, coercion, expertness, identification and legitimacy (e.g. Stern and El-Ansary, 1992). Main modes of VMS are administered systems, in which coordination is achieved through programmes developed by one or a few firms; contractual systems and corporate systems.

The final check of market channel performance deals with the required service output levels. For example, the lot size of the products supplied has to correspond to the needs of customers, the waiting time has to be acceptable, the density of buying locations has to be sufficiently high, and heterogeneity in the supply of products has to correspond to consumer preferences.

Institutional Economies

The theory of institutional economics is focused on the question how alternative sets of social rules (institutional structure, property rights) and economic organisations affect behaviour, allocation of resources, and equilibrium outcomes. This is highly relevant when the performance of food marketing systems in developing countries is studied.

Definitions of Institutions

Institutions are the organizing mechanisms of a society and property rights describe these institutions (Campbell and Clevenger, 1978). Property rights are the rights that individuals appropriate over their own labour and the goods and services they possess (Eggertsson, 1990). Institutions can be understood in two ways. The first is sociological: any behavioural regularity is an institution. The second is economic: institutions are the rules of the game in a society, or the humanly devised constraints that shape human interaction (North, 1990, p. 3).

Markets are institutions because they embody rules and regulations, formal or informal, which govern their operation. Contracts are institutions in that they lay down rules which govern activities of the contractual parties. Codes of conduct are institutions as well insofar as they can constrain the relationships between different individuals and groups (Nably and Nugent, 1989).

In the standard neoclassical general equilibrium model, commodities are identical, the market is concentrated in a single point in space, and the exchange is instantaneous; individuals are fully informed about the commodity and the terms of trade are known to both parties. Prices are a sufficient allocative devise to achieve highest value uses. The theory of price has provided valuable insights into the fundamental nature of exchange and resource allocation in decentralised markets (North, 1990, p. 30).

Coase (1960) made clear that it is only in the absence of transaction costs that the neoclassical paradigm yields the implied allocative results. With transaction costs, resource allocations are altered by property rights structures (North, 1990, p.28). In institutional economics, commodity markets are supposed to be imperfect and characterised by transaction costs which require institutions to regulate property rights and contracts, for example, marketing organisations, standardisation in grading or in contracts.

Institutional development should foster economic development by finding the mix of rules that facilitates or optimises the exchange process. Developing economies can be characterised by many cases of market failure and incomplete markets providing fruitful territory for institutional analysis (Thorbecke and Morrison, 1989). A specific role can be played by the state which can affect the wealth of a society by redefining the structure of property rights and by providing public goods which reduce the costs of transaction.

A major subject in transaction cost economics (TCE) is comparing the costs of transacting business 'across a market' with third parties with the costs of bringing the transaction in-house via vertical integration. High transaction costs can limit or prevent otherwise advantageous exchange. Historically, the state has lowered transaction costs by establishing and maintaining standards of measurement and by introducing and maintaining stable money.

We distinguish a broad and a narrow definition of transaction costs. The broad definition of transaction costs states that transaction costs originate in imperfect information, transportation, search, negotiation, recruitment, monitoring and supervision, motivation, enforcement, coordination, management, etc. (De Janvry and Sadoulet, 1994), or that transaction costs not only consist of measured costs (transportation, storage) but also of unmeasured costs such as waiting, bribery, uncertainty and governance (Lutz, 1994).

In the narrow definition, transaction costs are associated with the costs of both acquiring information about a potential exchange and completing the exchange. Eggertsson (1990, p. 15) distinguish: information costs (the search for information about price, quality, potential buyers and sellers), bargaining costs (the bargaining process), and enforcement costs (the making of contracts, the monitoring of contractual partners, the enforcement of the contract, the protection of property rights against third-party encroachment).

TCE starts from the assumption that the governance of contractual relations is primarily effected through institutions of private ordering rather than through legal centralism (Williamson, 1985). How and why are different transactions between parties created, carried out, or avoided? What types of bargaining strategies of parties can be expected? Is actor's behaviour characterised by bounded rationality (limited human capacity to anticipate or to solve complex problems) as a consequence of incomplete information and opportunistic behaviour?

A Framework of Analysis for a Food Commodity Market

A food commodity marketing system can be characterised by a flow chart representing product flows from producers to consumers passing through subsequent stages of the various marketing channels. At the commodity level, it is feasible to analyse the interaction between changes in the environment of the commodity system and the performance of the system (e.g. Campbell and Hayenga, 1978). Performance attributes regard flexibility of (marketing) institutions and organisations to adapt to changes in external factors such as government intervention, devaluation of the currency, the supply of public goods and services, consumer preferences and the supplying of substitutes.

A commodity marketing system may include several marketing channels consisting of actors supplying marketing services at different levels in the channel. At each level, spatially separated market-places[1] may be linked with each other through arbitrage. Arbitrage drives the prices in different markets, apart from transport and other transaction costs, into equality. Spatial networks of markets may have a horizontal (e.g. interlinked regional wholesale markets) or a vertical (e.g. markets in the assembly and distribution stage of agricultural products) dimension. The vertical dimension of a spatial network coincides with (a part of) a marketing channel.

Market performance can be assessed on different levels[2]. The analysis of market performance can be restricted to the functioning of a specific market-place where traders complete transactions (next sections). Market-places can be part of marketing channels as well as spatial market networks. A good performance and markets, as part of a marketing channel or a spatial network, is an essential condition for a proper functioning of food commodity marketing systems. In an evaluation of the performance of marketing channels, the services delivered to customers are assessed in relation to costs and profits in the subsection entitled 'Performance of Marketing Channels: the "vertical" Link Between Producers and Consumers'. In an evaluation of the performance of spatial networks it is assessed whether suppliers in different regions are equally compensated for the resources (product or value) they offer taking transport and other transaction costs into account (subsection headed 'Performance of Networks of Market-Places: 'Horizontal" Price Relations Between Markets').

The framework is built on the methodology (subsection headed 'Industrial Organisation Theory'), assuming a relationship between market structure, conduct and performance. Relevant elements of market structure, conduct and performance in developing countries are given in Table 12.1.

Performance Measurement on the Level of a Physical Market-Place

Structure

The following criteria are generally selected to evaluate market structure:

(1) Low degree of concentration of suppliers and buyers, unless economies of scale intervene ('sustainable industry structure');

(2) High degree of market transparency (information on supply, demand and prices);

(3) Absence of entry or exit barriers;

---

[1] A market-place is defined as a physical area where transactions are concluded. For example auctions, a village market, a regional market or an urban market. Others define it as an economic market (or economic region) where actors are competing with each other.

[2] The framework is confined to the functioning of the food marketing system. Performance is not measured at macro level (concerns the relative prices). Consequently, questions involving welfare economics go beyond the scope of the framework, which is directed at affectivity, equity and efficiency of inter-seasonal and inter-regional trade.

(4) Low transaction costs, both in a narrow sense (the costs of information, bargaining and contract enforcement) and a broad sense (costs of necessary marketing functions that are delivered); and

(5) Optimal institutional set-up (grades, standard contract, price information system, legal system, marketing institutions) to affect transaction costs.

TABLE 12.1: Elements of Structure, Conduct and Performance of Markets Within a Food Commodity Marketing System

Elements of structure	Elements of conduct	Elements of performance
-competition	-cooperation	-effectiveness of supply
-entry	-integration	-efficiency of supply
-market transparency	-strategies	-cost effectiveness
-infrastructure	-services	-market integration
-standardisation		-equity
-transaction costs		

Conduct

A market is not functioning properly when the distribution of market power between actors, as a result of cooperation or preferential access to resources and information, is such that one of the parties gains at the expense of other parties. The trader as entrepreneur has to compose an assortment of products and deliver the services that are desired by his customers. Traders may attempt to increase their profits by means of particular strategies:

— marketing strategies or policies to increase customer satisfaction or to decrease costs, e.g. innovations in products or distribution methods;

— strategies to enhance market power: collusion, mergers, rent seeking behaviour; the question is to know whether these strategies are in line with a sustainable industry structure (e.g. economies of scale).

— strategies to enhance competitive power by leadership in costs reductions.

Performance Measures: Within a Physical Market-Place

Affectivity is related to the degree to which actors or institutions are able to supply desired food products to the market-place. Are traders trying to maximise profits given a certain level of uncertainty, or, do they try to minimise risk given a satisfactory return on investments made?

Equity

This is related to the division of power among market parties in completing transactions. Is the market an equal opportunity institution? Are there, for example, any entry barriers for those who want to conclude a transaction?

Efficiency

If competition is (almost) perfect, workable or contestable, price levels are expected to be in line with costs implying that excessive profits are absent. Performance characteristics include level of transaction costs, returns on investments and level and distribution of profits among market participants. Do actors or marketing organisations aim at reducing transactions costs? Are new, adaptive and cost reducing institutions such as price information systems, standardisation in grades, measures or contracts implemented? Are returns on investment optimal in terms of asset turnover and profit margin?

Performance of Networks of Market-Places: 'Horizontal' Price Relations Between Markets

Structure

In the assembly stage of agricultural products, a spatial market network can be represented by a central market-place linked to a set of regional markets, which are surrounded by smaller satellite markets. A particular spatial market network may be linked to other networks. Integration among two or more market-places requires a proper physical, institutional and competitive environment, for example, an adequate infrastructure, absence of entry and exit barriers, standardisation of measures and common trade habits. In a competitive market, price integration is the outcome of an arbitrage process.

Conduct

The main behavioural aspect concerns the strategies of traders with respect to arbitrage. Traders aim to take advantage of price differences between markets that exceed transaction costs. They are supposed to buy with a view to selling elsewhere at a profit.

Performance Measure: Degree of Price Integration and Spatial Equilibrium

Effectiveness

This relates to trade flows generated when price differences between markets are greater than transaction costs, for example, the costs related to transport, storage and handling?

Equity

Are there any barriers for those who want to conclude a transaction? Is the spatial market system providing products and services to geographically dispersed consumers under the same conditions?

Are spatially dispersed markets governed by one and the same pricing system (law of one price) when taking transaction costs into account? Is a price shock in one of the central markets transferred to the satellite markets? Is there any delay in the transmision of this shock?

The degree of price integration among spatially dispersed markets must be assessed and related to market structure and behaviour of traders. Many recent studies analysed the performance of a network of spatially dispersed markets by measuring the degree of price integration among these markets. However, a high degree of price integration among markets does not necessarily imply that the market system is functioning properly. For example, one actor may be sufficiently powerful to fix prices in all markets under consideration. Co-integration analysis is a useful tool to measure the degree of price integration (e.g. Charemza and Deadman, 1992). Some recent applications concern Ghana (Alderman, 1993), Benin (Lutz, van Tilburg and van der Kamp, 1995) and India (Palaskas and Harris, 1993).

A disadvantage of price integration models is that product flows are not part of the analysis. Product flows between regions are usually difficult to measure. Spatial equilibrium models are suitable to approximate geographical flows of commodities in an indirect way. Assume that two or more regions with known supply and demand functions produce and consume certain commodities. Assume further that the regions are separated by known transfer costs. Given this knowledge, the problem is to determine the equilibrium levels of production, consumption and prices in each region and the equilibrium trade flows between regions (e.g. Takayama and Judge, 1971; Martin, 1981; Krishnaiah and Krishnamoorthy, 1988). It is a challenge to measure the effect of market imperfections with the help of this type of models (e.g. Hazell and Norton, 1986).

Performance of Marketing Channels: The 'Vertical' Link Between Producers and Consumers

Structure

Usually, a food commodity marketing system is composed of several parallel marketing channels. Is the institutional set-up optimal? Main structures of marketing channels that can be distinguished (section headed 'Approaches to Assess the Performance of Marketing Systems') are conventional marketing channel networks (CMC) and vertical marketing systems (VMS). Advantages and disadvantages of each of these types of governance structures in terms of transaction costs have to be assessed.

Conduct

A main aspect of conduct is the power used by channel members to achieve (vertical) coordination and cooperation in the marketing channel. To what extent is it allowed to use power to achieve optimal coordination? Is there a need for a channel leader?

Performance Measures: Service Outputs in Relation to Demand and Costs

Effectiveness

Regards the customer evaluation whether the final result of the flow of goods from producer to consumer through successive stages satisfies his needs at acceptable costs. The check concerns channel performance with respect to desired service output levels, for example, preferred lot size, quality, price, delivery time and location of the outlet.

Equity

Are there entry barriers for specific (groups of) actors? Is power used to the benefit of all participants in the marketing channel?

Efficiency

Are the transaction costs corresponding to the selected governance structure of the marketing channel optimal? Typical questions are: Is it possible that the marketing functions can be carried out more efficiently by, for example, reducing the number of intermediaries? Does (further) vertical integration reduce costs?

AN APPLICATION OF THE FRAMEWORK: THE FUNCTIONING OF THE MAIZE MARKET IN BENIN

Objective of the Case-Study

The main objective of the case study is to demonstrate the assessment of the efficiency of the spatial market network for maize in Benin (cf. section headed 'Performance of Networks of Market-Places: "Horizontal" Price Relations Between Markets'). The study is focused on spatial efficiency because the maize market can be characterised as a network of spot markets of relatively homogeneous goods where commercial services are supplied by flexible traders; it corresponds to a conventional marketing channel (cf. AGRER, 1986; and FAO, 1987). Moreover, an analysis of temporal price differences shows that storage is not an attractive policy for traders (cf. section headed 'Temporal Arbitrage, Costs of Storage and Rates of Return for Invested Capital'). Consequently, an analysis of spatial price integration seemed to be most interesting and as such only the relevant subset of the framework presented above is used in the analysis. The relevance of this objective was indicated in a comparative study of African and Asian countries by Ahmed and Rustagi (1987), indicating that especially in Africa marketing costs are high.

In the next section, elements of market structure and conduct are discussed. In the section following, the process of competition between different segments in a spatial marketing network is analysed. Subsequently, the role of traders with respect to temporal and spatial arbitrage is scrutinised. Finally, the efficiency of the spatial market network is discussed and some policy recommendations are derived.

Main Characteristics of Structure and Conduct of the Maize Market in Benin

Important differences exist between the south, where maize is the general food crop produced during two cropping seasons, and the north, where maize is a cash crop produced in only one cropping season. Because of this seasonality in production and because of variable climatic conditions, the marketing system needs to be flexible to meet the ever-changing supply and demand conditions at different market-places. The major maize granaries are located in the north of Benin where surpluses are important and storage conditions are favourable.

Towards the end of the 1988 lean season, consumers in Cotonou had to pay a price for maize three times higher than the price farmers in the south had received during the preceding harvest season; thus, two thirds of the consumer price was absorbed by commercial activities. This high margin between farmgate and consumer prices for the same unprocessed product is the origin of the negative connotation of commercial activities and exemplifies the importance of a thorough analysis of the efficiency of the marketing system.

In general, Benin is self-sufficient in maize, in fact even slightly overproducing. The distribution of maize is regulated by a private market system which is integrated into a larger international maize market (Nigeria, Niger and Togo[3]). The government has, only a limited role. Even during the Marxist/Leninist government from 1972 to 1990 the food market was controlled by private traders. The role of the state has been mainly concerned with the regulation of market transactions and the provision of the necessary physical infrastructure (indirect intervention).

Actors operate within a spatial network of both formal market-place (spot markets or periodic markets, subject to government regulations enforced by an official market organisation) and informal market-places (transactions effected in unregulated market-places or at the premises of one of the actors). The informal segment is important (see also Lutz, 1994) and is linked to a network of formal market-places. Usually traders are active in both segments simultaneously. The market structure is flexible and differs between regions and changes in time (seasons). Traders (wholesalers, retailers, brokers, assemblers) are flexible in the services they supply and, consequently, are difficult to categorise. The structure of the marketing channel through which the product passes from producer to consumer is also variable; the channels can be longer or shorter, depending on the behaviour of actors (traders, farmers, consumers) in the market. The existence of a number of alternative channels simultaneously is interpreted as a positive sign because it enables farmers and consumers to choose between the appropriate marketing services needed.

Market prices vary and are connected to a multitude of local units of measure. Price differences for homogeneous commodities may be explained by additional commercial services, but also by the typical characteristics of the market; bargaining may occur in two forms; bargaining about the price given the quantity or bargaining about the volume given a certain price. Three factors play a significant role in the bargaining process: prevailing supply and demand conditions on the market, individual bargaining capability and the various techniques of measures on a market.

Informal organisations are active in various spheres in the market serving the interests of their members. Representation of members and the provision of market services are their most common activities facilitating the functioning of the market. However, in some cases these organisations try to settle price levels and to control the entry of merchants. It is clear that these

---

[3] Despite the formal regulations on transborder trade, it can be shown that market integration is still retained by the informal market network (cf. Fanou/Lutz/Salami, 1991). Interventionist policies aiming to control trade flows failed in all these countries.

activities may hamper the functioning of the market and may be detrimental to the interests of the farmers. Non-resident wholesalers must respect the interests of local wholesalers (especially on the rural areas) as the latter's powers appeared to be of considerable significance.

Most traders have small financial resources and restrict their interventions to direct selling and buying. This explains the large number of retailers and petty wholesalers on the market. The opportunity costs of capital are high and this explains why most traders limit their involvement in credit arrangements or storage activities. The use of working capital with a high turnover rate facilitates the mobility of actors in the market: traders can easily shift among commodity markets and, consequently, potential competition is strong. Even large-scale wholesalers dealing in other commodities may decide to enter the maize market temporarily. This implies that the marketing channel can be characterised as a conventional marketing channel (CMC, cf. section headed 'Performance of Marketing Channels: the "Vertical" Link Between Producers and Consumers').

Wholesalers, in particular, buy a relatively important part of their produce on the informal maize market. The assembler, part of the wholesalers' network, guarantees a regular supply. The quantities handled per market day are relatively small, even for many of the wholesalers (less than 1000 kg of maize per market day, cf. Lutz, 1994, p. 96). Traders' gross margins (according to their own estimation) vary as a rule from 2 to 6 FCFA per kg.

Due to a lack of transparency, traders have to operate in networks in which information on prices, market regulations and activities of informal organisations is exchanged. Without contacts, it is difficult to develop a network and, consequently, it is risky to establish oneself in a new market. This may be perceived as an entry barrier. However, it is not set up artificially to protect certain interests, but for all new entrants in the channel it forms an obstacle to overcome. Assistance of friends and relatives is instrumental in this respect.

The application of formal regulations is hazardous, trading dues or buyer's licenses are determined by negotiation and only a fraction of the payments that could be collected are registered officially. This hampers the transparency of the market and causes inequality among the actors, especially since certain traders may strike a deal more easily with the inspectors than others. In addition, these discriminatory practices restrict entry to the market for those traders who are unable to participate due to lack of negotiating ability.

The interviews amongst farmers showed (Lutz, 1994) that they often sold their surplus in relatively small quantities in several transactions when household expenses needed to be financed. For many farmers maize stocks constituted a liquid resource, making the supply side of the market variable. Relations between farmers and merchants were flexible; the majority of farmers sought maximum prices, with the exception of some who preferred having a fixed relationship to assure access to various financing facilities. Implicitly, a certain contentment was expressed by the farmers with the commercial practices of the traders, because no major complaints were put forward in this respect.

For consumers, the same situation holds. Small quantities were bought regularly on nearby markets. Their storage capacity was limited because of lack of money. Generally, relations between consumers and traders were flexible; they sought the most attractive price and quality on the market. Credit did not play an important role, but in case maize was sold on credit the arrangement was often based on social rather than on economic consideration.

The Process of Competition in Market-Segments as a Result of Structure and Conduct Elements

To categorise competition between market-places in a spatial marketing network, surplus and deficit periods have to be taken into account. Seasonal fluctuations affect the structure of competition, as supply and demand conditions for maize vary considerably. Considering the flexible conduct of traders, competition cannot be assessed on the basis of the number of actors active on the market. It is potential competition that should be evaluated. This implies that a discussion focusing on entry/exit barriers is pertinent.

Four types of market-segments are distinguished in each regional or urban market-place: formal and informal, retail and wholesale, market-segments (cf. Lutz, 1994). In each segment, a different maize price exists depending on the commercial services delivered. However, as distances between the different segments are usually limited, the segments are strongly linked (cf. Table 12.5 for the linkage of prices between the formal market-segments).

In the informal retail market-segment, farmers and retailers sell small quantities, often on their premises, to consumers throughout the year. In informal wholesale market-segments, assembling traders buy from farmers on the road to the market or at the farmgate, especially during periods of abundant supply. Maize is to a lesser extent provided by resident wholesalers assuming the task of primary collection. This situation may change when regional surpluses dry up and wholesaler-distributors buy up maize in other surplus markets in order to supply retailers and assembling wholesalers. In the formal wholesale market-segment, wholesalers and farmers sell to retailers, assemblers or wholesalers during periods of abundance. The situation changes as described for the informal wholesale segment when deficits occur in the lean season. However, only minor quantities are traded in this segment as most actors prefer the informal wholesale segment in order to evade taxes. In the formal retail market-segment, retailers and some farmers sell to consumers on the periodic market.

Evidence was found for fierce competition in almost all the market-segments under study (cf. Lutz, 1994). When competition within a segment appeared to be rather weak, other segments in the market-place were able to contest the situation. This outcome gives rise to the expectation that prices in the spatial market network are integrated and that no profits in excess to normal rates of return can be realised by traders. This will be examined in the following sections. However, some artificial entry barriers — for farmers who wish to sell directly to non-resident wholesalers — were detected. These may influence prices for farmers negatively and, consequently, call for further research on assembly practices in the informal market-segment.

Temporal Arbitrage, Costs of Storage and Rates of Return for Invested Capital

When no entry barriers to trade exist, the degree of price integration depends on the quality of the physical and facilitating marketing infrastructure and the quality of market information. If temporal arbitrage is efficient, temporal price differences have to be smaller or equal to transfer costs. The storage costs can be subdivided into operating costs (fixed and variable) and quantitative and qualitative losses:

1. The variable costs involved in the storage activity consist mainly of capital costs (opportunity costs). As the banking system is not accessible to the majority of maize storing traders it is appropriate to base the estimate of opportunity costs on the profitability of other commercial services supplied by traders in the market. Two figures are crucial in such a calculation: an estimation of the average turnover rate of invested capital and the average gross margin traders realise.

   — A wholesaler selling (and buying) every market day 50 percent of his stock in a 4-day cycle, will realise a fourfold stock turnover per month (cf. Lutz, 1994, Table 5.8; this is in line with the estimation by traders that the average amounts sold equal the average amounts bought per market-day).

— A wholesaler accepts 5 FCFA/kg as a normal gross margin (cf. Lutz, 1994, Table 5.10). In Cotonou and Bohicon the average wholesale price is about 70 FCFA/kg (this corresponds to an average gross margin of = 7 percent). Taking into account fixed overhead costs (bags, bowls) and the labour supplied (opportunity cost for the trader's labour), the net margin is estimated to be somewhat lower and amounts to 5 percent.

This results in opportunity costs for capital equal 15 - 20 percent per month, depending on a three or fourfold stock turnover. If the turnover rate is lower and if the trader realises lower net margins this rate decreases. It is obvious from these calculations that a rate, for example, below 10 percent does not correspond with Beninois practice. Calculations reported by Alderman (1992, p. 6) provide comparable results for Ghana.

2. Quantitative pure 'weight' reduction occurs because the product loses some of its humidity during the storage period. Another source of loss is damage caused by insects and rats. The estimation of these percentages is based on the available literature (cf. Harnisch, 1986; Nago, 1981), which indicates that the average total loss for a 6-month storage period (the maximum length of the lean season in the south) approximates to a rate of 12 – 15 percent. We applied a growing rate to show increasing losses when the period of storage is extended and included a higher limit for the South to indicate favourable storage conditions in the north (cf. Lutz, 1994, Table 12.6).

Despite the fact that the average figures are quite correct, they may be only a rough indicator as deviations from the average may be significant. This makes storage a rather risky activity. The success of stock protection depends mainly on proper management. Has the maize been harvested without undue delay and is it well-dried to prevent insect infestation? Are the stores well-constructed? Information on such aspects is more difficult to collect for a trader than for the producer. So, from a management point of view one could say that a farmer has a comparative advantage over the trader to perform storage activities.

The costs mentioned in the last section were considered in order to estimate the rates of return for capital invested in a storage activity during the latter years of the eighties. It is an ex-post analysis to verify how arbitrage has been functioning to control price changes in the lean season.

Table 12.2 shows that the rate of return for invested capital is rather small, especially for the long-term storage activity (selling in May/June) as compared with the opportunity cost of capital as calculated above. Three characteristics make storage even more unattractive to traders:

(i) The risks arising from quantitative and qualitative losses during the storage period. Traders may be absent when the produce is harvested and stored and can, consequently, not assess the quality of the maize ex ante. Farmers do have a management advantage in this respect.

(ii) Risks resulting from price volatility. As was shown by Lutz (1994, Table 12.5), irregular price changes override short-run price trends. When maize is stored for a longer period, the seasonal factor becomes more important but even then the standard deviation for the rates of return remains high.

(iii) For traders with limited capital, depending to a large extent on trade profits for their income, a rate of return of 5 percent may not be sufficient for the most constraining production factor 'capital'. Capital is a scarce factor that limits the scale of the trade operations. Most traders prefer a high turnover rate since this leads to higher rates of return for the capital invested. Moreover, capital that is blocked in storage activities, may also block the returns that other factors (e.g. labour) produce. The opportunity costs for labour in the Beninois context is negligible because un-employment and under-employment are widespread.

Incidentally, it may still be attractive for traders to store maize. When no other profitable trading opportunities exist and when labour time has a high opportunity cost (e.g. traders who are also active as farmers) or when they lack time because of other obligations (illness, ceremonies, pregnancy etc.), they may engage in storage activities. Moreover, when traders are well-informed, short-run storage may be of interest in particular years. However, this is quite risky and will not fit into their average pattern of their business activities.

TABLE 12.2: Average Gross Margin per Sale for the Period 1985-1990: Rates of Return for Capital Invested in Storage Activities (Rate per Month and for Whole Storage Period)

		October	November	December	January	February	March	April	May	June	July
COTONOU	1*	8(21)	5(24)	0(17)	-(11)	9(16)	31(14)	30(8)	28(16)	17(15)	9(30)
		8.0	2.5	0.0	-0.5	1.8	5.2	4.3	3.5	1.9	0.7
	2					5(19)	24(6)	22(12)	19(20)	7(13)	-4(23)
						5.0	12.0	7.3	4.8	1.4	-0.7
BOHICON	1	17(34)	-1(27)	-2(28)	18(43)	8(30)	13(29)	20(31)	16(29)	11(25)	-1(29)
		17.0	-0.5	-0.7	4.5	1.6	2.2	2.9	2.0	1.2	-0.1
	2				13(29)	6(26)	8(15)	16(30)	10(26)	4(25)	-8(26)
					13.0	3.0	2.7	4.0	2.0	0.7	-1.1
PARAKOU	3			12(42)	23(70)	3(24)	11(29)	17(25)	16(31)	24(38)	30(41)
				12.0	11.5	1.0	2.8	3.4	2.7	3.4	3.8
	4						-3(4)	7(16)	5(16)	12(34)	16(34)
							-3.0	3.5	1.7	3.0	3.2
DJOUGOU	3			-3(8)	-3(18)	11(30)	15(23)	23(18)	33(20)	37(19)	14(20)
				-3.0	-1.5	3.7	3.8	4.6	5.5	5.3	5.9
	2					4(18)	7(10)	19(17)	28(19)	31(16)	38(21)
						4.0	3.5	6.3	7.0	6.2	6.3
AZOVE	1	4(18)	19(19)	17(18)	20(10)	35(21)	59(28)	56(24)	53(35)	45(34)	19(32)
		4.0	9.5	4.3	5.0	7.0	9.8	8.0	6.6	5.0	1.9
	2					6(9)	24(19)	20(19)	15(23)	8(23)	-11(27)
						6.0	12.0	6.7	3.8	2.0	-3.7
POBE	1	-8(4)	-5(8)	-3(13)	-2(11)	10(26)	6(19)	24(16)	30(25)	16(18)	-10(11)
		-8.0	-2.5	-1.0	0.5	2.0	1.0	3.4	3.8	1.8	-1.0
						4(18)	0(11)	14(12)	21(33)	5(16)	-19(9)
						4.0	0.0	4.7	5.3	1.0	-3.2
GLAZOUE	1	3(11)	3(14)	4(20)	5(18)	1(18)	24(25)	38(39)	47(43)	57(47)	24(34)
		3.0	1.5	1.3	1.3	0.2	4.0	5.4	5.9	6.3	2.4
	2					-10(6)	9(20)	17(15)	23(15)	30(15)	3(23)
						-10.0	4.5	5.7	5.8	6.0	0.5

* The storage strategy is based on the idea that the farmers/traders buy when prices attain minimum levels: September in the south = strategy 1, November in the north = strategy 3 (cf. Lutz, 1994, Table 6.5). In the south it may be difficult to stock in October/November because the short rainy season may hamper the drying of the product. The same calculation is done for actors that decide to stock only in the months of December/January (strategy 2). In Parakou a strategy (4) based on buying in February was distinguished.

* The first figure gives the net rent for the whole period. Between brackets the standard deviation is given, subsequently the rent is given per month. The storage costs are estimated at 4 FCFA per kg plus quantitative losses (cf. Lutz, 1994, Table 6.6). The profit is calculated as follows: first the buying price is augmented with the cost of storage (construction, preparation), subsequently the quantity sold is depreciated with the losses. Total invested capital and total revenues having been determined, the rate of return can be derived:

I = Capital Invested = (Storage Cost + Buying Price) x (Amount Bought)
R = Revenues   = (Selling price) x (Amount Bought - Losses)
  Rent (percent) = (R-I)/I) x 100

Source: Calculations based on prices (monthly averages) collected by the Ministry of Agriculture (MDR), for the period 1985 - 1990.

The demand for maize is relatively stable in Benin. Consequently, supply shocks are the main source of price variation. Seasonal price fluctuations are acceptable provided they remain within the limits set by storage and other transaction costs. The results of the price analysis show that storage is not an attractive activity to traders. This does not necessarily imply that the system cannot be made more efficient.

However, this is a question related to welfare economics and farm management economics, which needs serious examination but is beyond the scope of this chapter.

Spatial Arbitrage: Inter-regional Price Integration

Conditions for spatial price integration can be defined as follows (Tomek and Robinson, 1981, p. 151):

(i)   Price differences between any two regions (or markets) that trade with each other will just equal transfer costs; and

(ii)   Price differences between any two regions (or markets) that do not engage in trade with each other will be less than or equal to transfer costs.

Despite the anticipated contestability of markets, price differences between market-places are regularly larger than justifiable differences permit.[4] This indicates that some deficiencies in the system of market integration exist. Owing to the fact that traders buy on the basis of 'expected' prices, some observations exceeding justifiable price differences may be allowed. However, when this becomes a regular phenomenon and when certain markets are performing relatively worse than others, then it becomes interesting to analyse the price adaptation process between two markets after a price shock. Does lack of price information lead to short-term perturbations only, or does it affect price integration between markets also in the long run?

An Error Correction Model is derived to represent the process of price adaptation at the dependent market (price A) after a price shock at the central market (price B). In this equation the error correction term is interpreted as the mechanism adjusting for the disequilibrium in the period with lag (1-n).

$$A_1 = C + \delta(A_{1-n}\, B_{1-n}) + \sum_{j=o}^{n-1} \alpha_j + \sum_{j=o}^{n-1} \beta_j\, AB_{i-j} + TX_1 + e_1 \qquad (12.1)$$

The next question is how to choose the most suitable pairs of central and dependent markets. One may test for price integration between the regional markets of Azove and Nikki, but product flows indicate that direct transactions between these markets are rare and, when they do exist, they highly depend on prices in the urban markets of Bohicon and Cotonou. The choice of market pairs is based on the hierarchy in the maize market.

In line with the hierarchy of market-segments, retail markets are supposed to be directly related to the wholesale market in the same locality because the majority of the retailers in urban and regional markets buy their produce on the wholesale markets. We find the market of

---

[4]   Due to a lack of space this analysis will not be presented here. Lutz (1994, Section 6.3) compared price differences between market-segments with transaction costs (transport costs, taxes, etc.) for various types of traders.

Cotonou, the largest urban consumer market, which may be supplied by all other markets because price levels here are as a rule higher than elsewhere. Therefore, we start the analysis with the relationship between each of the wholesale markets and Cotonou as the reference market. The results of the tests for cointegration (Lutz, 1994, Table 7.6) indicate that there are no strong objections to this step.

The Retail-Wholesale Market Relationship in the Same Locality

The models that depict the relationship between wholesale and retail prices in the same locality have a relatively good fit (Table 12.3). The values for the coefficients δ and β_0 are all significant, indicating that a process of price integration exists. The test for market segmentation (no price integration between market pairs) is rejected for all pairs. However, market integration is less than perfect because the conditions for immediate or short-run integration are not satisfied. Some market pairs of wholesale and retail segments in the same locality (Glazoué, Savalou, Dassa, Hlassamé and Dogbo) are close to this situation: β_0 was larger than 0.80, which means that more than 80 percent of a price change on the wholesale market is transferred to the retail market on the same day. This result is not surprising because competition is strong in this market-segment.

The Wholesale Market Relationship (Tables 12.3 and 12.4)

The results for these models are less clear. The fit is much weaker for relations between wholesale markets: R^2 between 0.15 and 0.55, despite the price shocks that were taken into account by dummy variables. A large part of the variation is related to other variables. These may be related to market imperfections (e.g. a lack of information) and measurement errors, but also to substantial transaction costs and changing local supply conditions.

The main southern markets (Bohicon, Ketou, Azové and Glazoué) are relatively well-integrated with the reference market. The test for segmentation was rejected, although no immediate short-run integration could be observed. This also holds for Glazoué, where the error correction term in the relation was not significant. The linkages between the centre (Cotonou) and the other southern regional markets are less intensive. Price sluggishness is clearly demonstrated. However, a gradual process of integration takes place in the short run; the test for segmentation could not be rejected except for the market pair Dogbo-Cotonou (Table 12.3).

The northern markets are hardly integrated with the centre (Cotonou) in the south in the short run. The conditions for segmentation were rejected for Parakou but not for Tchaourou and Nikki. Even for Parakou the relation is weak because δ is only -0.07 (Table 12.3). This was to be expected since these markets are highly integrated with the maize market of Niamey (Niger). Moreover, transaction costs are high and prices follow a different seasonal harvest pattern.

Not all wholesale markets need to be integrated with the central reference market in the short run. When transaction costs with a central market are high and local surpluses are limited, arbitrage may be effective on an intra-regional level. Wholesale markets that are only weakly integrated with the national centre may be linked to the most nearby urban centre: Bohicon or Parakou. It is also useful to check if a segmented regional wholesale market is integrated with another regional surplus centre in the area. Thus, regional wholesale markets that are weakly integrated with one of the urban markets are linked to the most nearby regional centre.

The price relationships with an alternative reference market have been estimated (Table

TABLE 12.3: Error Correction Model Depicting the Short-Run Integration Process: All Wholesale Markets Are Related to Cotonou (W) and the Retail Markets are Linked to the Wholesale Markets in the Same Locality (R)*

	N	d^{***}	R^2	δ	α_1	α_2	α_3	α_4	β_0	β_1	β_2	β_3	β_4	β_5	F-seg**	F-short**	F-AR(4)
Cotonou (r)	200	0	0.39	-0.34*	-0.57*	-0.28*	-0.33*	-0.54*		-0.54*	0.60*	0.37*	0.57*	0.51*	18.2*	44.9*	2.4
Bohicon (r)	159	0	0.40	-0.32*						0.47*					53.1*	142.4*	1.1
Bohicon (w)	158	0	0.13	-0.10*	-0.03					0.19*	0.31*				7.4*	128.4*	1.5
Parakou (w)	140	0	0.13	-0.07*						0.02					11.5*	2292.8*	0.9
Azove (r)	182	0	0.48	-0.41*	-0.63*					0.65*	0.52*				53.2*	38.7*	1.4
Azove (w)	179	3	0.34	-0.04	-0.24*	-0.14*	-0.10			0.26*	0.27*				5.2*	257.3*	1.3
Dogbo (r)	183	0	0.57	-0.69*						0.81*					121.0*	9.0*	1.2
Dogbo (w)	183	3	0.35	-0.03						0.18*		0.02	0.30*		2.2	469.2*	1.8
Hlassamé (r)	135	0	0.72	-0.62*		-0.29*				0.83*	0.44*				176.2*	13.4*	1.4
Hlassamé (w)	151	2	0.14	-0.08	-0.10					0.18		0.05			3.2*	188.0*	0.9
Ketou (r)	139	0	0.36	-0.31*	-0.61*					0.56*					9.4*	28.0*	1.1
Ketou (w)	149	4	0.55	-0.16*		-0.40*	-0.51*	-0.55*		0.30*	0.58*	0.53*	0.43*	0.44*	13.1*	218.2*	0.9
Pobé (r)	110	0	0.43	-0.15*	-0.36*	-0.15	-0.34*	-0.37*		0.48*	0.28*	0.08	0.45*	0.31*	15.0*	145.1*	0.5
Pobé (w)	114	3	0.55	-0.06						0.26*					5.0*	379.4*	0.7
Dassa (r)	171	0	0.65	-0.19*	-0.55*	-0.39*				0.81*	0.50*	0.40*			79.2*	64.4*	0.6
Dassa (w)	172	1	0.32	-0.12*	-0.21*					0.18*	0.03				5.2*	370.7*	1.6
Savalou (r)	141	0	0.65	-0.26*	-0.46*					0.85*	0.46*				86.7*	60.5*	1.1
Savalou (w)	140	1	0.18	-0.10*	-0.06	-0.05				0.13	0.11	-0.13			4.3*	486.6*	0.1
Glazoué (r)	81	0	0.78	-0.54*						0.92*					140.2*	10.7*	1.0
Glazoué (w)	81	4	0.42	-0.11						0.33*					4.4*	107.1*	1.3
Nikki (w)	103	1	0.64	-0.05*						0.07					2.8	1053.5*	2.2
Tchaourou (w)	69	0	0.32	-0.04*	-0.11	-0.47*				0.05	-0.01	-0.09			1.8	1795.9*	1.1

* All coefficients marked with an asterisk are significant at the 5 percent significance level

** F-seg, test for segmentation. H_0: ν β=δ=o

F-short, test for short-run integration. H_1 : $\beta_0=1$ and δ=-1

F-AR(4), LM test for fourth order autocorrelation (see Lutz, 1994, Annex 7.1)

*** Number of dummies included for price shocks in the harvest season of June

Source: Survey Prices Wholesale/retail

12.4). The results indicate that some markets are more oriented to a centre lower in the hierarchy. The Mono markets of Azové and Hlassamé appear to have a stronger link with Hokicon than with Cotonou. This may be explained by the shorter distance between these markets. Dogbo is segmented from both urban centres (Cotonou and Bohicon) in the short run, but the results indicate that between Azové and Dogbo some degree of price integration exists and explains why Dogbo market prices are indirectly influenced by prices in the urban centres. The Dogbo market exports relatively small quantities of maize in the harvest season, but in the lean season maize has to be imported. Market relations between Dogbo and the urban centres are quite variable.

The same applies to Pobé; after the bad harvests of 1988, it was found that this market was weakly integrated with Cotonou, but the linkages with the nearby regional market of Ketou were more intensive, and they showed that this market was also integrated indirectly with the larger system.

The results for the markets in Zou Prefecture are somewhat disappointing. We expected a stronger linkage with Bohicon because this market is the most nearby urban centre. The results indicate the opposite: segmentation was not rejected for Glazoué/Bohicon, the coefficient $ß_0$ and the error correction term are not significant. However, in particular the regional market of Dassa, and to a lesser extent the market of Savalou, are integrated with Glazoué, lower in the market hierarchy but is the main market of the region.

Prices in Parakou are not linked more intensively to prices in Bohicon than to prices in Cotonou. The northern regional markets of Nikki and Tchaourou are not segmented from the Parakou market. However, also here price relations are quite weak. None of the $ß$ coefficients is significant and the process of price adaptation is mainly determined by a low absolute value of the error correction term. Tchaourou may seem to react to price changes in Bohicon ($ß_0$), but the effect is cancelled out when $ß_1$ and $ß_2$ are taken into account. Parakou is not a major consumer market for maize like Cotonou. The greater part of the regional surpluses are sold on the border market of Malanville where demand from Niamey determines the market conditions (cf. Fanou et al., 1991). Unfortunately we do not have any price series for this market.

Conclusions: The Speed of Price Change Transmission (Table 12.5)

When some evidence for long-run integration exists, but tests for 'immediate' short-run integration are rejected, the question arises of how long it will take before a price change in the reference market is transferred to the dependent market.

We found that price linkages between wholesale and retail markets in the same locality are quite intensive. Within 4 days at the least, 75 percent of a price change in the wholesale market is transferred to the retail market, and more than 90 percent after 20 days. Only for Pobé and Ketou are these percentages slightly smaller. Between wholesale markets the price linkages are less intensive. Price changes in Bohicon are more quickly transferred to Azové and Hlassamé than price changes in Cotonou. Table 12.5 confirms the results already presented in the other tables. A process of price integration is effective in the spatial marketing network. However, price sluggishness between several market pairs indicates that the market can perform better.

TABLE 12.4: Error Correction Model Depicting the Short-Run Integration Process: The Relationship With an Alternative Centre Lower in the Market Hierarchy

A_1 x B_1		N	d***	R^2	δ	α_1	α_2	α_3	α_4	B_0	B_1	B_2	B_3	B_4	F-seg**	F-short**	F-AR(4)
Pa x Bo	(w)	140	0		0.17	-0.09*				-0.00*					15.3*	2021.6*	0.4
Az x Bo	(w)	157	3		0.43	-0.13*	-0.37*	-0.17*		0.43*	0.41*	0.41*			10.8*	150.0*	2.2
Do x Bo	(w)	158	3		0.45	-0.04				0.15					1.8	481.9*	0.5
Hi x Bo	(w)	153	2		0.12	-0.11*				0.50*					6.4*	230.6*	1.1
Do x Az	(w)	182	3		0.42	-0.15*				0.28*					9.9*	175.1*	1.4
Hi x Az	(w)	153	2		0.07	-0.11*				0.10					2.6	172.2*	1.3
Po x Ké	(w)	113	3		0.57	-0.14*	-0.08			0.16*	0.27*				5.2*	137.3*	1.3
Da x Bo	(w)	158	2		0.39	-0.17*				-0.09					12.0*	351.5*	0.5
Sa x Bo	(w)	135	1		0.17	-0.09*				0.15					5.7*	465.9*	0.3
Gl x Bo	(w)	81	4		0.40	-0.15				0.20					2.6	67.9*	1.2
Da x Gl	(w)	81	1		0.60	-0.44*				0.61*					35.3*	30.2*	0.8
Sa x Gl	(w)	66	1		0.35	-0.12*	-0.11			0.37*	-0.09				5.5*	189.9*	0.3
Ni x Pa	(w)	102	1		0.67	-0.12*	0.09			0.15	-0.01				4.7*	324.2*	1.4
To x Pa	(w)	68	0		0.41	-0.08*	-0.18	-0.38*		0.21	-0.01	0.12			3.5*	498.9*	0.2
To x Bo	(w)	69	0		0.38	-0.05*	-0.10	-0.49*		0.20*	-0.04	-0.15			3.5*	1645.6*	0.5

* All coefficients marked with an asterisk are significant at the 5 percent significance level

** F-seg test for segmentation. H_s: v B=δ=o

F-short, test for short-run integration. H_s: B_o=1 and δ=1

F-AR(4), LM test for fourth order autocorrelation (see Lutz, 1994, Annex 7.1)

*** A_1 corresponds with the dependent market, B_1 is the centre.

Pa=Parakou, Bo=Bohicon, Az=Azové, Do=Dogbo, H1=Hlassamé, Po=Pobé, Ke=Ketou, Da=Dassa

Source: Survey prices wholesale/retail

TABLE 12.5: The Speed of Adjustment After a Price-change in the Reference Market

			Cumulative Partial Multiplier* After a Limited Number of Market Cycles					
A_1		B_1**	same Market Cycle	After 1 Cycle	After 2 Cycles	After 3 cycles	After 4 cycles	After 5*** cycles
Cotonou	(r) x	Cotonou (w)	0.54	0.83	0.88	1.16	1.11	1.04
Bohicon	(r) x	Bohicon (w)	0.47	0.64	0.76	0.83	0.89	0.92
Bohicon	(w) x	Cotonou (w)	0.19	0.31	0.39	0.46	0.52	0.57
Parakou	(w) x	Cotonou (w)	0.02	0.09	0.15	0.21	0.27	0.31
Azové	(r) x	Azové (w)**	0.65	0.76	0.83	0.89	0.92	0.91
Azové	(w) x	Cotonou (w)	0.26	0.47	0.38	0.64	0.60	0.60
Azové	(w) x	Bohicon (w)	0.43	0.68	0.93	0.87	0.89	0.91
Dogbo	(r) x	Dogbo (w)	0.81	0.94	0.98	0.99	1.00	1.00
Dogbo	(w) x	Azové (w)	0.28	0.39	0.48	0.56	0.63	0.69
Hlassamé	(r) x	Hlassamé (w)	0.83	0.94	0.98	0.99	1.00	1.00
Hlassamé	(w) x	Cotonou (w)	0.18	0.60	0.56	0.51	0.56	0.60
Hlassamé	(w) x	Bohicon (w)	0.50	0.56	0.61	0.65	0.69	0.72
Ketou	(r) x	Ketou (w)	0.56	0.80	0.96	0.90	0.89	0.85
Ketou	(w) x	Cotonou (w)	0.30	0.41	0.50	0.58	0.65	0.71
Pobé	(r) x	Pobé (w)	0.48	0.59	0.56	0.84	0.83	0.84
Pobeé	(w) x	Cotonou (w)	0.26	0.30	0.34	0.38	0.42	0.45
Pobeé	(w) x	Ketou (w)	0.16	0.42	0.52	0.59	0.65	0.71
Dassa	(r) x	Dassa (w)	0.81	0.86	0.91	0.90	0.92	0.93
Dassa	(w) x	Cotonou (w)	0.18	0.17	0.27	0.35	0.42	0.49
Dassa	(w) x	Glazoué (w)***	0.61	0.78	0.88	0.93	0.96	0.98
Savalou	(r) x	Savalou (w)	0.85	0.92	9.93	0.95	0.96	0.97
Savalou	(w) x	Cotonou (w)	0.13	0.23	0.08	0.17	0.25	0.36
Savalou	(w) x	Glazoué (w)***	0.37	0.24	0.33	0.41	0.48	0.54
Glazoué	(r) x	Glazoué (w)***	0.92	0.96	0.98	0.99	1.00	1.00
Glazoué	(w) x	Cotonou (w)***	0.33	0.40	0.47	0.53	0.58	0.63
Nikki	(w) x	Parakou (w)	0.15	0.15	0.25	0.36	0.46	0.55
Tchaourou	(w) x	Parakou (w)***	0.21	0.16	0.37	0.37	0.52	0.54
Tchaourou	(w) x	Bohicon (w)***	0.20	0.14	0.10	0.11	0.13	0.18

* The partial multiplier is derived from the ECM model. For a model with one lag the direct effect is equal to B_eAB_w, the second effect is $o(B_eAA_w)$, etc. (cf. Lutz, 1994, Annex 7.2).

** r = retail market, w = wholesale market

*** The market cycle of most market pairs correspond to a period of 4 days. Thus: 5 cycles of 4 days equals a period of 20 days Only in Glazoué and Tchaourou is the cycle longer; a period of 7 days. So, 5 cycles of 7 days equals a period of 35 days.

Source: Survey prices wholesale/retail

Results and Policy Recommendations

This study focused on the functioning of the maize market in Benin from a meso-economic perspective i.e. the functioning of the network of markets for maize. In this section an effort will be made to link the performance of the maize market in Benin to characteristics of its structure and conduct. The focus of the evaluation is on efficiency. In the final section some policy recommendations will be presented.

Efficiency of Market Performance

The assessment of economic efficiency of the market was focused on the question of whether services are supplied against acceptable gross margins. Efficiency of market functioning was operationalised as an analysis of gross margins. These were considered to be justified as long as services were supplied at normal rates of return. From the results it followed that seasonality in prices did not indicate that profits in excess of normal rates of return were realised by storage activities. On the contrary, for traders the storage service is a minor activity generating relatively small rents.

Despite the anticipated contestability, spatial price differences between market-segments are regularly larger than can be justified, indicating that there are some deficiencies in the system of market integration. The process of price adjustment and price integration between markets was scrutinised. Evidence is found for the existence of long-run price integration, indicating that prices at different market-places cohere in the long run and that a process of spatial arbitrage takes place. The condition of immediate, full integration is too stringent for the maize market in Benin. Serious price sluggishness is found, which is, as expected, more important when distances between market-places are larger. Hence, a lack of 'perfect' market integration is detected.

In order to understand this somewhat ambiguous result, the qualitative information that was collected on market structure and conduct components is helpful to explain the sluggishness of the price formation process. It is here that the SCP paradigm shows its suitability. Two aspects of the argument are distinguished: degree of competition, which is the major driving force in the market, and entry and exit barriers.

Competition

In order to characterise competition, the market was subdivided into market-segments and the results show that competition was fierce. When competition within one segment is limited, other segments may compete indirectly and preserve the conditions for a contestable market. Competitive strategies are based on prices and marketing costs and, as such, encourage traders to search for lower cost solutions. The structure of the marketing costs for different types of traders does not indicate that this leads to a process of concentration in the market (cf. Lutz, 1994, Section 6.3). On the contrary, small traders may successfully compete with large traders when distances are not too great. Consequently, artificial barriers seem to be ineffective as an instrument to block innovations; the flexible market will be competent to correct for deficiencies. Only at the local level, resident wholesalers may succeed in limiting competition from non-resident wholesalers. It is here that the development of a shorter, more efficient marketing channel may be hampered. Also the semi-illegal trade flows with neighbouring countries provide some scope for resident wholesalers to protect their market interests and consequently there is a call for further institutional supervision (cf. Fanou et al., 1991).

Entry and Exit Barriers

Thus far the conclusions have made it clear that the market structure is effective and competition within and between market-segments rather intense. However, evidence for immediate price integration was not found, so some form of entry barrier must exist to explain this result.

In Benin, the barriers to entry are not related to the traditional barriers discussed by Bain (1968) (cf. Lutz, 1994, Section 2.1.2). Economies of scale play a limited role because the cost structure of traders shows that only limited funds are needed to be able to start a viable commercial enterprise. Also sunk costs are limited and consequently exit barriers, which may arise when a firm's assets are specific to its production in progress, are negligible.

Product differentiation, another entry barrier stressed by Bain, does not constitute a barrier either. In our case, product differentiation should be interpreted as 'service' differentiation and plays a role in the marketing strategies of the traders. In the formal and informal market-segments a different mix of services (credit, transport) may be supplied occasionally. Generally, transactions are carried out in conventional marketing channels (CMC). As shifting costs are small, the mobility of farmers and consumers is high. Consequently, they may decide to sell or buy in the market-segment that offers the best prices (cf. Lutz, 1994, Sections 3.2, 5.2 and 6.3.1). Nevertheless, two types of entry barriers are relevant: local organisations regulating entry and market information.

Local Organisations Regulating Entry

This practice hampers the functioning of the market, because no cost argument justifies such regulation. However, not all activities of local organisations should be refused, as they might facilitate investments in infrastructure, the introduction of standards, and the process of consultation between traders and government in order to develop efficient trade policies.

Market information

A condition for successful market entry is access to adequate market information. This constitutes an entry barrier in particular for smaller traders. Obtaining information is costly in view of their small turnover (cf. Lutz, 1994, Section 6.3.1). The costs involved in the collection of information are a disincentive to traders who wish to extend their operations spatially. Market information may be seen as part of a trader's specific human capital. It explains why traders withdraw temporarily from the market, or stick to specific human capital. It explains why traders withdraw temporarily from the market, or stick to specific market-segments and continue with relatively less or even unprofitable activities. Others will stress that these barriers may be accepted as 'first-mover' advantages (cf. Stigler, 1968). However, even when lack of proper market information is not viewed as an entry barrier, it is clear that it is a catalyst in the process of competition and, as such, justifies our attention. This may explain the ambiguous result that markets are integrated in the long run but should do better in the short run. It is the lack of information on prices in alternative market-segments that makes the process of price adaptation.

Policy Recommendations

The question relevant for policy makers is how the institutional environment can improve the performance of the maize market in Benin. The emphasis is not on direct government intervention in the maize trade but on policies that reduce entry barriers and facilitate competition. According to the authors, improvements in the functioning of the maize market in Benin can be attained in the following areas:

— Standardisation of units of measurement and quality classes;
— Set-up of a reliable price information system in which information is given on supply and prices per quality class;
— Reduction in artificial entry barriers in formal market-segments;
— Encouragement of traders' organisations to participate in the development of market institutions to ameliorate market regulations and the market infrastructure;
— Decrease in regulatory uncertainty in the formal market-segments by enforcing

286

existing market regulations more rigorously;

—   Investment in, and maintenance of, market-places, roads and the telecommuni-
cation infrastructure; and

—   Provision of training and extension services adapted to the needs of farmers and
traders.

This chapter has illustrated that the development and application of a framework to assess the performance of the maize marketing system in Benin is a valuable instrument to find the weaknesses in the system and develop an optimal institutional market environment that fosters trade and economic growth.

REFERENCES

Abbott, J.C. ed., 1993. *Agricultural and Food Marketing in Developing Countries: Selected Readings.* C.A.B. International, Wallingford, UK.

Achrol R.S., T. Reve & L.W. Stern, 1983. "The Environment of Marketing Channel Dyads: a Framework for Comparative Analysis". *Journal of Marketing,* 547 (Fall), P. 55-67.

AGRER, 1986. Etude de la Commercialisation des Produits Vivriers au Benin, Volume 1, Bruxelles.

Ahmed, R. and N. Rustagi, 1987. Agricultural Marketing and Price Incentives: A Comparative Study of African and Asian Countries. Mimeograph (reprint), International Food Policy Research Institute, Washington.

Alderman, H., 1993. "Intercommodity Price Transmittal: Analysis of Food Markets in Ghana." *Oxford Bulletin of Economics and Statistics* 55, 1 p. 43-64.

Alderman, H., 1992. Food Security and Grain Trade in Ghana. Working Paper of the Cornell Food and Nutrition Policy Programme, 24 p, Washington.

Ardeni, P.G., 1989. "Does the Law of One Price Really Hold for Commodity Prices?" *American Journal of Agricultural Economics,* August 1989, p. 661-669.

Armah, P.W., 1989. Post-harvest Maize Marketing Efficiency: The Ghanaian Experience. Ph.D. Thesis, University College of Wales, Aberystwyth.

Badiane, O., 1989. Espace Regional Cerealier en Afrique de l'Ouest: Potentiel et Importance pour la Securité Alimentaire. Paper Presented at a Seminar on 'Regional Cereals Markets in West Africa'. CILLS/Club du Sahel, Lome.

Bain, J.S., 1968. *Industrial Organisation.* Second Edition, John Wiley, New York.

Bardhan, P., 1990. "Symposium on the State and Economic Development". *Journal of Economic Perspectives,* Vol. 4, No.3, p. 3-7

Bardhan, P., 198. *The New Institutional Economics and Development Theory: A Brief Critical Assessment.* World Development, 17, 9, p.1389-1395.

Baumol, W.J., J.C. Panzar and R.D. Willig, 1988. *Contestable Markets and the Theory of Industry Structure,* rev. ed. San Diego: HBJ Publishers.

Campbell, G.R. and Th. S. Clevenger, 1978. "An Institutional Approach to Vertical Coordination in Agriculture." N.C. Studies of the Organization and Control of the U.S. Food System, 17 p.

Campbell, G.R. and M.L. Hayenga, 1978. Vertical Organisation and Coordination on Selected Commodity Subsectors, AAEA Symposium, Blacksburg, Virginia, 122p.

Calton, D.W. and J.M. Perloff, 1994. *Modern Industrial Organisation*. Harper Collins College Publishers.

Caves, R. and M. Porter, 1976. "Barriers to Exit." In: Masson, R. and Qualls, P. (eds.), Essays in Industrial Organisation and vector autoregression. Edward Elgar.

Coase, R.H. ,1937. "The Nature of the Firm." *Economica* 4, 386-405.

Coase, R.H., 1960. "The Problem of Social Cost." *Journal of Law and Economics* 17, 1-44.

Cubbin, J.S. 1988. *Market Structure and Performance, the Empirical Research*. Harwood Academic Publishers.

Eggertsson, T., 1990. *Economic Behaviour and Institutions*. Cambridge University Press.

Fanou, L.K. C.H.M. Lutz and S. Salami, 1991. Les Relations entre les Marches de Mais du Benin et les Marches des Espace Avoisionants au Togo, au Niger et au Nigeria. FSA, Cotonou/Amsterdam.

FAO, 1987. Mission de Securite Alimentaire et de Commercialisation au Benin, Volume 1, Rome.

Harnisch, R., 1986. Situation du Stockage Traditionnel, Collectif et Industriel en Republique Populaire du Benin. GTZ, Eschborn.

Harriss, B., 1979. There is a Method in my Madness or Is It Vice Versa? Development Studies Discussion Paper 54, University of East Anglia, U.K.

Harriss, B., 1982. Agricultural Marketing in the Semi-arid Tropics of West Africa. East Anglia Development Studies Paper.

Harriss, B., 1993. Markets Within Markets and no Markets at all: Maps and Landscapes of Grain Markets in South Asia. Paper Presented at Workshop on Methods for Agricultural Marketing Research. March, 1993, CIP/LARI New Delhi, India, 49 p.

Hazell, P.B.R. and R.D. Norton, 1986. *Mathematical Programming for Economic Analysis in Agriculture*. MacMillan Publishing Company.

Hodgson, G.M., 1988. Economics and Institutions. Polity Press.

Hodgson , G.M., 1993. "Transaction Costs and the Evolution of the Firm." In: Christos Pitelis (ed.).

Hoff, K.A. and Braverman, J.A., 1993. The *Economics of Rural Organization:* Theory, Practice and Policy. Oxford University Press, New York.

Janvry, A. de, and E. Sadoulet, 1994. "Structural Adjustment under Transaction Costs." In: F. Heidhues and B. Knerr, *Food and Agricultural Policies Under Structural Adjustment*. Peter Lang Verlag, Frankfurt.

Killick, T., 1988. *A Reaction too Far: Economic Theory and the Role of the State in Developing Countries.* Heineman, London.

Kohls, R.L. and J.N. Uhl, 1990. *Marketing of Agricultural Products*, 7th ed. MacMillan, London.

Jones, W.O., 1984. "Economic Tasks for Food Marketing Boards in Tropical Africa." In J. Abbott 1993. *Agricultural and Food Marketing in Developing Countries, Selected Readings,* C.A.B. International, p.144-163.

Kohls, R.L. & J.N. Uhl, 1990. *Marketing of Agricultural Products*, 7th ed. MacMillan.

Kotler, P.H., 1994. *Marketing Management.* Prentice Hall.

Krishnaiah, J. and S. Krishnamoorthy, 1988. Interregional Allocation of Major Foodgrains in Andhra Pradesh: An Application of Spatial Equilibrium Model.

Lele, U., ed., 1992. *Aid to African Agriculture: Lessons from two Decades of Donors' Experience.* Johns Hopkins University Press, World Bank, Baltimore.

Lutz, C.H.M., 1994. *The Functioning of the Maize Market in Benin: Spatial and Temporal Arbitrage on the Market of a Staple Food Crop.* University of Amsterdam, Dep. of Regional Economics, 255p.

Lutz, C.H.M. & A. van Tilburg, 1992a. "Spatial Arbitrage Between Rural and Urban Maize Markets in Benin." In: L. Cammann (ed.) Traditional marketing systems. German Foundation for International Development (DSE), Feldafing, p. 90-100

Lutz, C.H.M. & A. van Tilburg, 1992b. Concurrentievormen op rurale en rubane maismarkten in Zuid-Benin, Tijdschrift voor Social Weternschappelijk Onderzoek van de Landbouw, Jarrgang 7, Nr. 3 p. 195-221.

Lutz, C.H.M., A. van Tilburg and B. van der Kamp, 1995. The Process of short- and long-term Integration in the Benin Maize Market. European Review of agricultural economics, Vol. 22-2, p. 191-212.

Marion, B.W. and W.F. Mueller, 1983." Industrial Organisation, Economic Power, and the Food". P.L. Farris (ed.), *Future Frontiers in Agriculture Marketing Research,* Iowa State University Press.

Martin, L.J., 1981. "Quadratic Single and Multi-commodity Models of Spatial Equilibrium: a Simplified Exposition." *Canadian Journal of Agricultural Economics* 29, 1, 21-48.

Matthews, R.C.O., 1986. "The Evolution of Agricultural Marketing Theory: Towards Better Coordination with General Marketing Theory." *Netherlands Journal of Agricultural Science* 34, 301-315

Nably, M.K. & J.B. Nugent, 1989. "The New Institutional Economics and its Applicability to Development." *World Development* 17, 9, p.1333-1347.

Nago, M.C. and V. Agbo, 1981. Stockage du Mais au Benin. Etude de cas du district d' Allada. FAO/Danida/African rural storage centre, Cotonou.

North, D.C., 1990. *Institutions, Institutional Change and Economic Performance*. Cambridge University Press.

Oladeji, S.O., 1987. Epargne et Credit en Milieu Rural (etude de cas de deux villages, Djobo et Gbedavo, dans la province du Zou). Faculte des Sciences Agronomiques, Abomey Calavi, Benin.

Palaskas, T.B. and B. Harriss, 1993. "Testing Market Integration: New Approahes with Case Material from the West Bengal Food Economy." *Journal of Development Studies*, Vol. 30, 1, p.1-57.

Ravallion, M., 1986. "Testing Market Integration." *American Journal of Agricultural Economics*, February, 102-109.

Ross, C.G., 1980. "Grain Demand and Consumer Preference in Senegal". *Food Policy 5*, p.273-281.

Schere, F.M. & D. Ross, 1990. *Industrial Market Structure and Economic Performance*, 3rd ed. Houghton Mifflin Company, Boston.

Sosnick, S.H., 1958. "A Critique of Concepts of Workable Competition." *The Quarterly Journal of Economics*.

Sterm, L.W. and A.I. El-Ansary, 1992. *Marketing Channels*, 4th ed. Prentice-Hall International Editions, London.

Stigler, G.J., 1968. "Barriers to Entry, Economies of Scale and Firm Size." In: Irwin, R.D. *The Organization of Industry*. Homewood, Illinois.

Takayama, T. and G.G. Judge, 1971. *Spatial and Temporal Price and Allocation Models*. North-Holland Publishing Company.

Thorbecke & Morrison, 1989. "Institutions, Policies and Agricultural Performance: a Comparative Analysis." *World Development*, 17, 9, p.1485-1498

Tilburg, A. van & C. Lutz, 1992. "Competition at Rural Markets in South Benin." In: L. Cammann (ed.) *Traditional Marketing Systems*. German Foundation for International Development (DSE) Feldafing, p.101-112.

Tilburg, A. van and C. Lutz, 1995. Framework to Assess the Performance of Food Commodity Marketing Systems, Conference of the Reseau SADAOC on "Sustainable Food Security in West Africa". Accra.

Timmer, C.P. 1986. *Getting Prices Right: The Scope and Limits of Agricultural Policy*. Cornell University Press, New York.

Tirole, J., 1988. *The Theory of Industrial Organisation*. The MIT Press.

Tomek, W.G. and K.L. Robinson, 1981. *Agricultural Production Prices*, 2nd ed. Cornell University Press, New York.

Williamson, O.E., 1985. *The Economic Institutions of Capitalism*. The Free Press.

AUTHOR INDEX

A

Abbott, J. C., 288, 290
Abott, P., 117
Abu, K., 143, 148
Achrol, R. S., 267, 288
Adolfse, L., 204
Agbessi Dos-Santos, H., 113
Agbo, V., 291
AGRER, 273, 288
Agripromo, 99, 100, 113
Ahmed, R., 234, 251, 273, 288
Alamgir, 53
Alan, W., 19, 33
Albersen, P., 36
Alderman, H., 135, 148, 188, 190, 204, 208,
          210, 212, 214, 217,
          229-231, 272, 277, 288
Al-Hassan, R., 79, 129
Amemiya, T., 211, 220, 229
Ancey, 84, 113, 256, 262
Andel, J., 115
Anderson, D., 21, 27, 33, 34, 42
Ankrah, E. K., 148, 152, 204
Ardeni, P. G., 288
Arecchi Alberto, 33
Arhin, K., 253, 262
Armah, P. W., 288
Arroyave, G., 210, 214, 230
Asante, F. A., 186, 187
Asenso-Okyere, W. K., 1, 2, 9, 13, 35, 36, 53,
          54, 174, 186, 187, 201,
          204, 205, 207, 229, 231
Askari, H., 155, 174
Atsma, P., 174, 233, 234
Attah, J. K., 229
Awusabo-Asare, K., 36, 53
Azam, J. P., 166, 172, 174

B

Badiane, O., 3, 13, 288
Bain, J. S., 265, 285, 288
Baker, 33, 91, 112, 114, 229
Bakker, 92, 93, 96, 117
Bakker-Frijling, M. J., 113
Baldus, R. D., 16, 33
Bapna, S. L., 156, 174
Bardhan, P., 264, 288
Barret, V., 30, 33

Basta, H. P., 149
Bateman, M. J., 159, 174
Baumol, W. J., 265, 288
Baur, H., 129, 133, 134, 138, 139, 143, 148
Behnke, R., 18, 33
Behrman, J. R., 13, 71, 156-159, 174, 204,
          227, 229, 231
Benneh, G., 1, 2, 134, 148, 174, 231
Benoit-Cattin, M., 115
Berkum, S., 258, 262
Besley, T., 166, 172, 174
Bevan, D., 166, 172, 174
Beynon, J. G., 153, 174
Bidani, B., 38, 54
Blackie, M. J., 115, 116
Blaikie, P., 24, 33
Bleiberg, F. M., 92, 113
Boateng, E. O, 36, 39, 53, 55, 60, 65, 71,
          134, 148, 190, 204, 209,
          229
Boffa, J. M, 115
Bond, M. E., 153, 154, 172, 174
Bonsu, M., 149
Boom, G.J. M., 9, 13, 35, 45, 46, 53-55, 186,
          204, 207, 231
Bos, A. K., 204, 233, 252
Bouis, H. E., 189, 204, 212, 214, 221, 224,
          229
Bound, J., 211, 218, 221, 229
Braverman, J. A., 289
Broekhuyse, J., 81, 86, 92, 113
Brown, C. K., 36, 53
Brun, A., 92, 113
Bryceson, D. F., 234, 241, 251
Buckwell, A. E., 105, 113
Bumb, B. L., 207-209, 229

C

Caldwell, 204
Calton, D. W., 265, 289
Campbell, G. R., 267, 268, 288
Cantrell, R., 115
Caves, R., 289
CEBV Study Team, 213, 229
CEBV, 254, 257, 262, 263
Chambers, R., 132, 148
Chhibber, A., 153, 174

293

SUBJECT INDEX

A

Absolute Poor, 27, 30
Active Population, 3, 7, 8, 30, 134
Acute, 188, 190, 191
Adjustment, 7, 8, 10-12, 30, 37, 53, 55,
    56, 71, 72, 79, 81, 153-157,
    160, 161, 163, 164, 166, 168,
    170-175, 204, 229, 252, 264,
    284, 285, 289
Adult Equivalent, 19, 197
AFS, 132, 134, 135, 137-140, 142-146,
    149, 152
Agricultural, 1, 3, 5-8, 10, 11, 13-15, 18,
    19, 21, 22, 24, 27,
    29-34, 53, 65, 69-71,
    75, 79-81, 84, 85,
    88-94, 96-98, 100,
    102, 104, 105,
    108-117, 121, 123,
    124, 126, 129-132,
    134, 140, 142,
    148-150, 153-155,
    157-159, 172-175,
    188, 204, 208-210,
    215, 224, 229-231,
    233-238, 240-242,
    244, 248-252, 258,
    264, 269, 271, 288-292
Agricultural Policy, 117, 131, 132, 291
Agriculture, 2-4, 8, 10, 13, 14, 16, 18,
    19, 27, 45, 51, 63, 64,
    66, 69-71, 73-75, 80,
    87, 88, 97, 100, 107,
    112, 115-118, 134,
    149, 153, 154, 156,
    166, 172, 174, 175,
    190, 204, 207-210,
    215, 216, 219,
    229-231, 234, 235,
    243, 247, 251, 262,
    278, 288-290
Agro-ecological, 36, 63, 65, 105, 106,
    188, 189, 209, 219,
    244, 248
Agro-forestry, 10, 97, 112
Agro-pastoralists, 107-109
Allocations, 84, 138, 268
Analysis of Farmers' Strategies, 80-83,
    132
Animal Husbandry, 97, 100, 104, 106,
    112, 136

Animal Traction, 33, 82, 87, 96, 108,
    112, 114, 115, 137
Anthropometric, 9, 42, 187, 192, 195,
    196, 202, 203
Anti-erosion, 88, 97, 112
Approach, 18, 19, 35, 36, 72, 79-81, 83,
    85, 87, 106, 111, 112, 117,
    130, 131, 154, 159, 174, 211,
    265, 266, 288
Assets, 2, 37, 43, 45-51, 75, 215, 285
Availability, 2, 24, 30, 32, 45-51, 81, 83,
    84, 86, 88-90, 97, 98,
    104, 105, 109, 111,
    125, 129, 130, 132,
    135, 138-140, 142,
    159, 172, 188-191,
    196-202, 207, 208,
    211, 212, 215, 219,
    241
Azove, 278, 279, 281

B

Barriers, 111, 254, 269, 271, 273, 276,
    285, 286
Benin, 5, 6, 25, 26, 28, 29, 233, 258,
    264, 265, 272-274,
    279, 284-291
Birth, 28, 30, 188, 192, 193, 195, 198,
    230
Bivariate, 189, 190
Body Mass Index, 43, 45-51, 198, 199,
    205, 215, 216, 231
Body Mass Indices, 9
Body Weight, 213-216, 218, 224, 227
Bohicon, 277-284
Breastfeeding, 192, 200-202
Bullocks, 136, 141, 237
Burkina Faso, 3-5, 25, 26, 28, 29, 79-82,
    84, 85, 99, 104, 108,
    110-118, 131, 132,
    149, 153, 154,
    157-160, 166-168,
    170-173, 175, 181,
    253-256, 258-260,
    262
Bush Farming, 132, 134-140, 142, 145,
    146
Business Income, 69, 75

C

D

Decision Variables, 87
Deforestation, 22, 23, 33
Degradation, 9, 10, 14, 15, 17, 23, 24,
    106, 134, 158, 173
Demand, 2, 4, 7-10, 16, 17, 19-22, 86,
    92, 93, 109, 111, 121,
    122, 128, 146, 189,
    196, 204, 205, 207,
    210, 211, 214, 217,
    218, 230, 237-239,
    257, 259, 261, 269,
    272-275, 279, 282,
    291
Demand Functions, 238, 272
Demand-side, 2
Demographic, 2, 42, 48, 53, 105, 106,
    157, 172, 187, 190,
    205, 215
Dependency Ratio, 8, 45, 48, 51, 129
Depreciation, 25, 27, 75, 259
Descriptive analysis, 83, 130, 131
Determinants, 18, 80, 83, 117, 157, 187,
    190, 201, 204, 209,
    229, 231
Devaluation, 5, 7, 27, 153, 252-254,
    256-259, 269
Deviations, 171, 190, 244, 249, 277
Discriminant Analysis, 42
Discriminatory, 275
Diseases, 14, 19, 21, 30, 49, 85, 188,
    192, 195, 197, 202
Dogbo, 280-284
Dual Monopoly, 266
Dummy Variable, 158, 159, 199, 215,
    216, 243, 244, 247,
    248

E

Ecological, 2, 14, 15, 24, 32, 33, 36, 63,
    65, 105, 106, 188,
    189, 209, 215, 219,
    244, 248
Economic Growth, 3, 160, 287
Economies of Scale, 60, 209, 269, 270,
    285, 291
Ecowas, 5
Education, 9, 31, 37, 42, 43, 46-51,
    63-66, 70, 188, 195,
    210, 215-217, 219,
    223, 229
Efficiency Wage, 211, 213, 218-220,
    230, 231

Efficient, 6, 8, 42, 212-214, 221, 252,
    264, 276, 279, 285,
    286
Elasticities, 153, 154, 161, 163-165,
    170, 172, 214, 223,
    224, 229, 240, 245,
    246, 248, 250
Electricity, 45-51, 198, 200, 202, 215,
    216, 219, 223
Employment, 2, 4, 18, 37, 42, 69, 70, 75,
    89, 129, 212, 215,
    216, 219, 221, 223,
    226, 227, 278
Endowment, 130, 211-213, 218
Energy, 21, 22, 38, 92, 93, 96, 113, 119,
    128, 134, 135, 145,
    146, 152, 201, 208,
    209, 213-215, 217,
    219, 223, 224, 227,
    230, 231
Energy Deficiencies, 214, 219, 224, 227
Entry, 265, 266, 269-271, 273-276, 285,
    286, 291
Environment, 7, 10, 14, 17-20, 22, 23,
    32, 33, 37, 43, 54, 99,
    129, 149, 229, 264,
    268, 271, 286-288
Environmental Degradation, 10, 134,
    158, 173
Equation, 38, 89-91, 93, 94, 97, 101,
    102, 155-157, 159,
    160, 164, 165, 171,
    197-199, 214,
    220-222, 225, 227,
    239, 240, 244, 248,
    279
Equity, 269-271, 273
Erosion, 20, 21, 34, 88, 97, 104-106,
    112
Error, 30, 156, 157, 211, 220, 221,
    279-283
Error Correction Model, 279, 281, 283
Europe, 5, 10, 13, 15, 256, 257
European Community, 252, 258
Exchange Rate, 5, 7, 19, 27, 57, 153,
    174, 175, 217, 252,
    254
Exit, 265, 266, 269, 271, 276, 285, 289
Exogenous, 84-87, 211, 212, 218, 220,
    223, 225-227
Expenditure, 3, 7, 13, 35, 36, 38-41, 46,
    47, 49-51, 54-57,
    59-63, 68-70, 75-78,
    113, 134, 146, 189,
    193, 196, 201, 215,
    230, 231

Extremely Poor, 62, 63

F

Fallow, 9, 84, 86, 88, 89, 97, 120, 121, 134, 138
Family Planning, 7, 51, 70
Farm, 8, 16, 30, 51, 69, 75, 76, 79-81, 83-85, 92, 100, 113, 117, 129-132, 135, 137-142, 145-150, 174, 175, 202, 209, 215, 216, 231, 234-237, 240-244, 246, 247, 250, 279
Farmers' Strategies, 80-83, 85, 87, 94, 105, 112, 130, 132
Farmers' strategy, 80
Farmgate Prices, 158, 250
Farming Systems Research, 81, 114, 117, 130, 149
Farm-household, 129, 131, 132, 137, 142, 145-147
Female, 43, 63-66, 70, 73-75, 113, 142, 151, 198, 215, 224, 226
Fertility, 8, 9, 20, 24, 51, 97, 105, 106, 109, 112, 116, 132, 149, 192, 204, 230
Fertilizer, 10, 82, 84, 86, 88, 97, 112, 130, 138, 51, 75, 139, 159, 201, 209, 210, 229, 237
Fiscal Policy, 7
Fish, 4, 134, 197, 199
Food Aid, 3, 5, 6, 24, 30
Food Availability, 2, 24, 188-191, 196-199, 202, 207, 208
Food Consumption, 2, 4, 25, 57, 71, 75, 76, 146-148, 210-213, 215-217, 224-227, 230, 234
Food Insecurity, 2, 3, 11, 24-26, 30, 69, 116, 117, 134, 147, 208
Food Intake, 137, 189, 190, 192, 197
Food Security, 1, 2, 4-6, 8-11, 13, 14, 24, 26, 27, 31, 32, 35, 53, 54, 70, 79-81, 96, 97, 105, 112, 115, 117, 129-132, 137, 145-147, 174, 186, 200, 204,

205, 231, 233, 250, 251, 288, 291
Food Security Policies, 4, 6
Food Supply, 2, 6, 24, 25, 27, 207, 208
Francophone Countries, 5
Full Information Maximum Likelihood (FIML), 211, 220, 221

G

Gambia, 25-30, 154, 175
Gap Index, 37-41
Gari, 5, 8
GATT, 10, 257, 258, 262
Gender, 2, 63, 64, 66, 71, 114, 139, 140, 142, 188, 198, 212, 217, 241
General Equilibrium Model, 268
Ghana, 3-7, 9, 10, 13, 22, 25-30, 34-43, 45, 48, 52-66, 69-74, 77-79, 106, 129-140, 142-150, 152-154, 157-161, 164-166, 171-176, 186-192, 195-197, 201, 203-205, 207-211, 214-217, 221, 223, 227-235, 240-244, 247-255, 258-260, 262, 272, 277, 288
Gini-coefficient, 39, 57, 61
GLSS, 38, 39, 42, 43, 45-51, 55, 56, 60-62, 73-75, 188-196, 204, 206, 208, 214, 215, 218, 225
Guinea Savannah, 134, 149
Guinea-bissau, 4, 25, 26, 28, 29

H

Hard Core, 36, 134
Head-count Index, 37, 39, 61, 62
Health, 2, 9, 13, 31, 37, 42, 49, 141, 186-190, 192, 195-205, 208, 209, 213, 215, 229, 230, 234, 240
Hectares, 22, 140, 157
Height-for-age, 9, 190-193, 195, 196, 198, 200-202, 206
Herbs, 85
Hierarchy, 279, 282, 283

O

P

Purchasing Power, 2, 4, 22, 24, 25, 27,
    57, 60, 62, 69, 111

Q

Quadratic, 157, 197, 218, 219, 227
Qualitative, 53, 276, 277, 285
Quantitative, 10, 24, 70, 86, 130, 131,
    174, 175, 253,
    276-278
Questionnaire, 55, 130, 131, 189, 241,
    244
Quintile, 57, 60, 61, 190, 191, 215

R

Rainfall, 85
Redistribution, 37, 225
Reduced-form, 155
Regional, 4, 5, 13, 24, 30-32, 53, 80-82,
    105, 110, 111, 129,
    130, 132, 134, 137,
    148, 149, 174, 175,
    233, 251-257, 260,
    262, 263, 269, 271,
    276, 279, 280, 282,
    288, 290
Regional cattle trade, 252, 255, 256, 260
Regression, 27, 42, 159-171, 173,
    176-185, 198, 200,
    210, 241-250
Remittances, 9, 45-51, 76, 129
Resilience, 17, 20, 43
Resource Use, 18, 138, 236
Retail, 20, 235, 256, 257, 276, 279-284
Risk, 2, 6, 37, 43, 94, 96, 114, 129, 156,
    157, 159, 171, 172,
    187, 192, 208, 266,
    270
Rural Poor, 16, 18, 30, 36, 71, 204
Rural-urban, 4, 8, 51, 202

S

SADAOC, 5, 11, 115, 131, 132, 135,
    137, 139, 140,
    142-146, 152, 204,
    251, 262, 291
Safe Water, 37
Safety Net, 9, 49
Safety Stock, 84, 86, 91, 93, 94, 120,
    122, 124, 147

Sahel, 16-18, 20-22, 24-30, 34, 115,
    118, 166, 167, 258,
    259, 261, 262, 288
Sahelian, 2, 4, 5, 14, 17, 21, 30, 33, 106,
    252, 256, 257,
    259-261
Sanitation, 2, 9, 37, 45, 188
SARI, 130-133, 139
Scenario, 85, 86
Seasonal Fluctuations, 275
Seasonal Poverty, 4
Seasonality, 4, 76, 210, 266, 274, 284
Sedentarisation, 16-18, 21
Self-employment, 69, 70, 75, 212, 215,
    216, 226
Self-sufficiency, 6, 17, 24, 147
Self-sufficient, 146, 147, 274
Semi-urban, 141, 189, 215
Senegal, 3, 4, 18, 22, 24-29, 34, 154,
    174
Sensitivity, 88, 96, 105, 109, 112, 221,
    261
Sensitivity Analysis, 96, 112
Services, 10, 11, 16, 19-21, 30, 31, 37,
    57, 63, 65, 66, 69, 73,
    74, 80, 131, 134, 137,
    141, 149, 153, 159,
    188, 203, 215, 219,
    251, 254, 266, 267,
    269-271, 273, 274,
    276, 284, 285, 287
Short-run, 146, 153, 154, 161, 163-165,
    167, 170, 265, 277,
    278, 280-283
Sierra Leone, 5, 6, 26, 28, 29, 224
Social Amenities, 4, 51, 202
Socio-cultural, 2, 14, 15, 24, 32
Socio-demographic, 105
Socio-economic, 14, 28, 55, 62-65, 69,
    81, 84, 105, 106, 112,
    129, 192
Soil Degradation, 9, 24
Source of Income, 45, 65, 76, 252
Spatial Arbitrage, 265, 273, 279, 285
Standard Deviation, 57, 156-159, 171,
    188, 215, 277, 278
Stocks, 2, 6, 21, 22, 91, 94, 95, 103,
    104, 121, 123, 135,
    142, 143, 147, 275
Strategies, 15, 17, 19, 20, 79-83, 85-88,
    92, 94, 97, 98, 104,
    105, 109-112,
    129-132, 136, 147,
    268, 270, 271, 285
Structure, 18, 53, 83, 109, 110, 134,
    253-257, 259,

T

U

V

W

Z